CHAP. 9
14, 16, 18

INTRODUCTION TO CONTROL SYSTEM TECHNOLOGY

Second Edition

Merrill's International Series in Electrical and Electronics Publications

SAMUEL L. OPPENHEIMER, *Consulting Editor*

INTRODUCTION TO CONTROL SYSTEM TECHNOLOGY

Second Edition

ROBERT BATESON

Charles E. Merrill Publishing Company
A Bell & Howell Company
Columbus   Toronto   London   Sydney

Published by Charles E. Merrill Publishing Co.
A Bell & Howell Company
Columbus, Ohio 43216

This book was set in Times Roman.
Cover Design Coordination: Will Chenoweth
Cover photo by Larry Hamill.

Library of Congress Catalog Card Number: 79–84687

International Standard Book Number: 0–675–08255–2

Printed in the United States of America

1 2 3 4 5 6 7 8 9 10—85 84 83 82 81 80

Preface

This text is intended for both two-year and four-year engineering technology programs. The approach is to promote learning by example and illustration, with numerous examples to describe the methods used to analyze and design control systems and a wide variety of illustrations of control systems and components.

Although a knowledge of calculus is not required, students should have a background in algebra, trigonometry, and, for Part Three, complex numbers. Derivatives and integrals are introduced and explained as they are required to describe a particular system. Some students will find this method of introducing calculus more interesting than the more conventional approach. In addition, SI units are used throughout the text in anticipation of eventual U.S. conversion to the metric system.

In response to suggestions from my colleagues, this edition has been *completely* reorganized to provide a more logical flow of information. The content is arranged to provide a natural progression from the simple to the complex and from components to systems. There is sufficient material for a full-year sequence. The first half is suitable for most two-year programs and the lower division of four-year programs. The second half is suitable for the upper division of four-year programs and the more advanced two-year programs.

Part One offers a discussion of the fundamentals of control that explains the concepts and terminology related to controls. The Laplace transforms are introduced at a level sufficient to allow students to obtain the transfer function of a component from the equation that defines the input/output relationship of the component.

In Part Two there is a complete discussion of control system components. Two chapters are devoted to the principles involved in the measurement of physical quantities and the characteristics of measuring means. A new chapter on microcomputers and solid-state components includes a review of digital fundamentals, a description of microprocessors and the microcomputers, a discussion of the four most common communications interfaces, the basic types of D/A and A/D converters, and a method of programming a microcomputer. The section on solid-state components includes examples of the use of SCRs, triacs, UJTs, diodes, and transistors. The remaining three chapters cover analog computers, controllers, control valves, and electric motors. The chapter on controllers includes new material on the PID control algorithm used in digital controllers.

Part Three covers the analysis and design of control systems. In the chapters on basic elements of a system and process characteristics, there are numerous step-by-step examples of the calculation of control parameters. These examples are quite helpful in developing calculation skills, and they make it possible for the student to carry out the complete analysis and design of a process control system. The chapter on frequency response features the development of BASIC programs to handle the tedious number crunching required to generate frequency response data, which greatly reduces the amount of class time needed to teach students how to obtain these data. The chapter on controller design, which features computer-aided design using a BASIC design program, emphasizes the design decisions rather than the details of calculation. The BASIC design program is the result of over five years of classroom experience in an engineering technology program.

The author is indebted to all who have worked in the field of control systems. I would like to thank those whose ideas have directly or indirectly contributed to this book, especially Dr. Stephen R. Cheshier, Purdue University, for his invaluable review. I would also like to thank my colleagues for their constant encouragement, and acknowledgement is due my students for their many helpful comments. I must also thank my children, Mark, Karen, and Paul, for their patience and understanding while I worked on the manuscript. Finally, I am grateful to my wife Betty for her encouragement, understanding, and her typing of the manuscript.

Robert Bateson

Contents

Contents

Part
One

INTRODUCTION

1

Basic Concepts and Terminology

1.1 Introduction

Control systems are everywhere around us and within us.[1] Many complex control systems are included among the functions of the human body. An elaborate control system centered in the hypothalamus of the brain maintains body temperature at 37 degrees Celsuis in spite of changes in physical activity and external ambience. In one control system—the eye—the diameter of the pupil automatically adjusts to control the amount of light that reaches the retina. Another control system maintains the level of sodium ion concentration in the fluid that surrounds the individual cells.

Threading a needle and driving an automobile are two ways in which the human body functions as a complex controller. The eyes are the sensor that detects the position of the needle and thread, or of the automobile and the center of the road. A complex controller, the brain, determines which actions must be performed to accomplish the desired result. The body implements the control action by moving the thread or turning the steering wheel; an experienced driver will anticipate all

---

[1] Instrument Society of America film, "Principles of Frequency Response," ISA, 1958.

3

types of disturbances to the system, such as a rough section of pavement, or a slow moving vehicle ahead. It would not be easy to reproduce in an automatic controller the many judgments that an average person makes daily and unconsciously.

Control systems regulate temperature in homes, schools, and buildings of all types. They also affect the production of goods and services by insuring the purity and uniformity of the food we eat, and by maintaining the quality of paper mill, steel mill, chemical plant, refinery, and all kinds of manufacturing plant products. Control systems help protect our environment by minimizing waste material that must be discarded, thus reducing manufacturing costs and minimizing the waste disposal problem. Sewage and waste treatment also requires the use of automatic control systems.

A control system is any group of components that maintains some desired result or value. From the previous examples, it is clear that a great variety of components may be a part of a single control system, whether they are electrical, electronic, mechanical, hydraulic, pneumatic, human, or any combination of these. The desired result may be, for example, the direction of an automobile, the temperature of a room, the level of liquid in a tank, or the pressure in a pipe.

The regulation of energy is the key to achieving control: control systems always involve changing conditions, and changing conditions always involve energy changes. For example, a system regulates the temperature in a room by controlling the energy input (fuel) supplied to the heater.

A control system is a group of components that maintains a desired result by regulating energy input.

1.2   Block Diagrams and Traveling Signals

Although it is not unusual to find several kinds of components in a single control system, or two systems with completely different kinds of components, all types of control systems have a common denominator: the mathematical equations that define the characteristics of each component. A wide range of control problems—including processes, machine tools, servomechanisms, space vehicles, traffic, and even economics—is analyzed by the same method. The important feature of each component is the effect it has on the system. The block diagram as a method of representing a control system retains only these important features of each component; signal paths connect the various blocks in the system. A single block is shown in Figure 1.1.

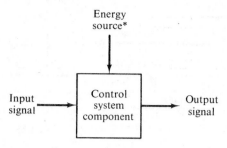

* The energy source is not shown on most block diagrams.
However, many components do have an external energy
source which makes amplification of the input signal possible.

FIGURE 1.1
A BLOCK REPRESENTATION OF A COMPONENT

Each component receives an input signal from some part of the system and thereby produces an output signal for another part of the system. The signals can be electric current, voltage, air pressure, liquid flow rate, liquid pressure, temperature, speed, acceleration, position, direction, etc. The signal paths can be electric wires, pneumatic tubes, hydraulic lines, mechanical linkages, or anything that transfers a signal from one component to another. The component may use some source of energy to increase the power of the output signal. Figure 1.2 illustrates block representations of different components.

The relationship between the input signal and the output signal is the most important characteristic of a component. The relationship is known as the *transfer function* of the component or its block representation. If the input signal is a combination of one or more sinusoidal signals, the component may change each sinusoid in two ways—size and timing.

A change in size is usually expressed as the change in the output signal corresponding to a unit change in the input signal, and is called the *gain* of the component.

$$\text{Gain} = \frac{\text{change in output size}}{\text{change in input size}} \left(\frac{\text{output units}}{\text{input units}} \right)$$

If the input and output signals have the same units, a change in size means that the output signal is smaller or larger than the input signal.

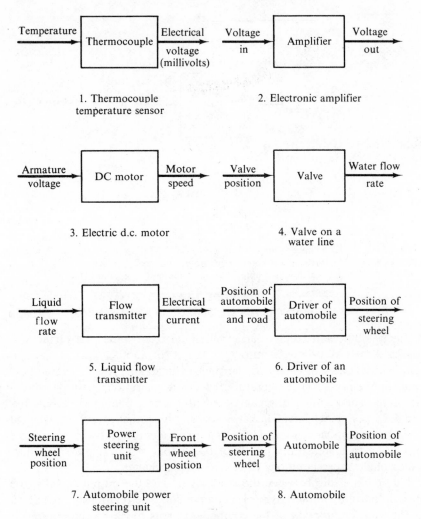

FIGURE 1.2

BLOCK REPRESENTATIONS OF CONTROL SYSTEM COMPONENTS

If the input and output signals have different units, the change in size must include the change in units. An amplifier that produces a 10 volt change in output for each 1 volt change in input has a gain of 10 volts per volt. A dc motor that produces a 1000 r/min change in speed for each 1 volt change in input has a gain of 1000 r/min per volt. A thermocouple that produces an output change of 0.06 millivolts for each 1° C change in temperature has a gain of 0.06 millivolts per degree Celsius.

A change in timing means that all or part of the signal is delayed so that the output signal occurs at a later time than the input signal. A component with a pure dead time characteristic will delay the entire signal by the same time period (called the dead time, t_d). Other characteristics will produce time delays that depend on the frequency of the signal. For example, a first-order lag characteristic with a corner frequency of 10 Hz will delay a sinusoidal signal of 1 Hz by about 6°, a signal of 10 Hz by 45°, and a signal of 100 Hz by about 84°. The exact nature of this timing change will be discussed in detail later.

> *The transfer function of a component or its block representation establishes the size and timing relationship between the output signal and the input signal.*

A block diagram consists of a block representing each component in a control system connected by lines that represent the signal paths. Figure 1.3 shows a very simple block diagram of a person driving an automobile. The driver's sense of sight provides the two input signals—the position of the automobile and the position of the center of the road. The driver compares the two positions and determines the position of the steering wheel that will maintain the proper position of the automobile. To implement the decision, the driver's hands and arms move the steering wheel to the new position. The automobile responds to the change in steering wheel position with a corresponding change in direction. After a short time has elapsed, the new direction moves the automobile to a new position. Thus there is a timing delay between a change in position of the steering wheel and the position of the automobile. This timing delay is included in the transfer function of the block representing the automobile.

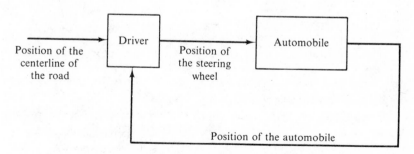

FIGURE 1.3

*SIMPLIFIED BLOCK DIAGRAM OF A
PERSON DRIVING AN AUTOMOBILE*

The loop in the block diagram indicates a fundamental concept of control. The actual position of the automobile is used to determine the correction necessary to maintain the desired position. This concept is called *feedback* and control systems with feedback are called *closed loop control systems*. Control systems that do not have feedback are called *open loop control systems* because their block diagram does not have a loop.

1.3 Open Loop Control

An open loop control system does not use a comparison of the actual result and the desired result to determine the control action. Instead, a calibrated setting—obtained by some sort of calibration procedure or calculation—is used to obtain the desired result.

The needle valve with a calibrated dial shown in Figure 1.4 is an example of an open loop control system. The calibration curve is usually obtained by measuring the flow rate for several dial settings. As the calibration curve indicates, different calibration lines are obtained for different pressure drops. Assume that a flow rate of F_2 is desired and a setting of S is used. As long as the pressure drop across the valve remains equal to $\triangle P_2$, the flow rate will remain F_2. If the pressure drop changes

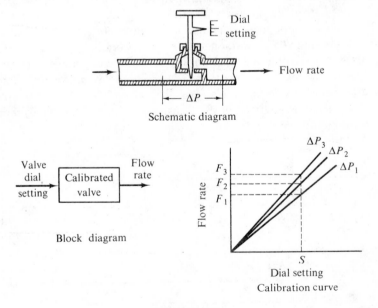

Schematic diagram

Block diagram

Dial setting
Calibration curve

FIGURE 1.4
A CALIBRATED NEEDLE VALVE

to $\triangle P_1$ or $\triangle P_3$, the flow rate will change to F_1 or F_3. The open loop control cannot correct for unexpected changes in the pressure drop.

The firing of a rifle bullet is another example of an open loop control system. A block diagram is shown in Figure 1.5. The desired result is to direct the bullet to the bull's-eye. The actual result is the direction of the bullet after the gun has been fired. The open loop control occurs when the rifle is aimed at the bull's-eye and the trigger is pulled. Once the bullet leaves the barrel, it's on its own: if a sudden gust of wind comes up, the direction will change and no correction will be possible.

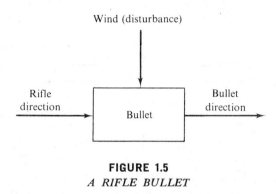

FIGURE 1.5
A RIFLE BULLET

The primary advantage of open loop control is that it is less expensive than closed loop control: it is not necessary to measure the actual result. In addition, the controller is much simpler because corrective action based on the error is not required. The disadvantage of open loop control is that errors caused by unexpected disturbances are not corrected. Often a human operator must correct slowly changing disturbances by manual adjustment. In this case, the operator is actually closing the loop by providing the feedback signal.

1.4 Closed Loop Control—Feedback

Feedback is the action of measuring the difference between the **actual** result and the desired result, and using that difference to drive the **actual** result toward the desired result. The term feedback comes from the direction in which the measured value signal travels in the block diagram. The signal begins at the output of the controlled system and ends at the input to the controller. The output of the controller is the input to the controlled system. Thus the measured value signal is fed back from the output of the controlled system to the input. The term *closed loop* refers to the loop created by the feedback path.

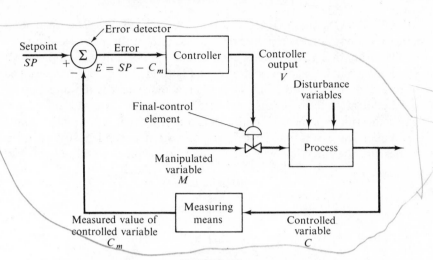

FIGURE 1.6

BLOCK DIAGRAM OF A CLOSED LOOP CONTROL SYSTEM

A block diagram of a closed loop control system is shown in Figure 1.6. The names of the components and variables in the diagram are used throughout this book. The student should be thoroughly familiar with these terms. The following four operations form the basis of the feedback control system.

1. Measurement of the controlled variable.
2. Computation of the difference between the measured value of the controlled variable and the desired value (the error).
3. Use of the error to generate a control action.
4. Use of the control action to drive the actual value of the controlled variable toward the desired value.

The *process* block represents everything performed in and by the equipment in which a variable is controlled. The process includes everything that affects the controlled variable except the automatic controller.

The *measuring means* senses the value of the controlled variable and converts it into a usable signal. Although the measuring means is considered as one block, it usually consists of a primary sensing element and a transmitter. The input-output curve of a temperature measuring means is shown in Figure 1.7. A typical temperature measuring means consists of a primary element such as a thermocouple or a resistance element and a transmitter. A thermocouple converts a temperature into a millivolt signal and the transmitter converts the millivolt signal into a usable electric current signal. A resistance element converts a tempera-

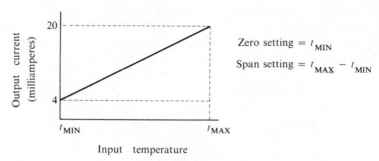

Zero setting $= t_{MIN}$

Span setting $= t_{MAX} - t_{MIN}$

FIGURE 1.7

INPUT-OUTPUT GRAPH OF A TEMPERATURE MEASURING MEANS

ture into a resistance value and the transmitter converts the resistance value into a usable electric current signal.

The *error detector* computes the difference between the measured value of the controlled variable and the desired value (or setpoint). The difference is called the *error* and is computed according to the following equation.

Error = (setpoint) − (measured value of controlled variable)

$$\text{or } E = SP - C_m$$

The *controller* acts on the input error signal to provide an output signal to the final control element that will tend to reduce the error signal. The control *mode* is the way in which a controller acts on an error signal to produce a control action. Several different control modes are used, and each mode produces a different controller output signal. The control modes include on-off, floating, proportional, integral, and derivative. The modes used in a particular controller depend on three factors.

1. How difficult the process is to control.
2. How accurately the process must be controlled.
3. How quickly the controller must respond to an upset.

The *final control element* is the component that uses the output signal from the controller to regulate the manipulated variable. Control valves and amplifiers are the most common final control elements. The solenoid gas valve on the furnace is the final control element of a household heating system. The controller output is an open or closed switch. A closed switch energizes the solenoid which opens the gas valve. An open switch de-energizes the solenoid which closes the gas valve. A power amplifier is the final control element of a typical dc motor speed control system.

A pneumatic control valve is often used as the final control element in processes (see Figure 1.8). An air pressure input signal acts on a diaphragm in opposition to a spring. An appropriate input signal can position the valve stem any place between open and closed.

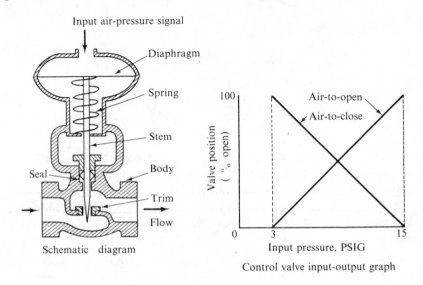

FIGURE 1.8
A PNEUMATIC CONTROL VALVE

The *setpoint* (*SP*) is the desired value of the controlled variable.

The *measured variable* (*C_m*) is the measured value of the controlled variable. It is the output of the measuring means and usually differs from the actual value of the controlled variable by a small amount.

The *error* (*E*) is the difference between the setpoint and the measured value of the controlled variable. It is computed according to the equation $E = SP - C_m$.

The *controller output* (*V*) is the control action intended to drive the measured value of the controlled variable toward the setpoint value. The control action depends on the error signal (*E*) and on the control modes used in the controller.

The *manipulated variable* (*M*) is the variable regulated by the final control element to achieve the desired value of the controlled variable. Obviously, the manipulated variable must be capable of effecting a change in the controlled variable. The manipulated variable is one of the input variables of the process. Changes in the load on the process necessitate changes in the manipulated variable to maintain a balanced condition. For this reason, the value of the manipulated variable is used as a measure of the load on the process.

The *disturbance variables* (*D*) are process input variables that affect the controlled variable but are not controlled by the control system. Disturbance variables are capable of changing the load on the process and are the main reason for using a closed loop control system.

The *controlled variable* (*C*) is the process output variable which is to be controlled. In a process control system, the controlled variable is usually an output variable which is a good measure of the quality of the product. The most common controlled variables are position, velocity, temperature, pressure, level, and flow rate.

The primary advantage of closed loop control is the potential for more accurate control of the process. There are two disadvantages of closed loop control: first, closed loop control is more expensive than open loop control. Second, the feedback feature of a closed loop control system makes it possible for the system to become unstable. An unstable system produces an oscillation of the controlled variable, often with a very large amplitude. (Stability will be studied in detail later on.)

A closed loop (or feedback) control system measures the difference between the actual value of the controlled variable and the desired value (or setpoint) and uses the difference to drive the actual value toward the desired value.

EXERCISES

1–1. Write a sentence that gives a general description of a control system.

1–2. List the two ways that a signal may be changed as it passes through a component.

1–3. A 20° C change in temperature resulted in a 5-milliamp change in the output current of a certain temperature transmitter. Determine the gain of the transmitter in milliamps per ° C.

1–4. Draw a block for each of the following control system components and name the input and output signals.
    a.   A household heating thermostat.
    b.   A solenoid valve for a household heating furnace.
    c.   A household heating furnace.
    d.   The inside of a house where the input signal is heat from the furnace.

1–5. A technician obtained the following calibration data for a needle valve similar to the one shown in Figure 1.4. Use the data to construct a calibration curve for use by the operators.

Valve settings	10	20	30	40	50	60	70	80	90	100
Flow rate, cc/min	24	47	71	95	120	145	168	188	215	238

1–6. Explain in writing the difference between open loop control and closed loop control and give the advantages and disadvantages of each.

1–7. List the four operations that form the basis of a feedback control system.

1–8. Draw the block diagram of a closed loop control system and indicate the process, the measuring means, the error detector, the controller, and the final control element. Also indicate the setpoint (SP), the measured variable (C_m), the error (E), the controller output (V), the manipulated variable (M), and the controlled variable (C).

1–9. The setpoint (SP) of a pressure control system is 32.3 psi, and the measured value (C_m) is 28.1 psi. What is the error (E)?

1–10. The setpoint (SP) of a temperature control system is 88° C, and the measured value (C_m) is 89.2° C. What is the error (E)?

1–11. The setpoint (SP) of a liquid flow control system is 0.100 cubic meters per minute, and the measured value (C_m) is 0.097 cubic meters per minute. What is the error (E)?

1–12. Give an example of an open loop control system.

1–13. Sketch the input-output graphs of a temperature transmitter, an air-to-open control valve, and an air-to-close control valve.

1–14. Draw the block diagram of a household heating control system. Use the four components described in Exercise 4.

    a. Identify the measuring means, the error detector, the controller, the final control element, and the process.

    b. Describe the operation of the controller. What is the relationship between the error signal and the controller output signal?

    c. If the desired room temperature is 72° F and the actual room temperature is 68° F, describe why the control system cannot immediately raise the actual temperature to 72° F.

2

Evaluation of Control Systems

2.1 Benefits of Automatic Control

Control systems are becoming steadily more important in our society. We depend on them to such an extent that life would be unimaginable without them. Automatic control has increased the productivity of each worker by releasing skilled operators from routine tasks and by increasing the amount of work done by each worker. Control systems improve the quality and uniformity of manufactured goods and services: many of the products we enjoy would be impossible to produce without automatic controls. Servo systems place tremendous power at man's disposal, enabling him to control large equipment such as jet airplanes and ocean ships.

Control systems increase efficiency by reducing waste of materials and energy, an increasing advantage as we seek ways to preserve our environment. Safety is yet another benefit of automatic control. Finally, control systems such as the household heating system and the automatic transmission provide us with increased comfort and convenience.

In summary, the benefits of automatic control fall into the following six broad categories.

1. Increased productivity
2. Improved quality and uniformity

3. Increased efficiency
4. Power assistance
5. Safety
6. Comfort and convenience

2.2 Load Changes

A control system must balance the material or energy gained by the process against the material or energy lost by the process in order to maintain the desired value of the controlled variable. Usually, the material or energy loss is the load on the process, and the manipulated variable must supply the balancing material or energy gain. However, sometimes the opposite condition exists and the manipulated variable must provide the material or energy loss.

A home heating system must balance the heat supplied by the furnace against the heat lost by the house in order to maintain the desired inside temperature. The heat loss is the load on the control system, and the energy supplied to the furnace is regulated by the manipulated variable.

A liquid level control system must balance the input flow rate against the output flow rate in order to maintain the level at the desired value. The output flow rate is the load on the system and the input flow rate is the manipulated variable.

The control of a variable speed motor driving a pump must balance the input power to the motor against the power delivered to the pump in order to maintain the desired pump speed. The power delivered to the pump is the load on the system, and the power input to the motor is regulated by the manipulated variable.

An air conditioning system must balance the heat removed by the air conditioner against the heat gained by the room in order to maintain the desired room temperature. The heat gained by the room is the load on the system, and the heat removed by the air conditioner is regulated by the manipulated variable.

The load on a process is always reflected in the manipulated variable. Therefore, the value of the manipulated variable is a measure of the load on a process. Every load change results in a corresponding change in the manipulated variable and, consequently, a corresponding change in the setting of the final control element. Consider a sudden increase in load for the pump control system described above. The increase in load tends to reduce the motor speed. The controller senses the reduced motor speed and produces a control action that increases the power input to the motor. In an ideal situation, the control action will cause the manipulated variable to match the increased load and the pump speed

will remain at the desired value. Within the control loop, the only variable that reflects the load change is the manipulated variable. For this
reason, it makes sense to define the load on the control system in terms
of the manipulated variable.

*The load on a control system is measured by the value of the
manipulated variable required by the process at any one time
in order to maintain a balanced condition.*

The load on a control system does not remain constant. Any uncontrolled variable that affects the controlled variable is capable of causing a load change. Each load change necessitates a corresponding
change in the manipulated variable in order to maintain the controlled
variable at the desired value. A closed loop control system automatically
makes the necessary change in the manipulated variable; an open loop
control system does not make the necessary change. Thus, a closed loop
control system is necessary if automatic adjustment to load changes is
desired.

There are usually several uncontrolled conditions in a process that
are capable of causing a load change. Some examples of load changes
are:

1.   A change in demand by the controlled medium. For example, opening the door of a house in winter necessitates more heat to keep the inside temperature at the desired value. Closing the door requires less heat.
Both are load changes. In a manufacturing process, a change in production rate almost always results in a load change: in a heat exchanger, a
flowing liquid is continuously heated with steam. A change in the liquid
flow rate is a load change because more heat is required.

2.   A change in the quality of the manipulated variable. For example, a
change in the heat content of the fuel supplied to a burner requires a
change in the rate at which the fuel is supplied to the burner. In a neutralizing process, a solution of sodium bicarbonate is used to neutralize
a fiber ribbon. A decrease in the concentration of sodium bicarbonate
is a load change because more neutralizing solution is required.

3.   Changes in ambient conditions. For example, if the outside temperature drops 20° F, more heat is required in order to maintain the desired
temperature in a house.

4.   A variable amount of energy absorbed or supplied within the
process. For example, using the range to prepare supper supplies the
house with a large quantity of heat. Less heat is required from the fur-

nace in order to maintain the desired temperature. Chemical reactions often generate or absorb heat as part of the reaction; these are load changes because, as the process generates or absorbs heat, less or more heat is required from the manipulated variable.

2.3 Objectives of a Control System

At first glance, the objective of a control system seems quite simple—to maintain the controlled variable exactly equal to the setpoint at all times, regardless of load changes or setpoint changes. In order to do this, the control system must respond to a change before the error occurs; unfortunately, feedback is never perfect because it does not act until an error occurs. A load change first must change the controlled variable: this produces an error. Then, the controller acts on the error to produce a change in the manipulated variable. Finally, the change in the manipulated variable drives the controlled variable back toward the setpoint.

It is more realistic for us to expect a control system to obtain as near perfect operation as possible. Since the errors in a control system occur after load changes and setpoint changes, it therefore seems natural to define the objectives in terms of the response to such changes. Figure 2.1 shows a typical response of the controlled variable to a step change in load.

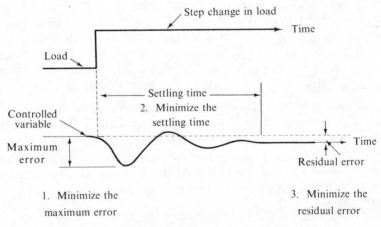

FIGURE 2.1

THE THREE OBJECTIVES OF A CLOSED LOOP CONTROL SYSTEM

One obvious objective is to minimize the maximum value of the error signal. Some control systems (with an integral mode) will eventually reduce the error to zero, while others require a residual error to compensate for a load change. In either case, the control system should eventually return the error to a steady, nonchanging value. The time required to accomplish this is called the settling time. A second objective of a control system is to minimize the settling time. A third objective is to minimize the residual error after settling out.

Unfortunately, these three objectives tend to be incompatible. For instance, the problem of reducing the residual error can be solved by increasing the gain of the controller so that a smaller residual error is required to produce the necessary corrective control action. However, an increase in gain tends to increase the settling time, and may also increase the maximum value of the error as well. The optimum response always is achieved through some sort of compromise.

Control Objectives

After a load change or setpoint change, the control system shall:

1. *Minimize the maximum value of the error.*
2. *Minimize the settling time.*
3. *Minimize the residual error.*

2.4 Damping and Instability

The gain of the controller determines a very important characteristic of a control system's response: the type of damping or instability that the system displays in response to a disturbance. The five general conditions are illustrated in Figure 2.2. As the gain of the controller is increased, the response changes in the following order: overdamped, critically damped, underdamped, unstable with constant amplitude, and unstable with increasing amplitude. Obviously, the unstable response does not satisfy the control objectives, nor does the overdamped response minimize the settling time. Therefore, the optimum response is either critically damped or slightly underdamped. Exactly how much damping is optimum depends on the requirements of the process.

Further insight about damping can be obtained by considering a familiar oscillating system—a child bouncing a ball. The ball will continue to bounce as long as the child pushes down when the ball is moving down—i.e., the force is in the same direction as the motion of the ball. The bouncing will die down quickly if the child pushes down when the ball is moving up—i.e., the force is in opposition to the motion of the ball. The oscillations of the ball are damped out by a force in opposition to the motion. Extending this concept to control systems, damping is a

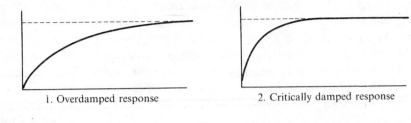

1. Overdamped response   2. Critically damped response

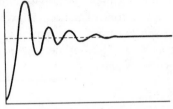
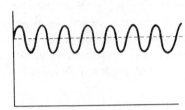

3. Underdamped response   4. Unstable-constant amplitude
                                                      response

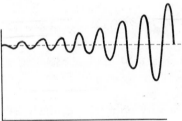

5. Unstable-increasing amplitude
             response

FIGURE 2.2

*THE FIVE TYPES OF RESPONSE TO A STEP CHANGE
IN LOAD OR SETPOINT*

force or signal that opposes the motion (or rate of change) of the controlled variable.

Several stabilizing techniques are used to increase the damping in a
system and thereby to allow a higher gain in the controller. The general
idea is to find a force or signal that will oppose changes in the controlled
variable. One such signal is the rate of change of the controlled variable.
In mathematics, the derivative of a variable is equal to its rate of change,
and this signal is referred to as the *derivative of the controlled variable*.
Damping is increased if the derivative of the controlled variable is subtracted from the error signal before it goes to the controller. This technique is sometimes called *output derivative damping*.

Another stabilizing signal is the derivative of the error signal. If the
setpoint is constant, this signal is equal to the negative of the derivative

of the controlled variable. Damping is increased if the derivative of the error is added to the error signal before it goes to the controller. This technique is usually called the *derivative control mode*.

Viscous damping is a stabilizing technique which is sometimes used in position control systems. It operates on the fact that frictional forces always oppose motion. A simple brake, a fluid brake, or an eddy current brake may be used to apply the damping force.

2.5 Criteria of Good Control

In order to evaluate a control system effectively, two decisions must be made: first the test must be specified. Second the criteria of good control must be selected. A step change in setpoint or load is the most common test. A typical step response test is illustrated in Figure 2.1. The three most common criteria of good control are: quarter amplitude decay, critical damping, and integral of absolute error. A discussion of each criterion follows.

1. *Quarter amplitude decay*—This criterion specifies a damped oscillation in which each successive positive peak value is one-fourth the preceding positive peak value. Quarter amplitude decay is a popular criterion because it is easy to apply in the field, and it provides a near optimum compromise of the three control objectives. Figure 2.3 illustrates the quarter amplitude decay response.

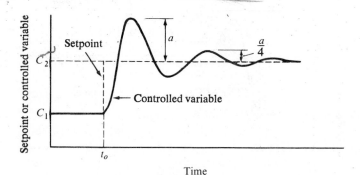

FIGURE 2.3

QUARTER AMPLITUDE DECAY RESPONSE TO A STEP CHANGE IN THE SETPOINT

2. *Critical damping*—This criterion is used when overshoot above the setpoint is undesirable. Critical damping is the least amount of damping that will produce a response with no overshoot and no oscillation. Elec-

trical instruments and some processes are critically damped. Figure 2.4
illustrates the critical damping criterion.

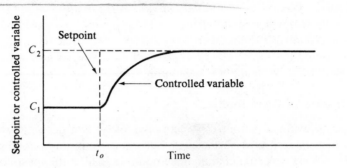

FIGURE 2.4
*CRITICAL DAMPING RESPONSE TO A STEP CHANGE
IN THE SETPOINT*

3.   *Minimum integral of absolute error*—This criterion specifies that
the total area under the error curve should be minimum. Figure 2.5
illustrates the minimum integral of absolute error criterion. The error is
the distance between C_2 and the controlled variable curve. The integral
of absolute error is the total shaded area on the curve. This criterion is
easy to use when a mathematical model is used to evaluate a control
system.

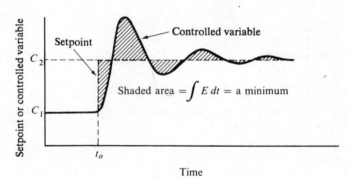

FIGURE 2.5
*MINIMUM INTEGRAL OF ABSOLUTE ERROR RESPONSE TO A
STEP CHANGE IN THE SETPOINT*

EXERCISES

2–1. Among the four types of load changes, which one is illustrated in each of the following examples.
  1. A change in demand by the controlled medium.
  2. A change in quality of the manipulated variable.
  3. A change in ambient conditions.
  4. A variable amount of energy absorbed or supplied within a process.

  a. A chemical process in plant A mixes two ingredients that combine to form a compound. Heat is generated by the reaction, and a control system is used to control the temperature of the mixture. What type of load change is a change in the amount of heat generated by the reaction?
  b. A process in plant B uses heated outside air to dry the product before packaging. A rainstorm raises the humidity of the outside air so that more heat is required to dry the product. What type of load change is this?
  c. A food process in plant C uses a dryer to toast corn flakes. What type of load change is a change in production rate from 200 pounds per hour to 300 pounds per hour?
  d. A food process in plant D uses a solution of sodium bicarbonate to neutralize synthetic meat fibers. The flow rate of sodium bicarbonate is the manipulated variable. What type of load change is a change in the concentration of sodium bicarbonate?

2–2. Which of the following is not an objective of a control system?
  a. To minimize changes in the manipulated variable.
  b. To minimize the maximum amplitude of the error signal.
  c. To return the error to a steady, non-changing condition in a minimum time.
  d. To minimize the residual error after stabilization.

2–3. List the five types of response to a step change in load and sketch the response curve of each one.

2–4. List the three criteria of good control and explain or illustrate each criterion.

2–5. In a chemical process, two components are blended together in a large mixer. The temperature of the mixture must be maintained between 110° C and 112° C. If the temperature exceeds 114° C, the finished product will not satisfy the specifications. Which of the three criteria of good control should be used for the temperature control system?

3

Types of Control

3.1 Introduction

Control systems are classified in several different ways. They are classified as closed loop or open loop, depending on whether or not feedback is used. They are divided into two broad categories—regulator systems and follow-up systems—depending on whether the setpoint is constant or changing. They are also classified according to the type of application, for example: process control, servomechanisms, sequential control, and numerical control. Finally, they are classified as analog or digital, depending on the nature of the signals. These general classifications are listed below.

Classifications of Control Systems

1. Feedback
   a. Not used—open loop
   b. Used—closed loop
2. Setpoint
   a. Seldom changed—regulator system
   b. Frequently changed—follow-up system

3. Application
   a. Process control
   b. Servomechanisms
   c. Sequential control
   d. Numerical control
4. Signal
   a. Analog
   b. Digital

Since Chapter One discussed open and closed loop control systems, this chapter will not deal with them, but will cover the other classifications.

3.2 Regulator and Follow-up Systems

Control systems are classified as regulator systems or follow-up systems, depending on how they are used. A *regulator system* is a feedback control system in which the setpoint is seldom changed, its prime function is to maintain the controlled variable constant despite unwanted load changes. A home heating system, a pressure regulator, and a voltage regulator are common examples of regular systems. Many process control systems are used to maintain constant processing conditions and, hence, are regulator systems.

A *follow-up system* is a feedback control system in which the setpoint is frequently changing. Its prime function is to keep the controlled variable in close correspondence with the setpoint as it changes. In follow-up systems, the setpoint is usually called the *reference variable.* A ratio control system, a strip chart recorder, and the antenna position control system on a radar tracking system are examples of follow-up systems. Many servomechanisms are used to maintain an output position in close correspondence with an input reference signal and, hence, are follow-up systems.

3.3 Process Control

Process control involves the control of variables in a manufacturing process—here, any combination of materials and equipment that produces or modifies a product, making it more useful and, hence, more valuable. A dairy, an automobile assembly plant, a refinery, an electric power plant, a steel mill, and a taconite plant are some examples of manufacturing processes. The most common controlled variables in a process are temperature, pressure, flow rate, and level. Others include

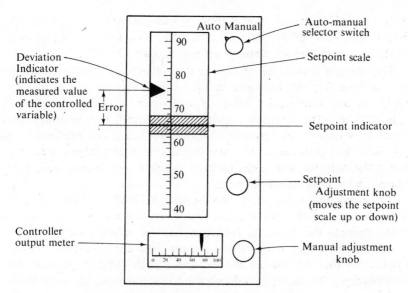

As shown above, the setpoint is 65, the measured
value of the controlled variable is 75, and the
controller output is about 76.

a) Schematic diagram

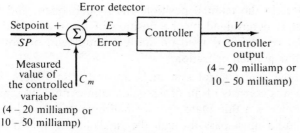

b) Block diagram

FIGURE 3.1
A TYPICAL PROCESS CONTROLLER

color, conductivity, pH, hardness, viscosity, and composition. Many
process control systems are used to maintain constant processing conditions and, hence, are regulator systems.

Process control systems may be either open loop or closed loop, but
closed loop systems are more common. The process control industry
has developed standard, very flexible controllers with one, two, or

three control modes. A closed loop process controller is illustrated in Figure 3.1. The error detector and controller are combined into a single instrument which displays all four signals (SP, C_m, E, and V).

The setpoint scale is a long tape or drum that travels behind a window with a fixed setpoint indicator in the center. The portion of the scale visible at any time varies from 20 to 50 per cent, depending on the manufacturer. The operator adjusts the scale until the desired setpoint value rests behind the setpoint indicator. Although the expanded scale just described is the most common method of indicating the setpoint, it is not the only method; some controllers use a galvanometer or a calibrated dial to indicate the setpoint.

The measured value of the controlled variable[1] is indicated on the setpoint scale by the deviation indicator. In this arrangement, the distance between the deviation indicator and the setpoint indicator is the deviation (or error) between the setpoint and the controlled variable. Controllers that do not use the expanded scale for the setpoint use a horizontal galvanometer to indicate the measured value of the controlled variable.

The other features shown in Figure 3.1 are available on most process controllers. The controller output is indicated by a horizontal galvanometer. The auto-manual selector switch allows transfer between automatic and manual operation: in the auto position, the loop is closed, and the controller output is determined by the error signal and the control modes. In the manual position, the loop is open, and the controller output is determined by the position of the manual adjustment knob. Manual operation is used on startup, shutdown, or anytime the operator prefers manual operation.

A liquid flow rate control system is illustrated in **Figure 3.2**. The measuring means consists of an orifice and a differential pressure transmitter. An orifice is a thin plate with a small hole positioned so that all the flowing liquid must pass through the small hole. The flow through the orifice is proportional to the square root of the pressure drop.

$$(\text{flow rate}) = (\text{constant}) \times \sqrt{(\text{pressure drop})}$$

The differential pressure transmitter converts the pressure drop across the orifice into a 4- 20-milliamp current signal proportional to the pressure drop. The current signal is transmitted to the controller as the measured value of the controlled variable. A square root scale in the controller allows the operator to read the flow rate directly. The con-

---

[1] The controlled variable is also referred to as the *process variable*.

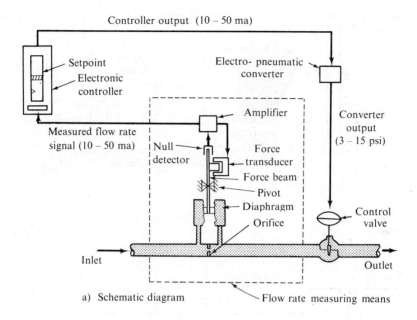

a) Schematic diagram

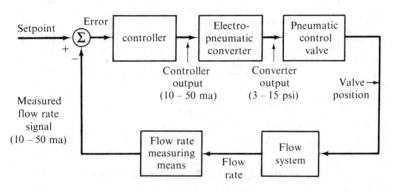

b) Block diagram

FIGURE 3.2

A LIQUID FLOW RATE CONTROL SYSTEM

troller output positions the control valve to maintain the desired flow rate.

The diaphragm arrangement converts the pressure difference across the orifice into a force on the lower end of the force beam. The force transducer at the other end of the beam produces the counterbalancing force. The null detector, amplifier, and force transducer make up a closed loop control system that maintains the force beam in the null

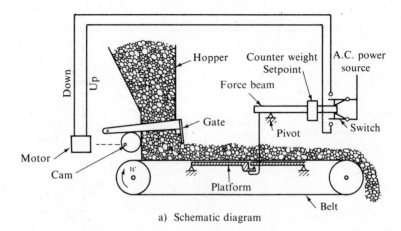

a) Schematic diagram

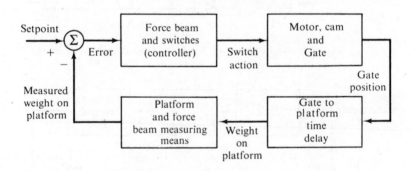

b) Block diagram

FIGURE 3.3

A SOLID FLOW RATE CONTROL SYSTEM

position. In this system, the null detector is the measuring means that senses the position of the force beam. The amplifier is the controller, and the force transducer produces a force proportional to the current supplied by the amplifier. This current is the source of the 4- 20-milliamp output signal from the flow rate measuring means.

The solid flow rate control system in Figure 3.3 uses a constant speed belt with a weigh platform. The solid flow rate is equal to the belt speed times the weight of material per unit length of belt. The weight of material on the platform applies a downward force on the left end of the

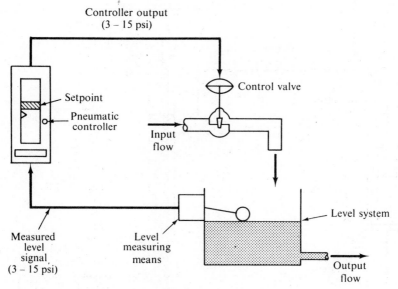

a) Schematic diagram

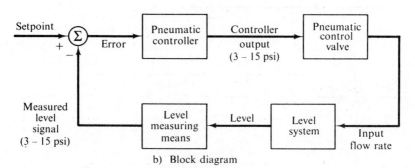

b) Block diagram

FIGURE 3.4

A LEVEL CONTROL SYSTEM

force beam; the counterweight supplies the counterbalancing force on the right side. The force beam operates the two limit switches which control the cam drive motor. If the material on the belt is too heavy, the beam will close the upper limit switch which drives the gate down; if the material is too light, the beam will close the lower limit switch which drives the gate up.

The tank level in Figure 3.4 will remain constant if the input flow rate is equal to the output flow rate. If the input flow rate exceeds the

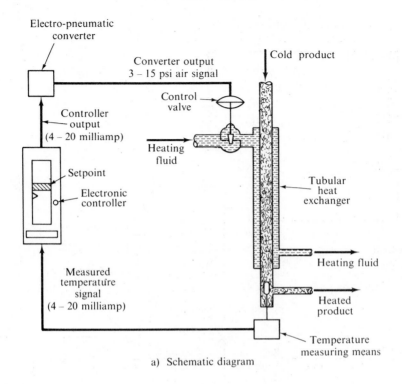

a) Schematic diagram

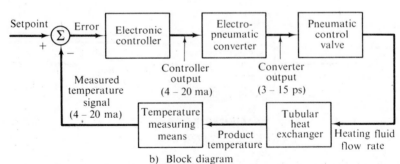

b) Block diagram

FIGURE 3.5

A TEMPERATURE CONTROL SYSTEM

output flow rate, the level will rise. If the input flow rate is less than the output flow rate, the level will drop. The controller uses the level signal to maintain the balance between the input and the output flow rates.

The tubular heat exchanger in Figure 3.5 is used to heat a liquid product—for example, the pasturization of milk. The heat exchanger consists of two concentric tubes: the product passes through the inner tube which is surrounded by the heating fluid contained in the larger tube. Steam is the most common heating fluid, but hot water and hot oil are also used. The control valve manipulates the heating fluid flow rate which determines the amount of heat transferred to the product. As it leaves the heat exchanger, the product temperature is measured. The controller compares the measured temperature with the setpoint, and manipulates the control valve to make sure that the two signals are equal.

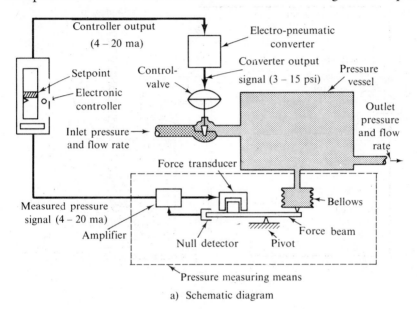

a) Schematic diagram

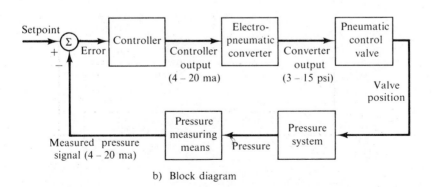

b) Block diagram

FIGURE 3.6
A PRESSURE CONTROL SYSTEM

The pressure control system of Figure 3.6 must maintain a balance between the input and output mass flow rates. The pressure measuring means is very similar to the differential pressure transmitter in the liquid flow control system. It uses a force balance principle to convert the process pressure into a 4- 20-milliamp current signal. The controller compares the measured pressure with the setpoint and manipulates the control valve to bring the two into correspondence.

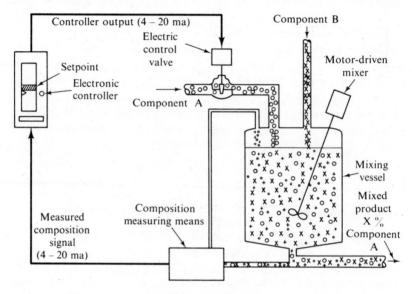

a) Schematic diagram

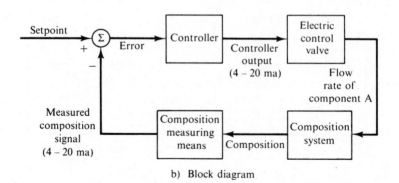

b) Block diagram

FIGURE 3.7
A COMPOSITION CONTROL SYSTEM

The composition control system in Figure 3.7 maintains the desired mixture of components A and B by manipulating the input of component B. The measuring means is some type of analyzer that measures the percentage of component A in the mixture. The mixing vessel blends the two components and smooths out the fluctuations in the flow rates of both components.

Manufacturing processes are often difficult to define mathematically. The process control engineer usually does not have a precise mathematical model of the process to work with, and the process model often changes with different operating conditions. Consequently, the process control industry has established standard signals for controllers, measuring means, and final control elements. The standard signals are: 3- 15-psi air pressure, and 4- 20 or 10- 50-milliamperes electric current. Although graphical or computer design methods may be used to obtain the approximate controller settings, final adjustments are almost always required to obtain the desired response of the actual system. Often, the control engineer's experience is sufficient to specify the controller without completing a formal design, and the controller adjustments are determined at startup.

3.4 Servomechanisms

A servomechanism is a feedback control system in which the controlled variable is mechanical position or motion. Many servomechanisms are used to maintain an output position in close correspondence with an input reference signal and, hence, are follow-up systems. Some process control systems are servomechanisms, while others may contain a servomechanism. There is no theoretical difference between a servomechanism and a closed loop process control system: the same mathematical elements are used to describe each system, and the same methods of analysis apply to each. However, because servo control and process control were developed independently of one another each has a different design method and a different terminology.

Servomechanisms usually involve relatively fast processes—the time constants may be considerably less than one second. Usually, the system is well-defined mathematically, and the controller may be designed to meet the system specifications by utilizing equipment which is most suitable for the application. Examples of servomechanisms are illustrated in Figures 3.8–3.12.

The hydraulic position control system in Figure 3.8 uses a lever to provide a mechanical feedback signal. The hydraulic valve is shown in the neutral position. If the setpoint lever is moved to the right, the valve spool also moves to the right, thus connecting the left side of the hy-

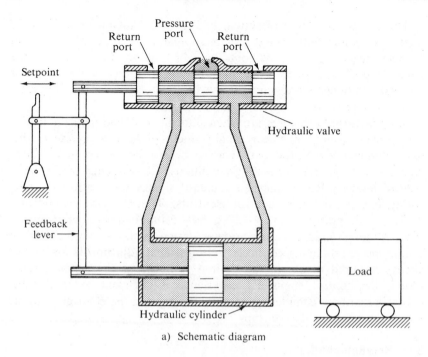

Setpoint

Return
port

Pressure
port

Return
port

Hydraulic valve

Feedback
lever

Load

Hydraulic cylinder

a) Schematic diagram

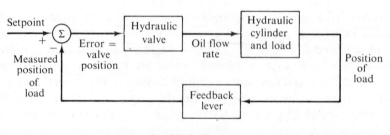

Setpoint

Σ

Error =
valve
position

Hydraulic
valve

Oil flow
rate

Hydraulic
cylinder
and load

Position
of
load

Measured
position
of
load

Feedback
lever

b) Block diagram

FIGURE 3.8

A HYDRAULIC POSITION CONTROL SYSTEM

draulic cylinder to the pressure port, and the right side to the return port. Hydraulic fluid will flow into the left side of the cylinder, moving the piston and load to the right until the valve is back in the neutral position. If the setpoint handle were moved to the left, the load would be moved to the left. For each position of the setpoint handle, there is a corresponding position of the load that will place the valve in its neutral position. Not much force is required to move the lever and control the great force exerted by the hydraulic cylinder.

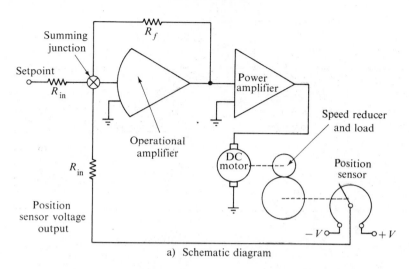

a) Schematic diagram

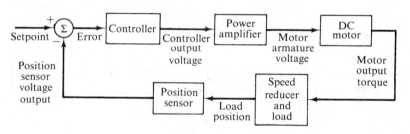

b) Block diagram

FIGURE 3.9

A DC MOTOR POSITION CONTROL SYSTEM

A dc motor position control system is illustrated in Figure 3.9. The position sensor is a 10K-ohm potentiometer with no stops and a 20° dead zone. The position sensor voltage output goes from $-V$ to $+V$ as the load rotates from its $+170°$ position to its $-170°$ position. The operational amplifier and the three resistors form the error detector and controller. The power amplifier is the final control element. The output of the controller is $-R_f/R_{in}$ times the sum of the setpoint voltage and the position sensor voltage.

$$\text{controller output} = (R_f/R_{in})\left(\text{setpoint volts} + \frac{\text{position}}{\text{sensor volts}}\right)$$

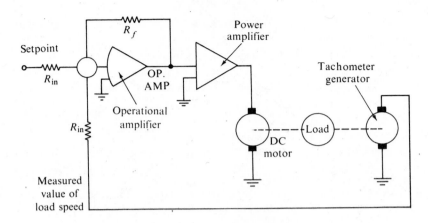

a) Schematic diagram

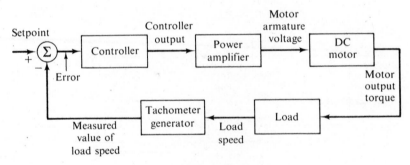

b) Block diagram

FIGURE 3.10
A DC MOTOR SPEED CONTROL SYSTEM

The power amplifier increases the voltage and the power of the controller output. The motor is a permanent-magnet field, armature-controlled, fractional-horsepower dc motor. The motor speed is proportional to the voltage applied to the armature by the power amplifier. When the sum of the setpoint and position sensor voltages is zero, the controller output voltage is zero and the motor speed is zero. When the above sum is not zero, the controller output voltage is not zero and the motor speed is not zero. The motor will always rotate in the direction that will drive the above sum toward zero. Notice that the negative feedback in this example is accomplished by making the sign of the position sensor voltage opposite the sign of the setpoint voltage.

The dc motor speed control system in Figure 3.10 uses the same controller, power amplifier, and motor as the position control system just discussed. The position sensor is replaced by a speed sensor: the tachometer generator. The speed sensor produces a voltage proportional to its speed. The negative feedback is accomplished by connecting the speed sensor so that its voltage is opposite the sign of the setpoint voltage. In this example, a small error signal is required to provide the armature voltage necessary to maintain the desired motor speed.

A mechanical speed control system or governor is shown in Figure 3.11. This device is used to control the speed of gasoline engines, gas

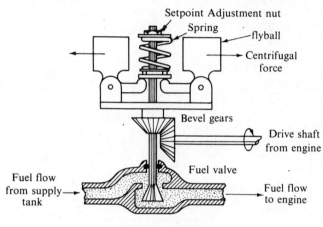

a) Schematic diagram

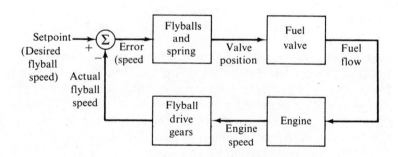

b) Block diagram

FIGURE 3.11
A MECHANICAL SPEED CONTROL SYSTEM

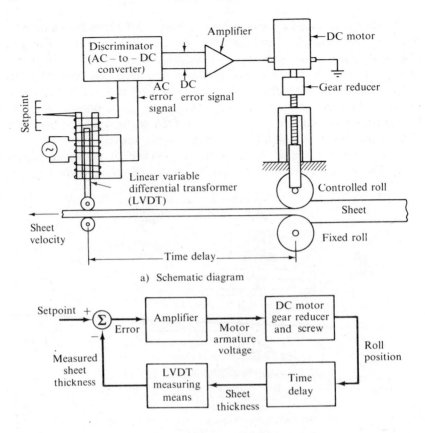

a) Schematic diagram

b) Block diagram

FIGURE 3.12
A SHEET THICKNESS CONTROL SYSTEM

turbines, and steam engines. A drive shaft from the engine rotates the flyball and spring assembly at a speed proportional to the engine speed. The rotation of the flyballs produces a centrifugal force that compresses the spring and raises the valve stem. Thus, the valve stem position is proportional to the engine speed. As the engine speed increases, the valve decreases the fuel flow. As the engine speed decreases, the valve increases the fuel flow. The engine will settle out at a speed that results in just enough fuel flow to balance the load on the engine.

The sheet thickness control system in Figure 3.12 uses a linear variable differential transformer as the thickness measuring means and error detector. The ac error signal from the LVDT is converted to a dc error signal by the *discriminator* (also called a *phase sensitive detector*). A discriminator is an ac-to-dc converter whose output magnitude varies linearly with the input magnitude, and whose output sign depends on the relative phase of the input. The dc error signal is amplified, and applied to the armature of the dc motor. The motor drives the upper roll which determines the sheet thickness. After a change in the upper roll position, the sheet must travel to the sensor before the change in thickness is measured. This time delay is represented by a block marked *time delay* in the block diagram.

3.5 Sequential Control

A sequential control system is one that performs a prescribed set of sequential operations. The automatic washing machine is a familiar example of sequential control: the control system performs the sequential operations of filling the tub, washing the clothes, draining the tub, rinsing the clothes, and spin drying the clothes. The automatic machining of castings for automobiles is another example of sequential control: a sequence of machining operations is performed on each casting to produce the finished part. A few basic concepts used in sequential control are explained in the examples illustrated in Figures 3.15 through 3.18.

Electromechanical components are frequently used in sequential control. Figures 3.13 and 3.15 contain schematic symbols of some of the most common sequential control components. The opening and closing of electric circuits occur in almost all sequential control systems. The *momentary contact pushbutton switches* are shown in Figure 3.13–a. A normally open switch will close the circuit path between the two terminals when the button is pushed and will open the circuit path when the button is released. A normally closed switch will open the circuit path when the button is pushed and will close the circuit path when the button is released.

The *limit switches* shown in Figure 3.13–b are actuated by a cam that operates the actuating arm. A normally open limit switch will close the circuit path between the two terminals when it is actuated and will open the circuit path when it is deactuated. A normally closed limit switch will open the circuit path when it is actuated and will close the circuit path when it is deactuated.

A *control relay* is a set of switches that are actuated when electric current passes through an electromagnetic coil. Figure 3.13–c illustrates a control relay with two switches. The relay coil is represented schematically by the circle with the designation 3CR. The CR signifies a control relay, and the 3 is a number designation to distinguish between two or more control relays. The relay is actuated or energized by com-

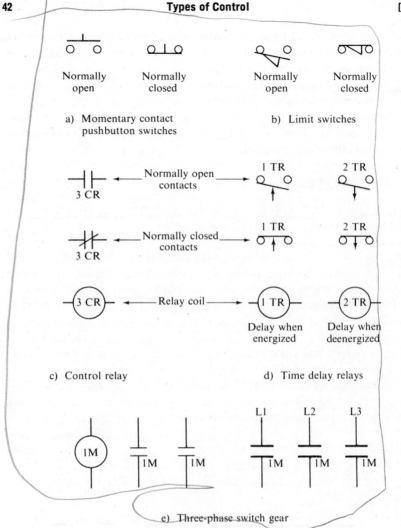

a) Momentary contact pushbutton switches

b) Limit switches

3 CR — Normally open contacts — 1 TR   2 TR

3 CR — Normally closed contacts — 1 TR   2 TR

3 CR — Relay coil — 1 TR   2 TR

Delay when energized   Delay when deenergized

c) Control relay

d) Time delay relays

1M   1M   1M   L1   L2   L3   1M   1M   1M

e) Three-phase switch gear

FIGURE 3.13
*SYMBOLS OF SWITCHES AND RELAYS
USED FOR SEQUENTIAL CONTROL*

pleting the circuit to the relay coil. The switches are represented by parallel lines with the 3CR designation. The designation is necessary because the relay contacts may occur anywhere in an electric circuit diagram. The designation identifies which relay actuates a particular set of contacts. Normally closed contacts are designated by a diagonal slash connecting the two contacts. A normally open contact will close the circuit path when the relay coil is energized and will open the circuit path when the relay is deenergized. A normally closed contact will open the circuit path when the relay coil is energized and will close the circuit path when the relay coil is deenergized.

Time delay relays are control relays that have provisions for a delayed switching action (see Figure 3.13–d). The delay in switching is usually adjustable, and it may take place when the coil is energized or when the coil is deenergized. The time delay relay is identified by a number followed by a TR. An arrow is used to identify the switching direction in which the time delay takes place. In relay 1TR, the delay occurs when the coil is energized. The contacts delay before they move up. There is no delay when relay 1TR is deenergized, and the contacts move down immediately. In relay 2TR, the delay occurs when the coil is deenergized. The contacts delay before they move down. There is no delay when relay 2TR is energized, and the contacts move up immediately.

In a time delay relay, the delay occurs in the direction indicated by the arrows.

Relays with heavy duty contacts are used to switch circuits that use large amounts of electric power. When the circuit load is an electric motor, the relay is called a *motor starter;* otherwise, it is called a *con-*

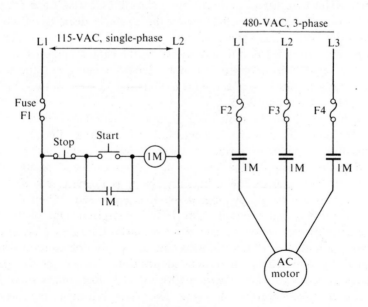

FIGURE 3.14
A CONTROL CIRCUIT FOR STARTING A LARGE AC MOTOR

tactor. An example of a relay for switching a three-phase system is shown in Figure 3.13–e. The three heavy contacts are used to switch the three lines supplying electric power to the load. The two light contacts are used in the control circuit.

A motor starter for starting and stopping a 480-VAC, three-phase motor clearly illustrates sequential control (Figure 3.14). Pushbutton switches in the 115-VAC control circuit are used to energize and de-energize the relay coil (1M). When the start button is pressed, coil 1M is energized, closing all four 1M contacts. The three large contacts connect the three 480-VAC lines to the motor. The small contact in parallel with the start switch is used to hold the circuit closed after the start button is released. When the stop button is pressed, the circuit breaks and all four 1M contacts open. The circuit remains deenergized after the stop button is released because the 1M holding contact is open.

Solenoid valves are used to control fluid flow in hydraulic or pneumatic systems. Symbols of common types of solenoid valves are illustrated in Figure 3.15. A solenoid valve is characterized by the number of ways that a line can be connected to it (two, three, or four). Each position is represented by a square. The flow paths are indicated by connecting lines or blocked lines (arrowheads indicate the direction of fluid flow). The valve is always shown in the deenergized position. The two, three, or four connecting lines (ways) are attached to the square which is in effect when the valve is deenergized. For a two-position valve, the square next to the spring is in effect when the valve is deenergized. For a three-way valve, the center square is in effect when the valve is deenergized. When a solenoid is energized, the square next to the solenoid is in effect. This is illustrated by the dotted symbols in Figures 3.15–a and 3.15–b.

A sequential control circuit for a pneumatic cylinder is shown in Figure 3.16. When the cylinder is connected to the exhaust line, the spring returns the piston to its retracted position. When the cylinder is connected to the air supply line, the piston is in its extended position. The solenoid valve exhausts the cylinder when it is deenergized, and supplies air pressure to the cylinder when it is energized. When the start button is pressed and released, relay 1CR is energized. One of the 1CR contacts holds the coil in the energized position; the other 1CR contact energizes solenoid *a,* which connects the air supply line to the cylinder. When the piston reaches its extended position, the end of the piston actuates limit switch 1LS. The actuation of 1LS deenergizes relay 1CR which, in turn, deenergizes solenoid *a.* The piston returns to the retracted position, ready for another cycle. In summary, each time the start button is pressed, the piston extends to the limit switch and then retracts

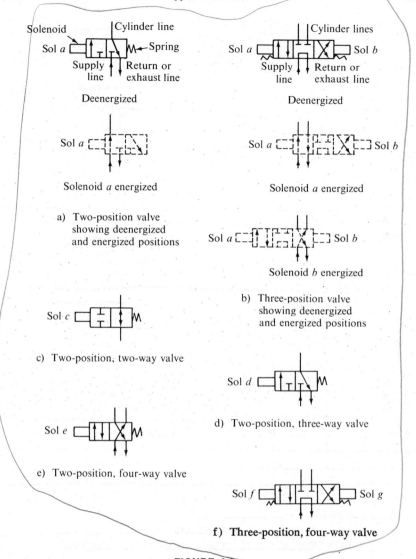

FIGURE 3.15

SYMBOLS OF SOLENOID VALVES USED FOR SEQUENTIAL CONTROL

to its starting position. The hydraulic equivalent of the above sequential system is shown in Figure 3.17.

A sequential control circuit for a hydraulic hoist is shown in Figure 3.18. This system uses a three-position solenoid valve which presents a special problem. Since it is not desirable to energize both solenoids at the same time, an *interlock* can be used to eliminate this possibility.

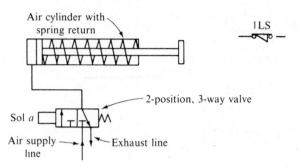

a) Schematic diagram

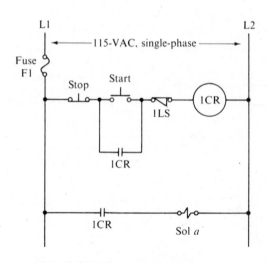

b) Electrical circuit diagram

FIGURE 3.16

A CONTROL CIRCUIT FOR A PNEUMATIC CYLINDER

The hydraulic hoist system is started and stopped by energizing relay 1M, which starts the pump motor and energizes the solenoid control circuit. When the up switch is pressed, relay 1CR is energized. Half of the interlock feature acts when the normally closed 1CR contact opens so that relay 2CR cannot be energized. The normally open 1CR contact closes and solenoid *a* is energized. This connects the hydraulic supply line to the bottom of the hydraulic cylinder, raising the platform. When the up button is released, relay 1CR is deenergized, and the solenoid

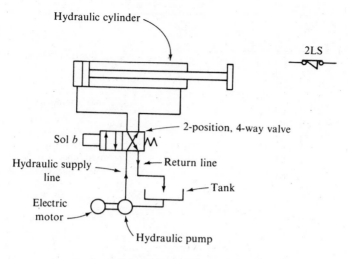

a) Schematic diagram

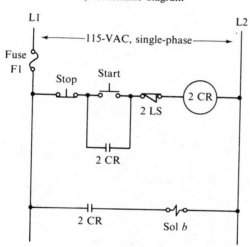

b) Electrical circuit diagram

FIGURE 3.17

A CONTROL CIRCUIT FOR A HYDRAULIC CYLINDER

valve goes to the neutral position. The cylinder is locked in its present
position as long as the solenoid is in neutral.

When the down switch is pressed, relay 2CR is energized. The normally closed 2CR contact opens so that relay 1CR cannot be energized.
This is the second half of the interlock feature. The normally open 2CR
contact closes and solenoid *b* is energized, connecting the hydraulic

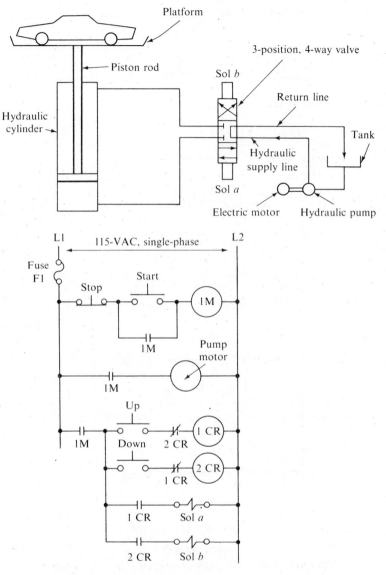

FIGURE 3.18

A CONTROL CIRCUIT FOR A HYDRAULIC HOIST

supply to the top of the cylinder and lowering the platform. When the down button is released, relay 2CR is deenergized, and the solenoid valve goes to the neutral position.

In summary: the platform goes up when the up button is pressed; down when the down button is pressed; and remains stationary when neither button is pressed. If the up button is pressed when the cylinder is at the top, a relief valve (not shown) provides a direct path from the pump outlet to the tank. The same is true if the down button is pressed when the cylinder is at the bottom. In addition, the system is interlocked so that both solenoids cannot be energized at the same time.

3.6 Numerical Control

Numerical control is a system that uses predetermined instructions to control a sequence of manufacturing operations. The instructions are usually coded in a symbolic program and stored on some type of storage medium, such as punched paper tape, magnetic tape, or punched cards. The term *control medium* is used to describe a storage medium which contains all the instructions required to accomplish a desired process or produce a desired part. The instructions consist primarily of numerical information—such as position, direction, velocity, cutting speed, etc.— hence the name numerical control. The manufacturing operations include: boring, drilling, grinding, milling, punching, routing, sawing, turning, winding (wire), flame cutting, knitting (garments), riveting, bending, welding, and wire processing. Although numerical control is not limited to the above operations, they do include most of the present applications.

The SLO-SYN system illustrated in Figure 3.19 uses open loop systems to control the x and y positions. The SLO-SYN motors rotate $1.8°$ for each pulse received from the control unit. If a 5-pitch lead screw is used, each step of the motor advances the table 0.001 inch. If a 10-pitch lead screw is used, each step of the motor advances the table 0.0005 inch.

Numerical control (N/C) has been referred to as flexible automation because of the relative ease of changing the control medium compared with changing cams, jigs, and templates. The same machine may be used to produce any number of different parts by using different control mediums. The numerical control process is most justified when a few parts are to be produced on a single setup or on a repetitive basis: it is seldom used to produce a single part continually on the same machine. N/C is ideal when a part is defined mathematically. With the increasing use of digital computers, more and more processes, designs, and products are being defined mathematically. Sometimes a drawing is unnecessary —a part that is completely defined mathematically can be described by equations instead.

A simplified explanation of the N/C process is illustrated in Figure 3.20. (a) The engineering drawing completely defines the desired part. (b) The part programmer uses the engineering drawing to determine the sequence of operations necessary to produce the part. He also speci-

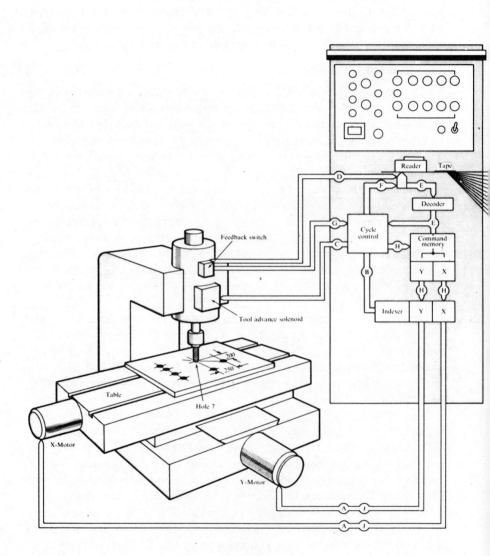

FIGURE 3.19

AN OPEN LOOP NUMERICAL CONTROL MACHINE TOOL
(Photo Courtesy Superior Electric Company)

The example below
illustrates how one cycle,
or block, of taped information
is processed and acted upon
by a tape controlled indexer.

A Assume the Table is in motion,
moving to position 6 as directed by the Indexer.

B When the Indexer completes both
the X and Y counts, it returns a signal to the Cycle Control.

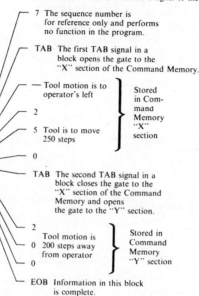

7 The sequence number is
for reference only and performs
no function in the program.

TAB The first TAB signal in a
block opens the gate to the
"X" section of the Command Memory.

— Tool motion is to   Stored
operator's left   in Command
2   Memory
"X"
5 Tool is to move   section
250 steps

0

TAB The second TAB signal in a
block closes the gate to the
"X" section of the Command
Memory and opens
the gate to the "Y" section.

2
Tool motion is   Stored in
0 200 steps away   Command
from operator   Memory
"Y" section
0

EOB Information in this block
is complete.

C The Cycle Control then energizes the
Tool Advance Solenoid.

D When the tool advances, the Feedback Switch closes, telling the Tape
Reader to read the next block of information.

E Tape information goes to the Decoder. In this case:

F EOB signal is relayed to Cycle Control, which stops Tape Reader before
reading next character.

G The Feedback Switch opens when
the tool is fully retracted, telling Cycle Control cycle is finished.

H Cycle Control, having received both
"EOB" signal from Decoder and
completion signal from the Feedback Switch,
tells the Command Memory to transfer to the Indexer
the motor stepping information
stored in Step E.

J Indexer starts the motors simultane-
ously, each stopping when it has
completed its required motion to
reach position 7. In this case, the
"Y" motor will stop first, having
moved 200 steps. The "X" motor
will continue until it has moved 250 steps.

K The cycle is complete, return to step B.

fies what tools will be used and how they will be used (cutting speeds and feed rates). (c) The part programmer records his decisions in the form of a symbolic program. APT (Automatically Programmed Tools) is one symbolic programming system used for this purpose. (d) The computer processes the symbolic program to produce the control medium (the tape used to control the manufacturing machines). The computer performs many of the calculations required to define each operation of the machine. (e) The machine operator prepares the

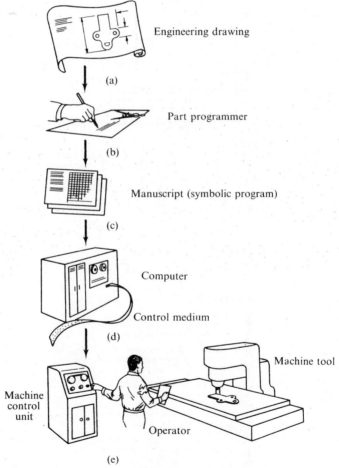

Engineering drawing

(a)

Part programmer

(b)

Manuscript (symbolic program)

(c)

Computer

Control medium

(d)

Machine tool

Machine control unit

Operator

(e)

FIGURE 3.20 [2]

A SIMPLIFIED EXPLANATION OF THE N/C PROCESS

[2] Olesten, *Numerical Control* (New York: John Wiley & Sons, Inc., 1970), pp. 3.

machine for operation, loads the control medium and starts the operation. The numerical control system follows the instructions on the control medium to produce the finished part. The operator monitors the operation and checks for possible malfunctions.

Numerical control systems use a rectangular cartesian coordinate system as the position reference. The N/C system controls the x, y, and z positions of the machine tool, according to the instructions in the control medium. Each axis has its own position control system, which may be open loop or closed loop (closed loop position control systems are called servomechanisms).

Figure 3.21 illustrates a numerical control machine tool that uses closed loop systems to control x, y, and z positions. The following steps are involved in changing the x-axis position.

Unit	Action Performed
1	The control unit reads an instruction on the control medium that specifies a $+0.004$-inch change in the x position.
2	The control unit sends a pulse to the machine actuator.
3	The machine actuator rotates the lead screw and advances the x-axis position $+0.001$ inch for each pulse received.

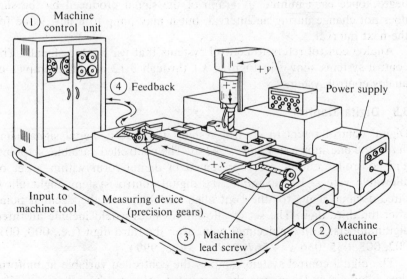

FIGURE 3.21 [3]

A CLOSED LOOP NUMERICAL CONTROL MACHINE TOOL

[3] *Ibid.*, pp. 12.

4   The measuring device measures the $+0.001$-inch change in the x-axis position and sends the information to the control unit.

1   The control unit compares the desired 0.004-inch motion with the actual 0.001-inch motion and sends another pulse.

—   The above sequence of operations is repeated until the measured motion agrees with the desired motion. The control unit then reads the next instruction on the control medium.

3.7 Analog Control

The signals in a control system are divided into two general categories: analog and digital. Graphs of an analog and a digital signal are shown in Figure 3.22.

An analog signal varies in a continuous manner and may take on any value between its limits. An example of an analog signal is a continuous recording of the outside air temperature. The recording is a continuous line (a characteristic of all analog signals). A digital signal varies in a discrete manner and may take only certain discrete values between its limits. An example of a digital signal is an outdoor sign that displays the outside air temperature to the nearest degree once each minute. A graph of the signal produced by the sign does not change during an interval, but it may jump to a new value for the next interval.

Analog control refers to control systems that use analog signals. The control systems shown in Figures 3.1 through 3.12 are all examples of analog control systems.

3.8 Digital Control

Digital control refers to control systems that use a digital signal similar to the one shown in Figure 3.22–b. The controlled variable, setpoint, and controller output are represented in digital form within a set of allowable values. For example, the digital control system might allow three decimal digits for the controlled variable, with the decimal point after the third digit. The set of allowable values would include all three digit numbers, with the decimal point after the third digit (i.e., 000, 001, 002, 003, 055, 056, . . . 267, 268, . . . 998, 999).

The digital control system samples the controlled variable at uniform time intervals and converts the measured value to the nearest smaller decimal number. The controlled variable is an analog signal which is converted to a digital signal, so the conversion is called an *analog-to-digital* conversion. As an example of analog-to-digital conversion, con-

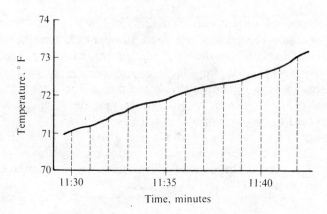

a) An analog signal of the outside air temperature

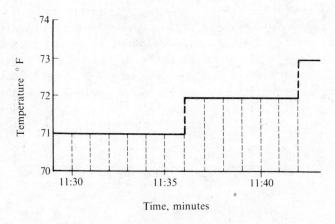

b) A Digital signal of the outside air temperature

FIGURE 3.22

*EXAMPLES OF ANALOG AND DIGITAL SIGNALS
OF THE SAME VARIABLE*

sider the sample taken at 11:31 in Figure 3.22. The analog value is
71.22° F, and it is converted into the digital signal of 71° F. The samples of the controlled variable are the only information about the
process used by the digital controller. Any changes that occur between
sampling intervals are not available to the controller. Obviously, the
sampling interval must be short enough to let the controller maintain
the error within acceptable limits. The output of the controller is also a
digital signal, which is changed only at uniform time intervals.

A digital control system uses a digital computer as the controller. All sizes of digital computers are used for control, but the type of application varies with the size and cost of the computer. Large expensive computers are used to handle a large number of control loops in a time-sharing arrangement to divide the cost of the computer among a number of control systems. In contrast, an inexpensive microcomputer may be devoted to the control of a single control loop.

An example of a time-sharing direct digital control system is shown in Figure 3.23. The process may be an entire manufacturing plant or a portion of a plant. The controlled variables are measured outputs

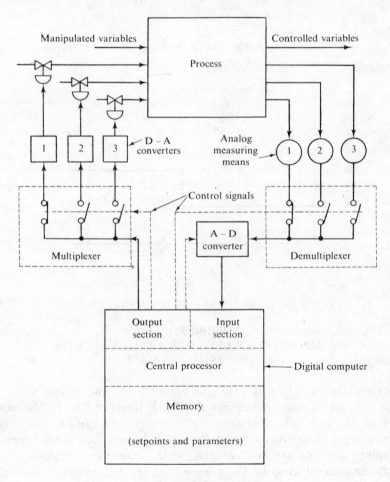

FIGURE 3.23

A DIRECT DIGITAL CONTROL SYSTEM

of the process that relate to the quality or production rate of the product. The manipulated variables are the inputs to the process that affect one or more of the controlled variables. The disturbances are unmeasured inputs to the process that affect the quality or production rate of the product.

In the operation of time sharing, the computer samples the controlled variables one at a time in a repetitive cycle. The central processor compares each sample of a controlled variable with its corresponding setpoint and determines the necessary corrective action in the form of an output signal. The output changes occur in a repetitive cycle that corresponds with the input sampling cycle.

The demultiplexer is a switching arrangement that selects each controlled variable for sampling. The A-D converter converts the analog signal into a digital signal which is received by the computer input section. The multiplexer is another switching arrangement that directs the computer output to the correct output device. The D-A converters convert the digital output signal into an analog signal and maintain the analog signal until it is changed by a subsequent output conversion.

An example of the application of a microprocessor in a sequential control system is shown in Figure 3.24. This system monitors the voltage and current supplied to critical (or emergency) loads, such as a life support unit in a hospital. If a power failure occurs, the control system starts the emergency generator and switches the generator output to the critical load. This assures a continuous source of power in situations where loss of power would have serious consequences.

The microprocessor consists of six units: the power supply, the central processor (cpu) and memory unit, the display and digital interface, the front panel, the analog interface, and the serial interface. The power supply provides the dc voltage necessary to operate the microprocessor. The ac line is the normal source of power, but an input from the engine cranking battery is provided to assure uninterrupted power to the microprocessor.

The cpu controls the interpretation and execution of instructions. It consists of an arithmetic section, working storage registers, logic circuits, timing and control circuits, decoders, and parallel buses for transfer of binary data and instructions. Memory includes random access memory (RAM) and read only memory (ROM). The RAM memory has both read and write capability and is used primarily as scratch pad memory for storage and retrieval of temporary data and instructions. The ROM is used to store the main program and permanent data.

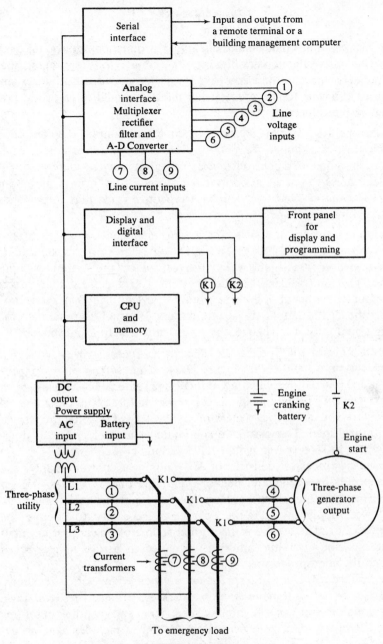

FIGURE 3.24 [4]

A MICROPROCESSOR CONTROL SYSTEM FOR AN EMERGENCY GENERATOR

[4] Schematic courtesy of ONAN Corporation.

The display and digital interface contains the electronic circuits that match the cpu to the two control relays and the front panel. The human interface provided by the front panel is one of the principal advantages of digital control. Relay K1 controls the emergency load switch. The actual switching action is performed by a linear motor (not shown

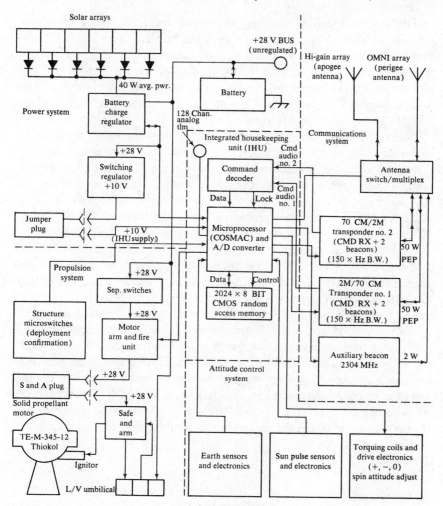

FIGURE 3.25 [5]

FUNCTIONAL BLOCK DIAGRAM, AMSAT-PHASE III

[5] Harris, Kearman, and King, Getting to Know Oscar from the Ground Up. *QST Series on Amateur Satellites*, pp. 35–36. American Radio Relay League, Inc., 1977.

in Figure 3.24), and the K1 relay operates the linear motor. The K2 relay is used to start and stop the generator engine.

The front panel contains a function switch, a digital display, and a keyboard. The function switch selects the signal that will appear on the digital display. The keyboard is used for manual input of data or instructions into memory.

The analog interface matches the cpu to the nine analog input signals. It includes a multiplexer, a rectifier, a filter, and an A-D converter. The multiplexer selects the input signals one at a time for input to the rectifier. The rectifier converts the ac analog signals into dc signals and the filter removes the ripple in the rectifier output. The A-D converter changes the dc analog signal into a digital signal (usually an 8- to 16-bit binary number).

The serial interface matches the cpu with an external terminal or a building management computer. The ability to communicate with a remote computer or terminal is another principal advantage of digital control. This feature provides the building manager with immediate information about the status of the various systems in the building.

Figure 3.25 shows the functional block diagram of the AMSAT-Phase III Amateur Satellite scheduled for launch in 1979. A CMOS microprocessor is utilized in a system known as the Integrated Housekeeping Unit (IHU). This system consists of a command decoder, analog-to-digital converter, microprocessor, and random access memory (RAM). The microprocessor is responsible for controlling virtually every function on board the spacecraft. Its duties include executing all telemetry and command requirements, monitoring the condition of the power and communications systems, providing the clocks needed for various spacecraft timing functions, and interacting with the attitude sensors and torquing magnet to adjust the orientation of the satellite in space. In addition, the IHU will make the final decision on whether all on-board systems are "go" for the kick-motor firing. If confirmed, the computer will send the command to fire, not a ground control station.

EXERCISES

3–1. Sketch the block diagram of an example of a regulator system and an example of a follow-up system. Use the block diagram to explain the difference between the two types of systems.

3–2.   Sketch a schematic diagram of a process controller. Include all essential knobs, scales, and indicators.

3–3.   Sketch the block diagram of a typical process controller. Indicate the error detector, the controller, and the following signals: SP, C_m, E, and V.

3–4.   Sketch the block diagram of an example of a process control system and explain its operation.

3–5.   Sketch the block diagram of an example of a servomechanism and explain its operation.

3–6.   Sketch the circuit diagram of a sequential control system with the following features.
     a.   Motor starter 1M is energized by pressing a momentary contact *start* switch.
     b.   Motor starter 1M can be deenergized only by pressing a momentary contact *stop* switch.

3–7.   Modify the circuit diagram of Exercise 6 to include the following additional features.
     a.   Limit switch 1L must be closed before the *start* switch can energize the motor starter 1M.
     b.   Either opening limit switch 1L or pressing the *stop* switch can deenergize motor starter 1M.

3–8.   Modify the circuit diagram of Exercise 6 to include the following additional features.
     a.   Limit switch 1L must close before the *start* switch can energize the motor starter 1M.
     b.   Opening limit switch 1L cannot deenergize the motor starter. The motor starter 1M can be deenergized only by pressing a momentary contact *stop* switch.

3–9.   Sketch the circuit diagram of a sequential control system with the following features.
     a.   A hydraulic cylinder is operated by a two-position, four-way valve as shown in Figure 3.17.
     b.   A 15-horsepower electric motor is controlled by a motor starter, as shown in Figure 3.15.
     c.   An interlock is provided so that the hydraulic cylinder cannot operate unless the 15-horsepower motor starter is energized.

3–10.   Numerical control is a system that:
     a.   Controls numbers in a manufacturing operation.
     b.   Uses predetermined numbers to control a sequence of manufacturing operations.

   c. Is most useful when a single part is produced continually on the same machine.

   d. Cannot be used to produce parts which are defined mathematically.

3–11. The control medium in an N/C system is:
   a. A fluid used to lubricate the controller.
   b. A written document prepared by the part programmer.
   c. The engineering drawing that completely defines the part.
   d. A storage medium which contains all the instructions to produce the part.

3–12. List the steps involved in a +0.003-inch change in the y-axis position of an N/C machine tool. A closed loop control system is used and each pulse advances the y-axis position 0.001 inches.

3–13. Analog control refers to control systems that use (continuous, discontinuous) signals.

3–14. A certain digital controller uses four binary digits for the controlled variable. A binary digit has only two possible values, 0 and 1. Thus a two-digit binary number has the following set of allowable values: 00,01,10,11. List the set of allowable values for the controlled variable when the four binary digits are used.

3–15. The digital signal illustrated in Figure 3.22 is obtained by eliminating the decimal part of the analog temperature signal. This is referred to as *truncating a signal*. Thus the signals 72, 72.3, 72.56, and 72.999 are all converted to 72 by the analog-to-digital conversion. An alternate method of conversion would be to round off to the nearest integer. In this method, 72 and 72.3 would be converted to 72, and 72.56 and 72.999 would be converted to 73. Redraw Figure 3.22-b using the nearest integer method of analog-to-digital conversion.

3–16. List one cycle of the sequence of operations performed by the DDC control system of Figure 3.23. Assume there are only three controlled variables and three control valves. The setpoint and control parameters for each controlled variable are stored in computer memory. The computer is programmed for a time-sharing operation, as described in Section 3.8.

3–17. List two principal advantages of a digital control system.

4

An Introduction to
Laplace Transforms

4.1 Introduction

The block diagram was introduced in Chapter One as a method of representing a control system. The essential feature of each block is the relationship between the input signal and the corresponding output signal. This input/output relationship is described mathematically by an equation that may include derivatives and integrals of a variable. The derivative is a mathematical expression of the rate of change of a variable. For example, the speed of an automobile is the derivative (or rate of change) of the distance traveled. The integral is a mathematical expression for the accumulation of an amount (or quantity) of a variable. For example, the integral of the speed of an automobile is the distance traveled. Figure 4.1 illustrates a typical control system component and its differential equation (an equation that contains a derivative). In simple terms, the equation states that the input signal (m) is equal to the output signal (h) plus the time constant (τ) multiplied by the rate of change of the output (dh/dt).

Although control systems are described by differential equations, most control system design and analysis does not use operational calculus. Instead, a transform is used to change the differential equation into an

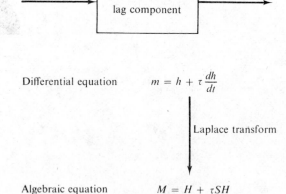

Differential equation $m = h + \tau \dfrac{dh}{dt}$

Laplace transform

Algebraic equation $M = H + \tau SH$

FIGURE 4.1

*THE BLOCK REPRESENTATION OF A TYPICAL
CONTROL SYSTEM COMPONENT*

algebraic equation. The transform is called the *Laplace transform.* In
Figure 4.1, the Laplace transformation of the differential equation
$m = h + \tau dh/dt$ produces the algebraic equation $M = H + \tau SH,$ where
S is a frequency parameter to be explained later. The algebraic equation
completely defines the size and timing changes between the input signal
(m) and the output signal (h). The frequency parameter (S) indicates
that the size and timing relationships depend on the frequency of the
input signal. This frequency dependency is a very important characteristic of control systems, and it will be discussed again in later chapters.

The transfer function of a component can be obtained by solving
its algebraic equation for the ratio of the output over the input. Although a complete discussion of the Laplace transform is beyond the
scope of this text, the steps necessary to obtain the transfer function
of a component from its differential equation are easy to master. Considering the widespread use of transfer functions in the control field,
some familiarity with the Laplace transform is a decided asset for a
technician.

4.2 Input-Output Relationships

The input-output equations of two typical components will be developed in this section. The two components (a liquid tank and an electrical RC circuit) are illustrated in Figures 4.2 and 4.3.

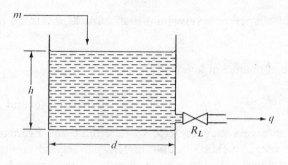

m = input flow rate, cubic meters/second

q = output flow rate, cubic meters/second

h = height of liquid in the tank, meters

d = diameter of the tank, meters

$A = \dfrac{\pi d^2}{4}$, area of the tank, square meters

R_L = flow resistance in the output line, $\dfrac{\text{newtons per square meter}}{\text{cubic meters per second}}$

FIGURE 4.2

A LIQUID TANK

Consider the liquid tank in Figure 4.2. The liquid level in the tank will not change as long as the inflow rate (m) and the outflow rate (q) are equal. If the inflow rate is greater than the outflow rate, the liquid level will rise. If the inflow rate is less than the outflow rate, the level will fall. During a certain time interval, (Δt), the amount of liquid in the tank will change by an amount (Δv) equal to the average difference between the inflow rate and the outflow rate multiplied by Δt.

$$\Delta v = (m - q)_{\text{avg}}\Delta t \text{ cubic meters} \qquad (4.1)$$

The change in the liquid level in the tank (Δh) is equal to the change in volume (Δv) divided by the cross-sectional area of the tank (A).

$$\Delta h = \frac{\Delta v}{A} = \frac{(m - q)_{\text{avg}}\Delta t}{A} \text{ meters} \qquad (4.2)$$

The average rate of change of the level in the tank is equal to the change in level Δh divided by the time interval Δt. For example, if the level changed 0.26 meters during a time interval of 100 seconds, then the average rate of change of level would be $0.26/100 = 0.0026$ meters per second.

$$\text{average rate of change of level} = \frac{\Delta h}{\Delta t}$$

$$= \frac{(m - q)_{\text{avg}}}{A} \text{ meters/second} \quad \textbf{(4.3)}$$

Different time intervals may be used to determine the average rate of change. The average of 0.0026 cubic meters per second could have resulted from a change of 0.026 meters during a time interval of 10 seconds or a change of 0.0026 meters during a 1-second interval. When the time interval diminishes to 0 seconds, we call it an instant of time. As the time interval Δt diminishes to an instant of time, the average rate of change becomes the instantaneous rate of change. In mathematics, the instantaneous rate of change of liquid level is called the derivative of level (h) with respect to time (t) and is designated by the symbol dh/dt. If the time interval (Δt) in Equation (4.3) diminishes to an instant, then the average rate of change of level becomes the instantaneous rate of change (dh/dt).

$$\text{As } t \rightarrow 0, \frac{\Delta h}{\Delta t} \rightarrow \frac{dh}{dt} = \frac{1}{A}(m - q) \text{ meters/second} \quad \textbf{(4.4)}$$

If the equation for flow out of the tank is linear (laminar flow), then the outflow rate (q) is given by the following equation (see Chapter 11).

$$q = \frac{\rho g h}{R_L} \text{ cubic meters/second} \quad \textbf{(4.5)}$$

where:

    $q =$ liquid flow rate, cubic meters/second
    $\rho =$ liquid density, kilograms/cubic meter
    $g =$ gravitational acceleration, meters/second^2
    $h =$ liquid level, meters
    $R_L =$ the laminar flow resistance, newton-second/meter^5

Substituting Equation (4.5) into Equation (4.4),

$$\frac{dh}{dt} = \frac{1}{A}\left(m - \left[\frac{\rho g}{R_L}\right]h\right)$$

or

$$R_L\left(\frac{A}{\rho g}\right)\frac{dh}{dt} = \left[\frac{R_L}{\rho g}\right]m - h$$

The term $(A/\rho g)$ is the capacitance (C) of the liquid tank (see Chapter 11), and the entire term $(R_L A)/(\rho g) = R_L C$ is called the time constant (τ) of the liquid tank. The term $R_L/(\rho g)$ is the steady-state gain (G) of the system. Substituting τ and G into the preceding equation gives us the final form of the differential equation for the liquid tank.

$$\tau \frac{dh}{dt} + h = Gm \tag{4.6}$$

Applying the Laplace transform to Equation (4.6),

$$\tau SH + H = GM$$

and solving for the ratio of H/M gives us the following transfer function for the liquid tank.

$$\frac{H}{M} = \frac{G}{\tau S + 1} \tag{4.7}$$

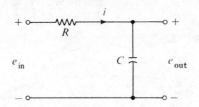

FIGURE 4.3

AN ELECTRICAL RC CIRCUIT

Consider the electrical circuit of Figure 4.3. If the output terminals are connected to a high impedance load, then the current through the resistor and the current through the capacitor are essentially equal. Let i represent the current that passes through the resistor and the capacitor. For the resistor, the current (i) is equal to the voltage difference $(e_{in} - e_{out})$ divided by the resistance (R).

$$i = (e_{in} - e_{out})/R \tag{4.8}$$

For the capacitor, the current (i) is equal to the capacitance (C) times the instantaneous rate of change of the capacitor voltage.

$$i = C \frac{de_{out}}{dt} \tag{4.9}$$

Equating the right-hand side of Equations (4.8) and (4.9) and setting $RC = \tau$ results in the following differential equation for the electrical circuit.

$$C \frac{de_{\text{out}}}{dt} = \frac{e_{\text{in}} - e_{\text{out}}}{R}$$

$$\tau \frac{de_{\text{out}}}{dt} + e_{\text{out}} = e_{\text{in}} \tag{4.10}$$

The Laplace transformation results in the following transfer function for the electrical circuit.

$$\frac{E_{\text{out}}}{E_{\text{in}}} = \frac{1}{\tau S + 1} \tag{4.11}$$

4.3 Logarithms—A Transformation

Logarithms are a familiar example of how a transformation makes problem solving easier. To make the study of control systems easier, and to clarify the Laplace transformation and its use, let us briefly examine the nature of the logarithmic transformation (see Figure 4.4).

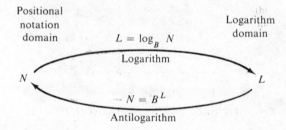

Positional notation domain

Logarithm domain

$$L = \log_B N$$

Logarithm

N L

$$N = B^L$$

Antilogarithm

Operations

Multiplication – – – – – – – – – Addition
Division – – – – – – – – – – – – – Subtraction
Exponentiation – – – – – – – – – Multiplication

Base e, $B = 2.71828\ldots$

Base 10, $B = 10$

Base 2, $B = 2$

FIGURE 4.4

THE LOGARITHMIC TRANSFORMATION

A number is usually represented by one or more decimal digits arranged in what is called a *positional notation*: the position of the digit determines whether the digit represents the number of ones, tens, hundreds, thousands, etc. For example: in the number 682, the digit 2 represents two times one, or two; the digit 8 represents eight times ten, or eighty; and the 6 represents six times one hundred, or six hundred.

We all know how to multiply and divide numbers in positional notation. We also know that it is usually easier to add and substract numbers than it is to multiply and divide them. The logarithmic transformation allows us to replace the operations of multiplication and division with the simpler operations of addition and subtraction. Most of us can square or cube a number, but what about raising a number to the 2.475 power? The logarithmic transformation allows us to replace the operation of exponentiation (raising to a power) with the simpler operation of multiplication.

If N, B, and L are three numbers such that
$$N = B^L$$

then the exponent (L) is the logarithm of the number (N) to the base (B). In other words, the logarithm of a number (N) to a given base (B) is the exponent (L) to which the base must be raised to produce the number (N). The logarithm is usually written as follows

$$L = \log_B N$$

We may think of the logarithm and antilogarithm as transformations between two regions or domains (see Figure 4.4). One domain contains all numbers (N) represented by positional notation using decimal digits. The other domain contains all the logarithms (L) of the numbers (N) in the positional notation domain. The logarithmic transformation transforms a number (N) into its corresponding logarithm (L), and is defined by the following relationship:

Logarithmic Transformation: $L = \log_B N$

The antilogarithmic transformation transforms a logarithm (L) into its corresponding number (N). It is defined by the following relationship:

Antilogarithmic Transformation: $N = B^L$

The three laws of exponents are the bases of the operations of multiplication, division, and exponentiation in the logarithmic domain. Consider two numbers, N_1 and N_2, with logarithms L_1 and L_2.

$$N_1 = B^{L_1} \qquad N_2 = B^{L_2}$$

Multiplication Law: $\quad N_1 \cdot N_2 = (B^{L_1}) \cdot (B^{L_2}) = B^{L_1 + L_2}$

Division Law: $N_1/N_2 = (B^{L_1})/(B^{L_2}) = B^{L_1 - L_2}$

Exponentiation Law: $N^x = (B^L)^x = B^{xL}$

OPERATION IN THE POSITIONAL NOTATION DOMAIN	CORRESPONDING OPERATION IN THE LOGARITHMIC DOMAIN
Multiplication, $N_1 \cdot N_2$	Addition, $L_1 + L_2$
Division, N_1/N_2	Subtraction, $L_1 - L_2$
Exponentiation, N^x	Multiplication, $x \cdot L$

Example 4.1   a. Compute $(24.3)(18.2)$ using logarithms to the base 10.   b. Compute $12.1^{2.475}$ using logarithms to the base 10.

Solution:

a. Complete the following four steps to find $(24.3)(18.2)$.

1   Transform 24.3 into the logarithmic domain.
    Transform 18.2 into the logarithmic domain.

2   Add the logarithms of the two numbers in the logarithmic domain to obtain the sum of 2.6457.

3   Transform 2.6457 into the positional domain to obtain 442.3 which is the product of 24.3×18.2.

4   Note that the answer could be obtained in the positional domain by the more difficult operation of multiplication.

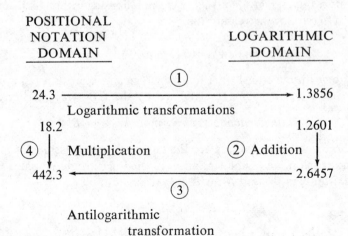

POSITIONAL NOTATION DOMAIN   LOGARITHMIC DOMAIN

24.3 ⟶ 1.3856
Logarithmic transformations
18.2   1.2601
④ Multiplication   ② Addition
442.3 ⟵ 2.6457
③
Antilogarithmic transformation

b. Complete the following four steps to find $12.1^{2.475}$.

1 Transform 12.1 into the logarithmic domain.

2 Multiply the logarithm of 12.1 by 2.475 to obtain the product 2.6799.

3 Transform 2.6799 into the positional domain to obtain 478.6; which is 12.1 raised to the 2.475 power.

4 Note that the answer could be obtained in the positional domain by the difficult operation of exponentiation. However, exponentiation by nonintegers is not done by hand calculation and therefore the logarithmic transformation is always used.

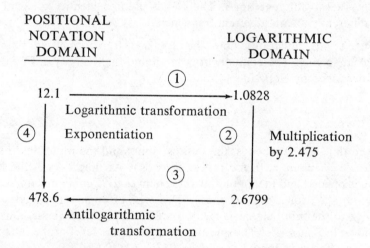

4.4 Laplace Transforms

The differential equations of typical control system components are developed in Section 4.2. The logarithm is used in Section 4.3 to illustrate how a transformation simplifies problem solving. The Laplace transform is introduced in this section to simplify the study of control systems.

The Laplace transform allows us to transform a differential equation into an algebraic equation—and since the latter equation is easier to solve, this is a step in the right direction. The differential equation of a component tells us the size of the output signal at each instant of *time* for any specified input signal. The input signal is determined by its size at each instant of *time*. Differential equations are said to be in the *time domain* because the signals are defined by their size at different times.

The algebraic equation tells us the changes in size and timing between the input signal and the output signal at each *frequency*. The input signal is specified by its frequency, size, and timing. The algebraic equations are said to be in the *frequency domain* because the size and timing changes are defined at different frequencies. The Laplace transform is illustrated in Figure 4.5.

Lower-case letters are used to represent signals in the time domain, and upper-case letters are used for signals in the frequency domain. In Figure 4.5 for example, the input signal is represented by $m(t)$ in the time domain and by $M(S)$ in the frequency domain. Similarly, the output signal is represented by $h(t)$ in the time domain and by $H(S)$ in the frequency domain. The (t) is used to identify a signal that has a different size at different times. The (S) is used to identify a signal that has different values at different frequencies.

Figure 4.6 illustrates how the Laplace transformation simplifies problem solving. The component is the liquid tank shown in Figure 4.2 and described by Equation (4.6).

$$\tau \frac{dh}{dt} + h = Gm \qquad \qquad \textbf{(4.6)}$$

Before time $t = 0$ seconds, the tank is empty and the input flow rate is zero (i.e., $h = m = 0$, for $t \leq 0$ seconds). At time $t = 0$, the input valve is opened and the input flow rate changes to K cubic meters/second (i.e., $m = K$, for $t > 0$ seconds). This type of change is called a *step change* in the input signal (m). A graph of input (m) versus time (t) is shown in Figure 4.6. The question is: what is the level of the tank after the step change in input? (i.e., $h = ?$, for $t > 0$).

Solution:

Transform the input (m) and Equation (4.6) into the frequency domain. A table of Laplace transform pairs may be used to make this transformation (see Table 1 in Appendix B).

TIME DOMAIN	FREQUENCY DOMAIN
$m = K$	$M = K/S$
$\tau \dfrac{dh}{dt}$	τSH
h	H
Gm	GM
$\tau \dfrac{dh}{dt} + h = Gm$	$\tau SH + H = GM$

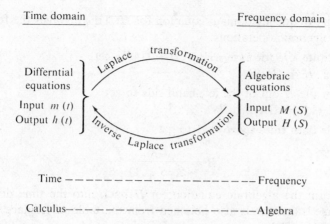

FIGURE 4.5

THE LAPLACE TRANSFORM

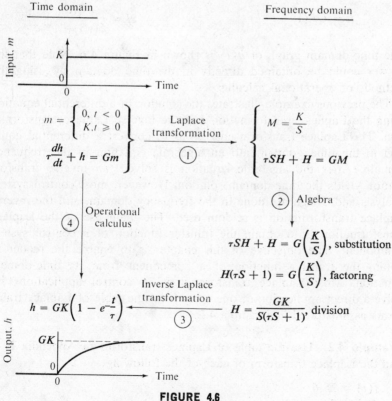

FIGURE 4.6

AN APPLICATION OF THE LAPLACE TRANSFORMATION

Solve the frequency domain equation for H. This involves the following three algebraic operations.

a. Substitute K/S for M in $\tau SH + H = GM$ to get
$\tau SH + H = GK/S$.

b. Factor the H out of the left-hand side to get
$H(\tau S + 1) = GK/S$.

c. Divide both sides by $\tau S + 1$ to get

$$H = \frac{GK}{S(\tau S + 1)}$$

Transform the algebraic equation for H back into the time domain. The inverse transform is also obtained from the table of Laplace transform pairs (see Table 1 in Appendix B).

TIME DOMAIN

$$h = GK(1 - e^{-\frac{t}{\tau}})$$

FREQUENCY DOMAIN

$$H = \frac{GK}{S(\tau S + 1)}$$

The time-domain graph of $h(t)$ is shown in Figure 4.6. Note that the answer could be obtained directly in the time domain by using the methods of operational calculus.

The previous example illustrates the solution of a differential equation using the Laplace transformation and the inverse Laplace transformation. The Laplace transform enables us to convert a differential equation in the time domain into an algebraic equation in the frequency domain. After the algebraic equation is solved, an inverse transformation yields the time-domain solution. However, most control system analysis and design are done in the frequency domain, and the inverse Laplace transformation is seldom used. The main use of the Laplace transformation is to obtain the transfer function of a control system component. The objective of this chapter is to enable the reader to obtain the transfer function of a component from its time-domain equation. Most Laplace transformations in control applications involve a direct application of one entry from the table of Laplace transforms as illustrated in Example 4.2.

Example 4.2   Use the table of Laplace transforms in Appendix B to find the Laplace transform of each of the following:

a. $f(t) = 25.6$
b. $f(t) = 5.6 \, dx/dt$
c. $f(t) = 12.2 \, d^2x/dt^2$

d. $f(t) = 7.1 \int x \, dt$
e. $f(t) = 18t^4$
f. $f(t) = 4.2 \sin 100t$

Solution:

a. From Entry No. 2, $F(S) = 25.6/S$.
b. From Entry No. 5, $F(S) = 5.6SX(S)$.
c. From Entry No. 6, $F(S) = 12.2S^2X(S)$.
d. From Entry No. 8, $F(S) = 7.1X(S)/S$.
e. From Entry No. 3, $F(S) = (18)(4!)/S^5 = 432/S^5$.
f. From Entry No. 12, $F(S) = (4.2)(100)/(S^2 + 100^2)$.

Some Laplace transformations require the use of more than one entry from the table for the solution. These problems require considerably more insight in the use of functional notation than the previous examples. Fortunately, most transfer functions can be obtained by the simple methods illustrated in Example 4.2. The following examples are included for those readers interested in the more difficult transformations.

Consider the following two functions:

$$g(t) = t^3 \text{ and } f(t) = 32(t - a)^3$$

According to the rules of functional notation, $g(t - a) = (t - a)^3$ and

$$f(t) = 32(t - a)^3 = 32g(t - a)$$

By Entry No. 3, the Laplace transform of $g(t)$ is

$$G(S) = 3!/S^4 = 6/S^4$$

By Entry No. 4, the Laplace transform of $f(t)$ is

$$F(S) = 32e^{-aS}G(S) = 32e^{-aS}(6/S^4) = 192e^{-aS}/S^4$$

Note that both Entry No. 3 and Entry No. 4 were used to obtain the final expression for $F(S)$. Example 4.3 illustrates a procedure for finding the Laplace transform when two entries are involved.

Example 4.3 Find the Laplace transform of each of the following:

a. $f(t) = 9.2e^{-2(t-4)}$
b. $f(t) = 16.1 \sin 100(t - 0.3)$

Solution:

a. Let $g(t) = e^{-2t}$ and $g(t - 4) = e^{-2(t-4)}$.
    From Entry No. 9, $G(S) = 1/(S + 2)$.
    By functional notation, $f(t) = 9.2g(t - 4)$.
    From Entry No. 4, $F(S) = 9.2e^{-4S}G(S) = 9.2e^{-4S}/(S + 2)$.

b.   Let $g(t) = \sin 100t$ and $g(t - 0.3) = \sin 100(t - 0.3)$.
    From Entry No. 12, $G(S) = 100/(S^2 + 100^2)$.
    By functional notation, $f(t) = 16.1g(t - 0.3)$.
    From Entry No. 4, $F(S) = 16.1e^{-0.3S}G(S)$
    $$= 1610e^{-0.3S}/(S^2 + 10{,}000).$$

4.5 Transfer Functions

The major use of the Laplace transformation in the study of control
systems is to obtain the transfer function of a component. The transfer
function is obtained by solving the frequency-domain algebraic equation for the ratio of the output signal over the input signal. As an example, consider the liquid system described by Equation (4.6).

$$\tau\frac{dh}{dt} + h = Gm \tag{4.6}$$

The frequency-domain algebraic equation is obtained by applying the
Laplace transformation to each term in Equation (4.6) and solving
for the ratio H/M.

$$\tau SH + H = GM$$
$$H(\tau S + 1) = GM$$
$$H = \left[\frac{G}{\tau S + 1}\right]M$$
$$\frac{H}{M} = \left[\frac{G}{\tau S + 1}\right] = \text{the transfer function} \tag{4.7}$$

The transfer function is defined by Equation (4.7). Notice that the
output (H) can be obtained by multiplying the input (M) by the transfer function (H/M).

> *The transfer function of a component is the ratio of the fre-
> quency domain output* (H) *over the input* (M).
> *The frequency domain output* (H) *of a component is equal
> to the product of the input* (M) *times the transfer function.*

If a sinusoidal signal is applied to the input of a linear component,
the output signal is a sinusoidal signal with the same frequency as the
input signal. However, the output signal usually has a different amplitude and phase from the input signal as illustrated in Figure 4.7. The
transfer function of the component gives us the amount of change in the

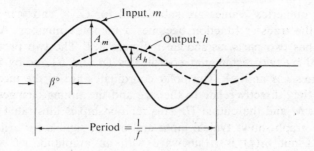

For $S = j\omega = j2\pi f$, the transfer function has a

magnitude (M) and an angle $\theta°$ such that

$M = \dfrac{A_h}{A_m}$ = amplitude ratio (change in size)

$\theta° = \beta° =$ phase difference (change in timing)

FIGURE 4.7

*THE RELATIONSHIP BETWEEN THE TRANSFER FUNCTION
AND THE CHANGE IN SIZE AND TIMING*

amplitude and phase between the sinusoidal input and the sinusoidal output signals.

The transfer function of the liquid system given by Equation (4.7) will be used to illustrate how the transfer function gives us the amount of the amplitude and phase changes. Equation (4.7) contains three parameters: G, τ, and S. The first two parameters, G and τ, depend on the resistance (R_L), capacitance (C), and liquid density (ρ) of the liquid system. For a given system, G and τ are constants. The third parameter (S) is a frequency parameter whose value depends on the frequency of the input signal (M). For sinusoidal inputs with a frequency of f Hertz, the frequency parameter is equal to $j2\pi f$ or $j\omega$ where $\omega = 2\pi f$. The symbol j represents the square root of -1.

$$S = j\omega$$
$$\omega = 2\pi f \text{ radians/second}$$
$$j = \sqrt{-1}$$

Substituting $j\omega$ for S in Equation (4.7), we have

$$\frac{H}{M} = \left[\frac{G}{j\omega\tau + 1}\right] \tag{4.12}$$

When numerical values are substituted for G, τ, and ω in Equation (4.12), the transfer function becomes a complex number. A complex number has two parts, as shown in Figure 4.8. The two parts may be expressed by two rectangular coordinates (a and b), or by two polar coordinates (M and β). The polar coordinates are of the most interest because they directly relate to the size and the timing changes between the input M and the output H. This relationship is illustrated in Figure 4.7. The graph shows typical input and output signals for a liquid tank system. Input $m(t)$ is a sinewave with an amplitude of A_m and a frequency of f. The output $h(t)$ is also a sinewave with a frequency of f and an amplitude of A_h. The output (h) lags the input (m) by β degrees. The magnitude of the transfer function when $S = j\omega$ is equal to the ratio of the output amplitude (A_h) over the input amplitude (A_m). That is

$$M = A_h/A_m$$

The angle of the transfer function (θ) is equal to the phase difference (β) between the input and the output signals.

$$\theta = \beta$$

The above statements are true for the transfer function of any linear system. The changes in size and timing at a given frequency are referred to as the *frequency response of the component*. Frequency response is a very important characteristic of control system components—a subject to which Chapter Thirteen is devoted.

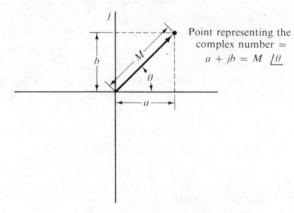

Point representing the complex number =
$a + jb = M \ \underline{/\theta}$

FIGURE 4.8

A GRAPH OF A COMPLEX NUMBER

The transfer function of a component defines the changes in size and timing between the input and the output for each input frequency (f).

Example 4.4 A liquid tank has a transfer function described by Equation (4.7). The input flow rate (m) varies sinusoidally about an average value with an amplitude of 0.0002 cubic meters/second and a frequency (f) of 0.0001 Hertz. The time constant (τ) is 1,590 seconds and the gain (G) is 2,000 seconds. Determine the amplitude and phase of the output (h).

Solution:

The transfer function gives us the amplitude ratio and the phase difference, so the first step is to use Equation (4.12) to determine the magnitude and angle of the transfer function.

$$\frac{H}{M} = \frac{G}{j\omega\tau + 1} = \frac{G}{1 + j2\pi f\tau}$$

$$= \frac{2,000}{1 + j2\pi(0.0001)(1,590)} = \frac{2,000}{1 + j1} = \frac{2,000}{1.414\underline{|45°}}$$

$$= 1,414\underline{|-45°}$$

The output amplitude is equal to the input amplitude times the magnitude of the transfer function.

$$\text{output amplitude } h = 1,414 \times 0.0002 = 0.283 \text{ meters}$$

The phase difference is −45°, so the output (h) lags the input (m) by 45°.

Example 4.5 A component is described by the following equation. Determine the transfer function.

$$m = 8\frac{dh}{dt} + 4h + 3\int h \, dt$$

Solution:

Determine the frequency-domain equation by a Laplace transformation.

$$M = 8SH + 4H + \frac{3}{S} H$$

$$SM = (8S^2 + 4S + 3)H$$

$$\text{Transfer function} = \frac{H}{M} = \frac{S}{8S^2 + 4S + 3}$$

EXERCISES

4-1. The liquid tank shown in Figure 4.2 has the following parameter values.

$R_1 = 6.51 \times 10^6$ newton-seconds/meter^5
$C = 2.68 \times 10^{-4}$ cubic meter/newton per square meter
$\rho = 1000$ kilograms/cubic meter (water)
$g = 9.81$ meters/second

Equation (4.6) is the differential equation that describes the relationship between the input flow rate (m) and the output level (h). Determine the numerical value of the coefficients of dh/dt, h, and m, and write Equation (4.6) using the numerical values.

4-2. Repeat Exercise 1 for the electrical circuit shown in Figure 4.3 and described by Equation (4.10). The electrical circuit has the following parameter values.

$R = 10,000$ ohms
$C = 25 \times 10^{-6}$ farad

4-3. Use logarithms to the base 10 to complete each of the following exercises.

a. 4.62×8.73 b. 12.8×482 c. 0.945×6.38
d. $55.5/22.7$ e. $4.62/37.8$ f. $0.735/7.02$
g. $(8.62)^4$ h. $(4.68)^{3.2}$ i. $(75.3)^{0.82}$

4-4. Determine the frequency-domain expression, $F(S)$, for each of the time-domain expressions, $f(t)$, (i.e., find the Laplace transform of each expression).

a. $f(t) = 7.8$

b. $f(t) = 3.2\, dx/dt$

c. $f(t) = 18.3\, t^3$

d. $f(t) = 5.3 \sin(377t)$

e. $f(t) = -37.2 \dfrac{d^2x}{dt^2}$

f. $f(t) = 9.2 \cos(1000t)$

h. $f(t) = 2.3t\sin(100t)$

g. $f(t) = 7.3\, g(t - a)$

i. $f(t) = 4.8t\cos(377t)$

j. $f(t) = 5.7 \dfrac{d^3x}{dt^3}$

k. $f(t) = 6.2\, e^{-8t}$

l. $f(t) = 77.8 \int x\, dt$

4-5. Determine the time-domain expression for each of the following frequency-domain expressions (i.e., find the inverse Laplace transform of each expression).

a. $800/S$

b. $7.3\, S\, X(S)$

c. $(1/S)X(S)$

d. $-3.2/(S + 8)$

e. $4.3\, S^3\, X(S)$

f. $\dfrac{8.2}{S(2.3S + 1)}$

g. $3.7\left(\dfrac{4!}{S^5}\right)$

h. $5.6\, S^2\, X(S)$

i. $42\left(\dfrac{3!}{(S+2)^4}\right)$ j. $77\left(\dfrac{10}{S^2+10^2}\right)$

k. $-36\left(\dfrac{S}{S^2+7^2}\right)$ l. $102\left(\dfrac{S^2-12^2}{(S^2+12^2)^2}\right)$

4–6. The liquid tank system shown in Figure 4.2 is described by Equation (4.6). When $R_LC = 2,000$ and $R_L/(\rho g) = 1,000$, Equation (4.6) becomes

$$2,000\frac{dh}{dt} + h = 1,000m$$

where:

h = the liquid level, meters
m = the input flow rate, cubic meters/second
t = time, seconds

Determine the transfer function, $H(S)/M(S)$, from the differential equation.

4–7. The electric RC circuit shown in Figure 4.3 is described by Equation (4.10).

$$(RC)\frac{de_o}{dt} + e_o = e_i$$

Determine the transfer function, $E_o(S)/E_i(S)$, from the differential equation for the following values of R and C.

a. $R = 10,000$ ohms and $C = 10^{-5}$ farad
b. $R = 1,000,000$ ohms and $C = 10^{-4}$ farad
c. $R = 100$ ohms and $C = 10^{-6}$ farad

4–8. The dead-time process shown in Figure 11.10 is described by the equation

$$f_o(t) = f_i(t - t_d)$$

where:

f_o is the output signal, kilograms/second
f_i is the input signal, kilograms/second
t_d is the dead time lag, seconds

Determine the transfer function, $F_o(S)/F_i(S)$, from the above equation if the dead time, t_d, is 310 seconds.

4–9. A typical temperature measuring means is described by the following differential equation.

$$6\frac{di}{dt} + i = (0.4)\theta$$

where:

$$i = \text{the output current signal, milliamperes}$$
$$\theta = \text{the input temperature signal, } ° \text{ Celsius}$$

Determine the transfer function, $I(S)/\theta(S)$, from the differential equation.

4–10. A process-control valve/electropneumatic converter is used in the control systems illustrated in Figures 3.2, 3.5, and 3.6. A typical converter and valve combination is described by the following differential equation.

$$0.0001 \frac{d^2x}{dt^2} + 0.02 \frac{dx}{dt} + x = 0.3i$$

where:

$x = \text{the valve stem position, inches}$
$i = \text{the current input signal to the converter, milliamperes}$

Determine the transfer function, $X(S)/I(S)$, from the differential equation.

4–11. A tubular heat exchanger similar to the one shown in Figure 3.5 is described by the following differential equation.

$$25 \frac{d^2\theta}{dt^2} + 26 \frac{d\theta}{dt} + \theta = 125x$$

where:

$x = \text{the valve position, inches}$
$\theta = \text{the temperature of the fluid leaving the heat exchanger, } ° \text{ Celsius}$

Determine the transfer function, $\theta(S)/X(S)$, from the differential equation.

4–12. A manufacturing plant uses a liquid surge tank to feed a positive displacement pump. The pump supplies a constant flow rate of liquid to a continuous heat exchanger. The surge tank is described by the following differential equation.

$$h(t) = 0.5 \int m(t) \ dt$$

where:

$h(t) = \text{the level of liquid in the surge tank, meters}$
$m(t) = \text{the difference between the input flow rate and the output flow rate, cubic meters per second}$
$t = \text{time, seconds}$

Determine the transfer function, $H(S)/M(S)$, of the surge tank.

4–13. The spring-mass-damping system shown in Figure 11.9–a is described by the following differential equation.

$$M \frac{d^2x}{dt^2} + R \frac{dx}{dt} + K x = F$$

where:

M = the mass, kilograms
R = the dashpot resistance, newtons/meter per second
K = the spring constant, newtons/meter
x = the position of the mass
F = the external force applied to the mass, newtons

Determine the transfer function, $X(S)/F(S)$, if

M = 3.2 kilograms
R = 2.0 newtons/meter per second
K = 800 newtons per meter

4–14. An armature-controlled dc motor is sometimes used in speed- and position-control systems (see Figures 3.9 and 3.10). The dc motor operation is described by the following set of equations.

a. $e = R i + L \frac{di}{dt} + K_e w$

b. $T = K_t i$

c. $T = J \frac{dw}{dt} + b w$

where:

e = the armature voltage, volts
i = the armature current, amperes
w = the motor speed, radians per second
T = the motor torque, newton-meters
J = the moment of inertia of the load, kilograms-meter^2
b = the damping resistance of the load, newton-meters/radian per second
R = the armature resistance, ohms
L = the armature inductance, henrys
K_e = the back emf constant of the motor, volts/radian per second
K_t = the torque constant of the motor, newton-meters/ampere

A small permanent-magnet dc motor has the following parameter values.

$J = 8 \times 10^{-4}$ kilogram-meter^2
$b = 3 \times 10^{-4}$ newton-meter/radian per second
$R = 1.2$ ohms
$L = 0.020$ henrys
$K_e = 5 \times 10^{-2}$ volts/radian per second
$K_t = 0.043$ newton-meter/ampere

Substitute the above parameters into the preceding equations to obtain the exact differential equations of the dc motor. Determine the transfer function, $W(S)/E(S)$, by transforming all three equations into frequency-domain algebraic equations. Use algebraic operations to obtain the ratio of W/E, which is the desired transfer function.

4–15. A proportional-plus-integral controller is described by the following differential equation.

$$i = K e + \frac{K}{T_i} \int e \, dt$$

where:

$i =$ the current output of the controller, milliamperes
$e =$ the input error signal, volts
$K =$ the gain setting of the controller
$T_i =$ the integral setting of the controller

Determine the transfer function, $I(S)/E(S)$, if K is 10 and T_i is 0.08.

4–16. A proportional-plus-derivative controller is described by the following differential equation.

$$i = K e + K T_d \frac{de}{dt}$$

where T_d is the derivative setting of the controller and the other parameters are those defined in Exercise 15. Determine the transfer function, $I(S)/E(S$, if K is 1.4 and T_d is 0.01.

4–17. A three-mode controller is described by the following differential equation.

$$i = K e + K T_d \frac{de}{dt} + \frac{K}{T_i} \int e \, dt$$

The parameters are defined in Exercises 15 and 16. Determine the transfer function, $I(S)/E(S)$, if K is 3.6, T_d is 0.008, and T_i is 2.2.

Part Two

CONTROL SYSTEM COMPONENTS

5

Measuring Means
Components

5.1 Introduction

Position, displacement, velocity, acceleration, and force are the most common controlled variables in servomechanisms. Temperature, flow rate, pressure, and level are the most common controlled variables in industrial processes. The purpose of this chapter is to illustrate the basic principles involved in the measurement of these important process variables. No attempt is made to include a complete treatment of these transducers; however, the examples selected do represent a reasonable cross section of the almost endless variety of transducers.

5.2 Transducers

A transducer is a device that receives a signal in one form and converts it to a signal in a different form. The American Standards Association (ASA) definition is: "A transducer receives information in the form of a physical quantity and transmits this information in the form of another physical quantity."[1] According to this definition, every component in a control system is a transducer. In a more limited usage, the term refers to

[1] (ASA, 85.1—1963). "Terminology for Automatic Control" ASA Bulletin C85.1, American Society of Mechanical Engineers, United Engineering Center, 345 East 47th Street, New York, N.Y. 10017, 1963.

those devices which are used to measure some physical quantity, such as temperature, pressure, force velocity, etc. In this sense, only components which provide the measuring means in a control system are considered transducers.

The basic elements of a transducer are illustrated in Figure 5.1. Most transducers consist of two distinct parts; a primary element and a secondary element. The primary element is a device in which some physical quantity is dependent upon the value of the measured variable. The dependent physical quantity may be a displacement, a force, an electrical voltage, or an electrical resistance. The secondary element may be called an amplifier, a transmitter, converter, discriminator, or detector. Its purpose is to convert the output of the primary element into a useable electric or pneumatic signal. Typical input-output curves of the primary element, secondary element, and transducer are included in Figure 5.1.

The different types of primary and secondary elements are too numerous to mention here. Examples of several types of transducers are included for each of the more common controlled variables. The purpose is to illustrate the basic principles of transducers with a series of specific examples. In a control system, the function of the transducer is to provide a signal which is a measure of the controlled variable. For this reason, the examples are classified by the variable that is measured rather than by the manner in which the output is developed.

5.3 Position and Displacement Measurement

General Considerations. Position and displacement measurement is divided into two general types: linear and angular. Linear position or displacement is measured in units of length. Angular position or displacement is measured in radians or degrees. The primary standard of length is the international meter, which is defined in terms of the wavelength of the red-orange line of krypton 86. All other units of length are defined in terms of the primary standard. The meter is the basic unit of length in the International System of Units (SI), and the radian is the unit of angle. A radian is equal to $180/\pi$ degrees (approximately 57.3 degrees). Linear and angular displacement are illustrated in Figure 5.2.

Linear displacement can be converted into angular displacement of a roller as illustrated in Figure 5.2-c. The angular displacement in radians is equal to the linear displacement divided by the radius of the wheel $\triangle \theta = \triangle S / R$.

Thickness is a linear dimension which is frequently measured in terms of a displacement. Some type of mechanical linkage is used to convert variations in thickness into a corresponding displacement.

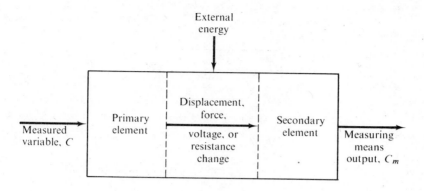

a) Block diagram

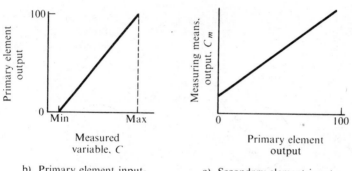

b) Primary element input-output curve

c) Secondary element input-output curve

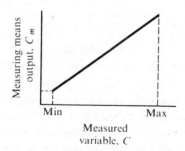

d) Measuring means input-output curve

FIGURE 5.1

BASIC ELEMENTS OF A MEASURING-MEANS COMPONENT
(TRANSDUCER)

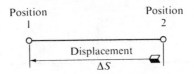

a) Linear displacement

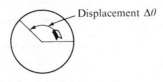

b) Angular displacement

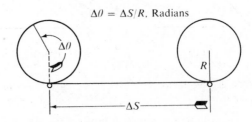

c) Conversion of linear displacement to
angular displacement

FIGURE 5.2

LINEAR AND ANGULAR DISPLACEMENT

Potentiometers. A potentiometer consists of a resistance element with a sliding contact which can be moved from one end to the other. Potentiometers are used to measure both linear and angular displacement, as illustrated in Figure 5.3. The wire-wound resistance element produces a uniform drop in the applied voltage, E_s. As a result, the voltage of the sliding contact is directly proportional to its distance from the reference end.

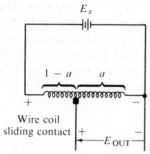

a) A linear displacement potentiometer

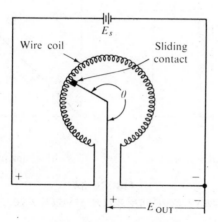

b) An angular displacement potentiometer

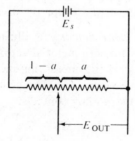

c) Schematic representation of a potentiometer

FIGURE 5.3

POTENTIOMETERS

Potentiometers may be used with single or double excitation. *Single excitation* of a potentiometer is illustrated in Figure 5.4. The output changes from 0 to E_s volts as the sliding contact moves from the reference end to the opposite end. The parameter a in Figure 5.4 is the proportionate position of the sliding contact. That is, a is the distance from the reference end to the sliding contact divided by the total length of the resistance element.

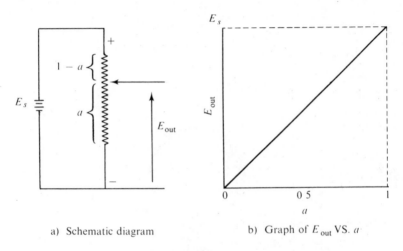

a) Schematic diagram   b) Graph of E_{out} VS. a

FIGURE 5.4

A POTENTIOMETER WITH SINGLE EXCITATION

A potentiometer with *double excitation* is illustrated in Figure 5.5. The output voltage changes from $-E_s$ to $+E_s$ volts as the sliding contact moves from the reference end to the opposite end. The voltage is 0 when the sliding contact is in the center of the resistive element.

A *two-potentiometer error detector* is illustrated in Figure 5.6. The position of the setpoint potentiometer determines the setpoint voltage signal, SP. The position of the measuring-means potentiometer determines the measured value signal, C_m. The error signal is determined by the voltage between the two sliding contacts.

$$E = SP - C_m$$

If both potentiometers are in the same position, then the sliding contacts are at the same voltage and the error signal is equal to 0 volts. When the two potentiometers are not in the same position, the error voltage is equal to $SP - C_m$ in both sign and magnitude.

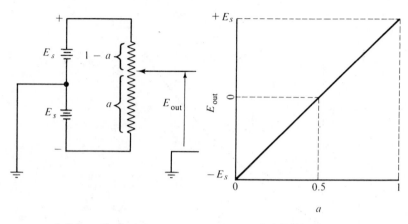

a) Schematic diagram   b) Graph of E_{out} VS. a

FIGURE 5.5

A POTENTIOMETER WITH DOUBLE EXCITATION

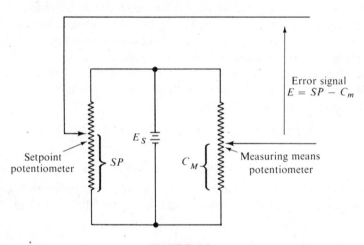

FIGURE 5.6

THE TWO-POTENTIOMETER ERROR DETECTOR

The *non-linearity* of a potentiometer is the amount of deviation of the output graph from a straight line. It is expressed by stating the maximum deviation as a percentage of full scale. For example, consider the non-linearity illustrated in Figure 5.7. If the maximum non-linearity is 0.1 volt and the full-scale range is 10 volts, then the non-linearity is 100 $(0.1)/10 = 1$ per cent.

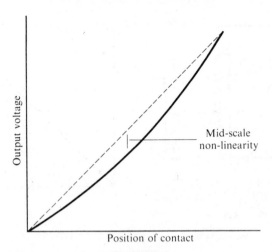

FIGURE 5.7

NON-LINEARITY

The *resolution* of a potentiometer is the smallest possible change in the output signal, expressed as a percentage of the full-scale output. The smallest change is determined by the voltage step between adjacent loops in the wire-wound element, as shown in Figure 5.8. If there are N turns in the element, then the voltage step between successive turns is $E_T = E_S/N$, where E_S is the full-scale voltage. The percentage resolution is given by the following relationship.

$$\text{Resolution, } \% = 100E_T/E_S = 100(E_S/N)/E_S$$

or

$$\text{Resolution, } \% = 100/N \qquad \textbf{(5.1)}$$

Potentiometers are subject to a non-linearity error whenever a current passes through the lead wire connected to the sliding contact. This

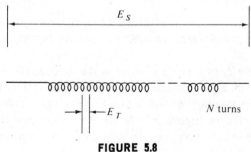

FIGURE 5.8

RESOLUTION OF A POTENTIOMETER

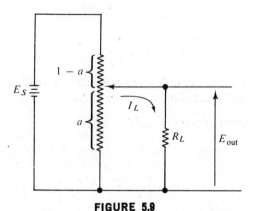

FIGURE 5.9
LOADING ERROR IN A POTENTIOMETER

error is called a *loading error* because it is caused by the load resistor connected between the sliding contact and the reference point. A potentiometer with a load resistor is illustrated in Figure 5.9. If R_P is the resistance of the potentiometer and a is the proportionate position of the sliding contact, then aR_P is the resistance of the portion of the potentiometer between the sliding contact and the reference point. The load resistor, R_L, is connected in parallel with the aR_P portion of the potentiometer. The equivalent resistance of this parallel combination is

$$\frac{aR_LR_P}{R_L + aR_P}$$

The resistance of the remaining portion of the potentiometer is equal to $(1 - a)R_P$, and the equivalent total resistance is the sum of the last two values.

$$R_{EQ} = (1 - a)R_P + \frac{aR_LR_P}{R_L + aR_P}$$

The output voltage, E_{OUT}, may be obtained by voltage division as follows.

$$E_{OUT} = \left[\frac{\dfrac{aR_LR_P}{R_L + aR_P}}{(1 - a)R_P + \dfrac{aR_LR_P}{R_L + aR_P}}\right]E_S$$

$$E_{OUT} = \left[\frac{aR_L}{R_L - aR_L + aR_P - a^2R_P + aR_L}\right]E_S$$

$$E_{OUT} = \left[\frac{a}{1 + ar - a^2r}\right]E_S$$

where

$$r = \frac{R_P}{R_L}$$

The loading error is the difference between the loaded output voltage (E_{OUT}) and the unloaded output voltage (aE_S).

$$\text{Loading error} = aE_S - E_{OUT}$$

$$= aE_S - \left[\frac{a}{1 + ar - a^2r} \right] E_S$$

$$= \left[\frac{a^2r - a^3r}{1 + ar - a^2r} \right] E_S, \text{ volts}$$

The loading is usually expressed as a percentage of the full-scale range, E_S.

$$\text{Loading error, } \% = 100 \left[\frac{a^2r(1 - a)}{1 + ar(1 - a)} \right] \qquad \textbf{(5.2)}$$

Example 5.1   The potentiometer in Figure 5.9 has a resistance of 10,000 ohms and a total of 1000 turns. Determine the resolution of the potentiometer and the loading error caused by a 10,000-ohm load resistor when $a = 0.5$.

Solution:

The resolution is given by Equation (5.1).

$$\text{Resolution} = 100/N = 100/1000 = 0.1 \text{ per cent}$$

The loading error is given by Equation (5.2).

$$\text{Loading error} = 100 \left[\frac{a^2r(1 - a)}{1 + ar(1 - a)} \right]$$

$$a = 0.5$$

$$r = 10,000/10,000 = 1$$

$$\text{Loading error} = 100 \left[\frac{(0.5)^2(1)(0.5)}{1 + (0.5)(1)(0.5)} \right]$$

$$\text{Loading error} = 10 \text{ per cent}$$

Linear Variable Differential Transformers. The linear variable differential transformer (LVDT) is a rugged electromagnetic transducer used to measure linear displacement. A diagram of an LVDT is shown in Figure 5.10. It consists of a single primary winding and two secondary windings on a common form. A movable magnetic core provides a variable coupling between the windings.

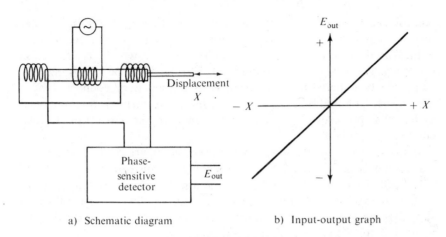

a) Schematic diagram   b) Input-output graph

FIGURE 5.10

A LINEAR VARIABLE DIFFERENTIAL TRANSFORMER

The primary winding is excited by an ac voltage. The secondary windings are cross-connected, with the two output lines connected to the input of a phase-sensitive detector. When the core is centered, the output of the secondary is 0 volts. A displacement on either side of the null point produces an ac voltage in the secondary winding. The amplitude of the ac voltage is proportional to the amount of displacement, and the phase angle depends on the direction of the displacement. A positive displacement produces a 0° phase angle. A negative displacement produces a 180° phase angle.

The phase-sensitive detector converts the ac secondary voltage into a dc voltage, E_{OUT}. The magnitude of the dc voltage is proportional to the amplitude of the ac voltage. The sign of the dc voltage is positive if the ac phase angle is 0°, and the negative if the ac phase angle is 180°. The result is the overall input-output graph illustrated in Figure 5.10-b.

In some ac systems, the ac secondary voltage is used as the error signal. The phase-sensitive detector is not required in these systems. Two identical LVDTs may be wired in opposition to form an error detector similar to the potentiometer error detector shown in Figure 5.6.

Synchro Systems. A synchro is a rotary transducer that converts augular displacement into an ac voltage, or an ac voltage into an angular displacement. Three different types of synchros are used in angular displacement transducers: the control transmitter, control transformer, and control differential.

Synchros are used in groups of two or three to provide a means of measuring angular displacement. For example, a control transmitter and control transformer form a two-element system that measures the angular displacement between two rotating shafts. The displacement measurement is then used as an error signal to synchronize the two shafts. The term *electronic gears* is sometimes used to describe this type of system, because the two shafts are synchronized as if they were connected by a gear drive. The addition of a control differential forms a three-element system that provides adjustment of the angular relationship of the two shafts during operation.

A two-element synchro system is shown in Figure 5.11. The control transmitter is designated CX and the control transformer is designated CT. Both the transmitter and the transformer have an H-shaped rotor with a single winding. Connections to the rotor winding are made through slip rings on the shaft. The stators each have three coils spaced 120° apart and connected in a Y configuration.

An ac voltage is applied to the rotor winding of the control transmitter. This voltage induces ac voltage in the three stator windings which are uniquely determined by the angular position of the rotor. The voltages induced in the transmitter are applied to the transformer stator windings which, in turn, induce a voltage in the transformer rotor winding. The ac voltage induced in the transformer rotor winding is uniquely determined by the relative position of the two rotors, as shown in Figure 5.11. A graph of the transformer rotor voltage (e_{OUT}) is included for each of the three relative rotor positions shown. The amplitude of e_{OUT} is a maximum when the angular displacement of the two rotors is 0° or ± 180°, and zero when the angular displacement is ± 90°. Notice the change in sign of the ac voltage between the 0° and 180° positions.

A graph of the amplitude of the output voltage (e_{OUT}) versus the angular displacement (θ) is shown in Figure 5.12. The output voltage is described by the following mathematical relationship.

$$e_{OUT} = (E_m \cos \theta) \sin \omega t \tag{5.3}$$

where:

E_m = The maximum amplitude

θ = the angular displacement

ω = the radian frequency of the ac voltage applied to the transmitter rotor.

The sign and magnitude of the amplitude term ($E_m \cos \theta$) is uniquely determined by the angular displacement θ, as shown in Figure 5.12. The graph is reasonably linear for values of θ from 20° to 160°. The operating point is located at the center of this linear region (i.e., $\theta = 90°$). The

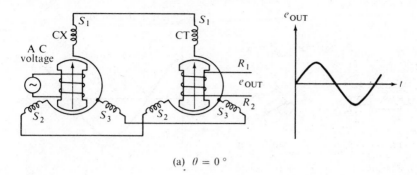

(a) $\theta = 0°$

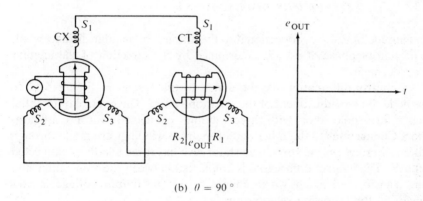

(b) $\theta = 90°$

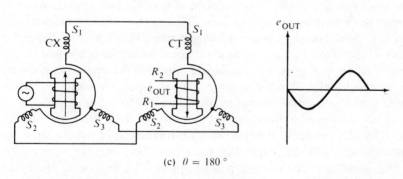

(c) $\theta = 180°$

FIGURE 5.11
A SYNCHRO ANGULAR DISPLACEMENT TRANSDUCER

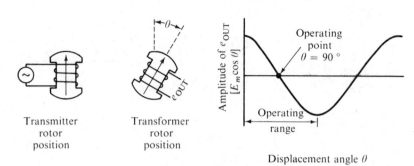

$$e_{\text{OUT}} = [E_m \cos \theta] \sin \omega t$$

FIGURE 5.12

SYNCHRO DISPLACEMENT TRANSDUCER OUTPUT AMPLITUDE VERSUS DISPLACEMENT ANGLE θ

magnitude of e_{OUT} is proportional to the amount of angular displacement. The sign or phase of e_{OUT} is determined by the direction of the angular displacement.

A control differential may be added to the system in Figure 5.11 to provide remote adjustment of the operating point. The control differential has a three-pole rotor, with three windings connected in a Y configuration. Connections to the other end of the roto windings are made through three slip rings on the shaft. The stator also has three windings connected in a Y. The control differential is connected between the transmitter and the transformer, as shown in Figure 5.13. The output voltage is now given by the following relationship.

$$e_{\text{OUT}} = [E_m \cos (\theta + \theta_d)]\sin \omega t \tag{5.4}$$

The angle θ_d is the relative displacement of the differential rotor. When θ_d is zero, the three-element system is no different from the two-element system (the dotted output curve). However, when θ_d is not zero, the entire curve is displaced by angular amount equal to θ_d. In other words, the control differential provides a means of adjusting the operating point by simply moving the differential rotor. This is especially useful when the transmitter and transformer rotors are connected to rotating shafts.

An application of a three-element synchro system is illustrated in Figure 5.14. The process consists of two rotating rolls which must be synchronized with the capability to adjust their angular relationship during operation. The control is accomplished by regulating the speed of the slave roll to maintain the desired angular relationship. The three-element synchro system acts as the measuring means, setpoint, and error

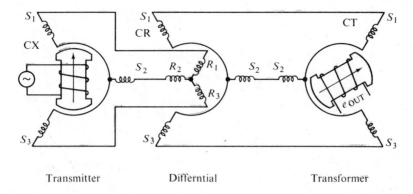

Transmitter   Differntial   Transformer

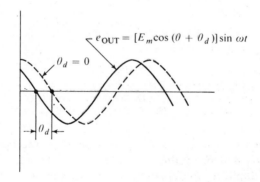

FIGURE 5.13

A THREE-ELEMENT SYNCHRO SYSTEM

detector. The setpoint is the angular position of the differential rotor. It represents the desired angular relationship between the two driven rolls. The transmitter and transformer combination compares the angular relationship of the two rolls with the desired relationship, and produces an ac output signal proportional to the difference (i.e., the error signal). The converter-amplifier acts as a high-gain proportional controller. The gain is sufficiently high that the proportional offset is maintained within acceptable limits.

The fourth type of synchro should be mentioned for the sake of completeness. This is the control receiver (CR). It is similar in construction to the control transmitter; the main difference lies in how it is connected and used. The difference in connection consists of connecting the ac voltage to the receiver rotor as well as to the transmitter rotor. The transmitter sche-

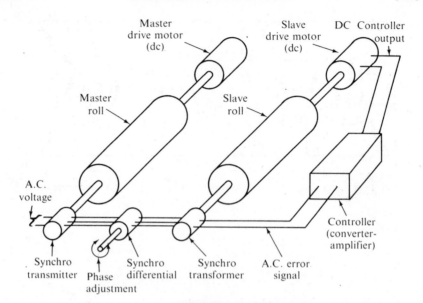

FIGURE 5.14

A THREE-ELEMENT SYNCHRO SYSTEM USED TO SYNCHRONIZE
TWO SHAFTS WITH ADJUSTMENT OF THEIR ANGULAR
DISPLACEMENT DURING OPERATION

matic in Figure 5.11 could be converted to a receiver schematic by connecting R_1 and R_2 to the ac voltage. The function of the control receiver is to duplicate the position of the transmitter. If the angular displacement between the two rotors is not zero, a torque is developed that tends to reduce the displacement to zero. If the load on the receiver rotor is slight, it will duplicate the position of the transmitter rotor. A synchro transmitter-receiver system is used for remote indication of angular position. It is also used for remote adjustment of angular position when the load is slight and high accuracy is not required.

Example 5.2　The synchro system in Figure 5.14 operates at a frequency of 400 Hz. The maximum amplitude of the transformer rotor voltage is 22.5 volts. Determine the ac error signal produced by each of the following pairs of angular displacements.

a.　$\theta = 90°, \quad \theta_d = 0°$

b.　$\theta = 60°, \quad \theta_d = 0°$

c.　$\theta = 135°, \quad \theta_d = -15°$

d.　$\theta = 100°, \quad \theta_d = -45°$

Solution:

Equation (5.4) gives the relationship between θ and θ_d and the error signal, e_{OUT}.

$$e_{OUT} = E_m \cos(\theta + \theta_d) \sin \omega t$$

At 400 Hz, $\omega = 2\pi(400) = 2,570$ radians/second. The maximum amplitude, $E_m = 22.5$ volts.

a. $\cos(\theta + \theta_d) = \cos(90 + 0) = \cos 90 = 0$

$$e_{OUT} = 0 \text{ volts}$$

b. $\cos(\theta + \theta_d) = \cos(60 + 0) = \cos 60 = 0.5$

$$e_{OUT} = (22.5)(0.5) \sin 2570t$$
$$e_{OUT} = 11.25 \sin 2570t \text{ volts}$$

c. $\cos(\theta + \theta_d) = \cos(135 - 15) = \cos 120 = -0.5$

$$e_{OUT} = (22.5)(-0.5) \sin 2570t$$
$$e_{OUT} = -11.25 \sin 2570t \text{ volts}$$

or

$$e_{OUT} = 11.25 \sin(2570t + 180°) \text{ volts}$$

d. $\cos(\theta + \theta_d) = \cos(100 - 45°) = \cos 55° = 0.574$

$$e_{OUT} = (22.5)(0.574) \sin 2570t$$
$$= 12.9 \sin 2570t$$

Pulse Generator. A pulse generator displacement transducer is illustrated in Figure 5.15. The roller converts linear displacement into a rotation of the pulse generator. The pulse generator converts this rotation into a sequence of pulses. The number of pulses is proportional to the linear displacement. A pulse counter or some type of digital controller is used to receive the pulse output of the displacement transducer.

The total number of pulses generated by a pulse generator displacement transducer is given by the following equation.

$$N_T = N_R X/(\pi D) \tag{5.5}$$

where:

$N_T =$ the total number of pulses

$N_R =$ the number of pulses per revolution

$X =$ the total displacement

$D =$ the diameter of the roller

A simple algebraic manipulation of Equation (5.5) results in the following useful form.

$$X = \left(\frac{\pi D}{N_R}\right) N_T \qquad (5.6)$$

The term $(\pi D/N_R)$ in Equation (5.6) is the displacement per pulse.

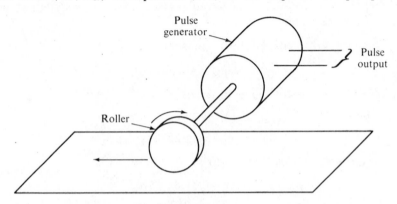

FIGURE 5.15

A PULSE GENERATOR DISPLACEMENT TRANSDUCER

Example 5.3   The pulse generator displacement transducer in Figure 5.15 has the following specifications.

$$N_R = 180 \text{ pulses/revolution}$$
$$D = 5.91 \text{ centimeters}$$

Determine the linear displacement per pulse and the displacement measured by each of the following total pulse counts.

a. $N_T = 700$

b. $N_T = 2220$

Solution:

Equation (5.6) provides the necessary relationship.

$$X = \left(\frac{\pi D}{N_R}\right) N_T$$

The linear displacement per pulse is given by $\pi D/N_R$.

$$\pi D/N_R = \pi(5.91)/(180)$$
$$= 0.103 \text{ centimeter/pulse}$$
$$= 1.03 \text{ millimeter/pulse}$$

a. $N_T = 700$

$$X = (0.103)(700) = 72.2 \text{ centimeters}$$

b. $N_T = 2220$

$$X = (0.103)(2220) = 229 \text{ centimeters}$$
$$= 2.29 \text{ meters}$$

5.4 Velocity Measurement

General Considerations. Velocity is the rate of change of displacement or distance. It is measured in units of length per unit time. Velocity is a vector quantity that has both magnitude (speed) and direction. A change in velocity may constitute a change in speed, a change in direction, or both.

Angular velocity is the rate of change of angular displacement. It is measured in terms of radians per unit time or revolutions per unit time. Angular velocity measurement is much more common in control systems than linear velocity measurement. Even when linear velocity is measured, it is usually converted into an angular velocity and measured with an angular velocity transducer. Two angular velocity transducers are considered: tachometers and pulse generators.

Tachometer. A tachometer is an electric generator used to measure angular velocity. A dc tachometer is illustrated in Figure 5.16. The coil is mounted on a metal cylinder called the *armature*. The armature is free to rotate in the magnetic field produced by the two permanent-magnet field poles. The two ends of the coil are connected to opposite halves of a seg-

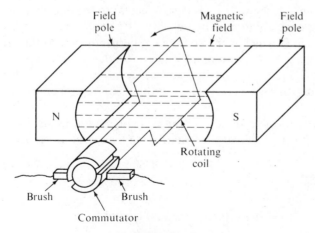

FIGURE 5.16
A TACHOMETER GENERATOR

mented connection ring called the *commutator*. There are two segments
on the commutator for each coil on the armature (only one is shown in
Figure 5.16). For example, an armature with 11 coils would have a
commutator with 22 segments.

The two carbon *brushes* connect the lead wires to the commutator segments. The brushes and commutator act as a reversing switch that reverses the coil connection once for each 180° rotation of the armature.
This switching action converts the ac voltage induced in the rotating coil
into a dc voltage. In other words, the commutator and brush constitute an
ac-to-dc converter.

The dc tachometer produces a dc voltage which is directly proportional
to the velocity of the armature. This voltage is based on the following fact:
A voltage is induced in a conductor when it moves through a transverse
magnetic field. If the conductor, magnetic field, and the velocity are
mutually perpendicular, the induced voltage is given by the following
equation.

$$E_L = LBV \tag{5.7}$$

where:

E_L = the induced voltage, volts

L = length of the conductor, meters

B = flux density of the magnetic field, webers
per square meter

V = velocity of the conductor, meters/second

In a tachometer, the velocity is perpendicular to the magnetic field only
twice during each rotation. The velocity (V) in Equation (5.7) is replaced by $V \sin \theta$, which is the component perpendicular to the magnetic
field. This means that a sinusoidal voltage is induced in each coil. However, when several coils are spaced evenly around the armature, the
rectified voltage is very nearly equal to the maximum obtained in Equation (5.7).

The velocity of the conductors may be expressed in terms of the average radius (R), and the angular velocity (S) in revolutions per minute.

$$V = 2\pi RS/60 \tag{5.8}$$

where:

V = the velocity, meters/second

R = the average radius, meters

S = angular velocity, revolutions/minute

Finally, the tachometer has N conductors of length L connected in series. The total voltage is the sum of the identical voltages induced in each conductor as given by the following equation.

$$E = NE_L \qquad\qquad (5.9)$$

Combining Equations (5.7), (5.8), and (5.9) produces the following equation for the voltage output of a tachometer.

$$E = K_E S = 30K_E\omega/\pi \qquad\qquad (5.10)$$

$$K_E = \frac{2\pi RBNL}{60} \qquad\qquad (5.11)$$

where:

$E =$ the tachometer output voltage, volts

$K_E =$ the EMF constant, volts/rpm

$S =$ the angular velocity, revolutions per minute

$\omega =$ the angular velocity, radians per second

$R =$ the average radius, meters

$B =$ the flux density of the magnetic field, webers/square meter

$N =$ the effective number of conductors

$L =$ the length of each conductor, meters

Example 5.4   A dc tachometer has the following specifications.

$R = 0.03$ meter

$B = 0.2$ weber/square meter

$N = 220$

$L = 0.15$ meter

Determine K_E and the output voltage at each of the following speeds.

$S = 1000, 2500,$ and 3250 rpm

Solution:

$$K_E = \frac{2\pi RBNL}{60}$$

$$K_E = \frac{2\pi(0.03)(0.2)(220)(0.15)}{60}$$

$$K_E = 0.0207 \text{ volt/rpm}$$

For $S = 1000$ rpm,

$$E = (0.0207)(1000) = 20.7 \text{ volts}$$

For $S = 2500$ rpm

$$E = (0.0207)(2500) = 51.8 \text{ volts}$$

For $S = 3250$ rpm

$$E = (0.0207)(3250) = 67.3 \text{ volts}$$

Pulse Generator and Frequency Converter. A pulse generator velocity transducer is illustrated in Figure 5.17. The magnetic pickup is a variable inductor that produces a pulse each time a gear tooth passes by. The gear tooth produces the pulse by changing the reluctance of the flux path as it passes by the magnetic pickup. If there are N teeth on the gear, the pick-up will produce N pulses per revolution. The pulse frequency (f) is the number of pulses per second. The relationship between the angular velocity, pulse frequency, and number of pulses per revolution is given by Equation (5.12).

$$S = 60f/N \qquad\qquad \textbf{(5.12)}$$

where:

$S = $ angular velocity, revolutions/minute

$f = $ pulse frequency, Hz

$N = $ number of pulses per revolution

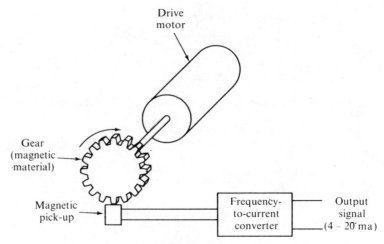

FIGURE 5.17

*A MAGNETIC PICKUP AND FREQUENCY CONVERTER
USED AS A VELOCITY TRANSDUCER*

The frequency-to-current converter converts the pulse frequency into an electric current signal suitable for use in electronic controllers and indicators .

Example 5.5 A magnetic pickup velocity transducer uses a 32-tooth gear. Determine the pulse frequency if the gear is rotating at 850 revolutions/minute.

Solution:

Equation (5.12) must be solved for f.

$$f = SN/60$$
$$f = (850)(32)/60$$
$$f = 453 \text{ Hz}$$

5.5 Acceleration Measurement

General Considerations. Acceleration is the rate of change of velocity. The measurement of linear acceleration is based on Newton's law of motion: $F = Ma$—i.e., the force (F) acting on a body of mass (M) is equal to the product of the mass times the acceleration (a). The acceleration is measured indirectly by measuring the force required to accelerate a known mass (M). The units of linear acceleration are meters/second².

Angular acceleration is the rate of change of angular velocity. The measurement of angular acceleration is usually obtained by differentiating the output of an angular velocity transducer. Angular acceleration is expressed in terms of radians/second² or revolutions/second².

Accelerometers. A schematic diagram of a linear accelerometer is shown in Figure 5.18. The accelerometer is attached to the object whose acceleration is to be measured and undergoes the same acceleration as the measured object. The mass M is supported by cantilever springs attached to the accelerometer frame. Motion of the mass M is damped by a viscous oil surrounding the mass.

The accelerometer is basically a spring-mass-damping system similar to the second-order processes covered in Section 12.5 (Figure 12.16). A second-order system is characterized by its resonant frequency (f_o) and its damping ratio (ζ), as determined by the following equations.

$$f_o = \frac{1}{2\pi} \sqrt{\frac{K}{M}} \qquad \qquad \textbf{(5.13)}$$

$$\zeta = \sqrt{\frac{b^2}{4KM}} \qquad \qquad \textbf{(5.14)}$$

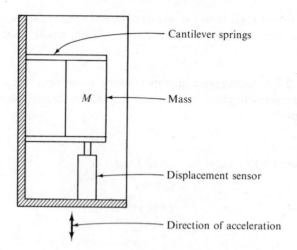

FIGURE 5.18

AN ACCELEROMETER

where:

f_o = resonant frequency, Hz

ζ = damping ratio

K = spring constant of the cantilever springs, newtons/meter

M = mass, kilograms

b = damping constant, newton-seconds/meter

Consider the situation in which the accelerometer frame in Figure 5.18 is accelerated upward at a constant rate. The mass M will deflect the cantilever springs down until. the springs exert a force large enough to accelerate the mass at the same rate as the frame. When this occurs, the spring force (KX) is equal to the accelerating force $(F = Ma)$.

$$KX = Ma$$

or

$$X = \frac{M}{K}a \qquad (5.15)$$

where:

X = the displacement of the mass, meters

M = the mass, kilograms

K = the spring constant of the cantilever springs, newtons/meter

a = acceleration, meters/second^2

Equation (5.15) indicates that the displacement X is proportional to the acceleration and may be used as a measure of the acceleration.

However, most accelerometers are used to measure accelerations that change with time. The response of the accelerometer depends on the frequency with which the measured acceleration changes. If the frequency is well below the resonant frequency (f_o), then Equation (5.15) is accurate and the displacement may be used as a measure of the acceleration. At frequencies well above the resonant frequency, the mass remains stationary. The displacement is equal to the displacement of the accelerometer frame. It is not a measure of the acceleration. At frequencies near the resonant frequency, the displacement of the mass is greatly exaggerated. Again, the displacement is not a measure of the acceleration. In conclusion,

An accelerometer must have a resonant frequency considerably larger than the frequency of the acceleration it is measuring.

Thompson[2] has shown that the error in Equation (5.15) is less than 0.5 per cent if the resonant frequency (f_o) is at least 2.5 times as large as the measured frequency when the damping ratio is 0.6.

Example 5.6   The accelerometer in Figure 5.18 has the following specifications .

$$M = 0.0156 \text{ kilogram}$$
$$K = 260 \text{ newtons/meter}$$
$$b = 2.4 \text{ newton-seconds/meter}$$
$$X_{max} = \pm 0.3 \text{ centimeters}$$

Determine the following.
a.   The maximum acceleration that can be measured.
b.   The resonant frequency, f_o.
c.   The damping ratio, ζ.
d.   The maximum frequency for which Equation (5.15) can be used with less than 0.5 per cent error.

Solution:

a.   The maximum acceleration may be determined by using Equation (5.15).

---

[2] W. T. Thompson, *Mechanical Vibrations* (Englewood Cliffs, N.J.: Prentice-Hall, 1948).

$$X = \frac{M}{K}a$$

$$a_{max} = X_{max} K / M$$
$$a_{max} = (0.003)(260) / (0.0156)$$
$$a_{max} = 50 \text{ meters} / \text{second}^2$$

b.   The resonant frequency is given by Equation (5.13).

$$f_o = \frac{1}{2\pi}\sqrt{\frac{K}{M}} = \frac{1}{2\pi}\sqrt{\frac{260}{0.0156}}$$
$$f_o = 20.6 \text{ Hz}$$

c.   The damping ratio is given by Equation (5.14).

$$\zeta = \sqrt{\frac{b^2}{4KM}} = \sqrt{\frac{2.4^2}{4(260)(0.0156)}}$$
$$\zeta = 0.595$$

d.   The maximum frequency for an error less than 0.5 per cent is $f_o/2.5$.

$$f_{max} = \frac{20.6}{2.5} = 8.25 \text{ Hz}$$

5.6   Force Measurement

General Considerations. Force is a physical quantity which produces or tends to produce a change in the velocity or shape of an object. It has a magnitude and a direction which are both defined by Newton's Second Law of Motion.

$$F = Ma \qquad\qquad\qquad \textbf{(5.16)}$$

where:

F = force applied to mass M, newtons
M = mass, kilograms
a = acceleration of mass M, meters/second^2

The magnitude of force F is equal to the product of the magnitudes of M and a. The direction of force F is the same as the direction of acceleration a.

Equation (5.16) does not mean that there are no forces on a body if it is not accelerating, only that there is no net unbalanced force. Two equal and opposite forces applied to a body will balance; and no acceleration will result. All methods of force measurement use some means of producing a measurable balancing force. Two general methods are used to produce the balancing force: the null balance method and the displacement method.

A beam balance is an example of a null balance force measuring means. The unknown mass is placed on one pan. Accurately calibrated masses of different sizes are placed on the other pan, until the beam is balanced. The unknown mass is equal to the sum of the calibrated masses in the second pan.

A spring scale is an example of a displacement type of force measuring means. The unknown mass is placed on the scale platform which is supported by a calibrated spring. The spring is displaced, until the additional spring force balances the force of gravity acting on the unknown mass. The displacement of the spring is used as the measure of the unknown force.

Three force transducers are covered in this section. The strain-gage load cell and the volumetric load element are displacement types. The unknown force is applied to an elastic member. The displacement of the elastic member is converted to an electric signal proportional to the unknown force. The pneumatic force transmitter is a null balance type. The unknown force is balanced by the force produced by air pressure acting on a diaphragm of known area. The air pressure is proportional to the unknown force and is used as the measured value signal.

Strain-Gage Force Transducers. Strain is the displacement per unit length of an elastic member. For example, if a bar of length L is stretched to length $L + \triangle L$, the strain, ϵ, is equal to $\triangle L/L$. A strain gage is a means of converting a small strain into a corresponding change in electrical resistance. It is based on the fact that the resistance of a fine wire varies as the wire is stretched (strained).

There are two general types of strain gages: bonded and unbonded. *Bonded strain gages* are used to measure strain at a specific location on the surface of an elastic member. The bonded strain gage is cemented directly onto the elastic member at the point where the strain is to be measured. The strain of the elastic member is transferred directly to the strain gage, where it is converted into a corresponding change in resistance. *Unbonded strain gages* are used to measure small displacements. A mechanical linkage causes the measured displacement to stretch a strain wire. The change in resistance of the strain wire is a measure of the displacement. The displacement is usually caused by a force acting on an elastic member. The unbonded strain gage measures the total displacement of an elastic member. The bonded strain gage measures the strain at a specific point on the surface of an elastic member.

The *gage factor* of a strain gage is the ratio of the unit change in resistance to the strain.

$$G = \frac{\triangle R/R}{\triangle L/L}$$ (5.17)

where:

G = gage factor

$\triangle R$ = change in resistance, ohms

R = resistance of the strain gage, ohms

$\triangle L$ = change in length, meters

L = length of the strain gage, meters

The gage factor is usually between 2 and 4. The effective length (L) ranges from about 0.5 centimeter to about 4 centimeters. The resistance (R) ranges from 50 ohms to 5000 ohms.

The *stress* on an elastic member is defined as the applied force divided by the unit area. If F is the applied force and A is the cross-sectional area, then the stress is equal *to F/A*. In elastic materials, the ratio of the stress over the strain is a constant called the *modulus of elasticity (E)*.

$$E = S/\epsilon$$ (5.18)

where:

E = modulus of elasticity, newtons/square meter

S = stress, newtons/square meter

ϵ = strain, meters/meter

An example of a strain-gage force transducer is illustrated in Figure 5.19. The cantilever beam is the elastic member (a diving board is a familiar example of a cantilever beam). The unknown force is applied to the end of the beam. The strain produced by the unknown force is measured by a bonded strain gage cemented onto the top of the beam. The center of the strain gage is located a distance of L units from the end of the beam.

The cantilever beam assumes a curved shape that approximates a semicircle. The top surface is elongated, while the bottom surface is compressed. Halfway between these two surfaces is a neutral surface, in which there is no displacement. The stress at any point on the top surface of the cantilever beam is given by the following equation.

$$S = \frac{6FL}{bh^2}$$ (5.19)

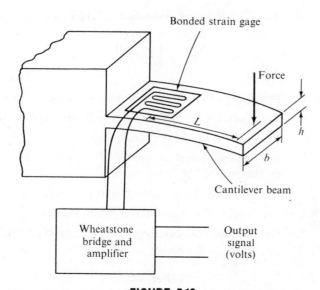

Bonded strain gage

Force

L

h

b

Cantilever beam

Wheatstone
bridge and
amplifier

Output
signal
(volts)

FIGURE 5.19

A STRAIN-GAGE FORGE TRANSDUCER

where:

S = stress, newtons/meter

F = the applied force, newtons

L = the distance from the point to the end of the
beam, meters

b = the width of the cantilever beam, meters

h = the height of the cantilever beam, meters

Equations (5.17), (5.18), and (5.19) can be combined to obtain an expression for the unit change in resistance produced by the unknown force F. We will begin with Equation (5.17).

$$G = \frac{\triangle R/R}{\triangle L/L} = \frac{\triangle R/R}{\epsilon}$$

$$\triangle R/R = G\epsilon \qquad\qquad \textbf{(5.20)}$$

However the strain (ϵ) can be expressed in terms of the stress and the modulus of elasticity (E) by solving Equation (5.18) for E.

$$\epsilon = S/E \qquad\qquad \textbf{(5.21)}$$

Substitute the *RHS* of Equation (5.21) for ϵ in Equation (5.20).

$$\triangle R/R = GS/E \qquad\qquad \textbf{(5.22)}$$

Finally, substitute the *RHS* of Equation (5.19) for *S* in Equation (5.22).

$$\frac{\triangle R}{R} = \left[\frac{6GL}{bh^2E}\right]F \qquad (5.23)$$

where:

$\triangle R$ = the change in resistance of the strain gage, ohms

R = the unstrained resistance of the strain gage, ohms

G = the gage factor of the strain gage

L = the distance from the center of the strain gage to the end of the beam, meters

b = the width of the cantilever beam, meters

h = the height of the cantilever beam, meters

E = the modulus of elasticity of the cantilever beam, newtons/square meter

TABLE 5.1
MODULUS OF ELASTICITY OF COMMON METALS

Material	E (newtons/square meter)
Steel	1.8×10^{11} to 2.1×10^{11}
Aluminum	5.5×10^{10} to 7.0×10^{10}
Copper-copper alloys	9.5×10^{10} to 1.25×10^{11}

Example 5.7 The strain-gage force transducer in Figure 5.19 has the following specifications.

Cantilever beam

    Material: steel

    $E = 2 \times 10^{11}$ newtons/square meter

    Maximum allowable stress $= 3.5 \times 10^8$ newtons/square meter

    $b = 1$ centimeter

    $h = 0.2$ centimeters

    $L = 4$ centimeters

Strain gage

    Gage factor $= 2$

    Nominal resistance $= 120$ ohms

Determine the maximum force which can be measured and the change in resistance produced by the maximum force.

Solution:

The maximum force is obtained by substituting the maximum allowable stress into Equation (5.19).

$$S = \frac{6FL}{bh^2}$$

$$F_{max} = \frac{S_{max}\, bh^2}{6L}$$

$$F_{max} = \frac{(3.5 \times 10^8)(0.01)(0.002)^2}{(6)(0.04)}$$

$$F_{max} = 58.3 \text{ newtons}$$

The change in resistance is obtained from Equation (5.23).

$$\triangle R = R\left[\frac{6GL}{bh^2E}\right]F$$

$$\triangle R = 120\left[\frac{(6)(2)(0.04)}{(0.01)(0.002)^2(2 \times 10'')}\right]58.3$$

$$\triangle R = 0.42 \text{ ohm}$$

Pneumatic Force Transducers. A pneumatic force transducer is illustrated in Figure 5.20. The unknown force F is balanced by the force of the air pressure against the effective area of the diaphragm. The ball and nozzle is arranged such that the balance of the two forces is automatic. For example, suppose the force F increases. The force rod moves upwards, reducing the opening between the ball and nozzle. The pressure

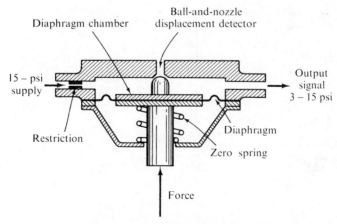

FIGURE 5.20

A PNEUMATIC FORCE BALANCE FORCE TRANSDUCER

in the diaphragm chamber increases and restores the balanced condition. The air pressure signal (P) in the diaphragm chamber is determined by the following equation.

$$F = (P - 3)A \qquad\qquad \textbf{(5.24)}$$

where:

F = the unknown force, pounds

P = air pressure in the diaphragm chamber, pounds/square inch

A = effective area of the diaphragm, square inches

The $(P - 3)$-term simply indicates that 3 psi corresponds to a force of zero. The signal range is from 3 to 15 psi.

Example 5.8   The pneumatic force transducer in Figure 5.20 has an effective area of 2.1 inches². Determine the force range of the transmitter.

Solution:

Equation (5.24) may be used.

$$F = (P - 3)\, A$$
$$F = (15 - 3)\,(2.1) = 25.2 \text{ pounds}$$

Volumetric Load Elements. An application of a volumetric load element is illustrated in Figure 5.21. The weight of the tank applies a force on the

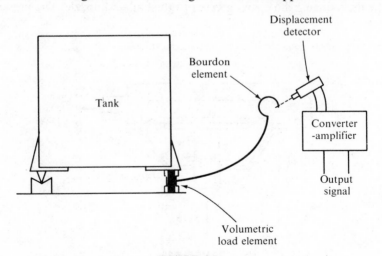

FIGURE 5.21
A VOLUMETRIC FORCE TRANSDUCER USED TO MEASURE THE WEIGHT OF A TANK

volumetric load element, producing a fluid pressure in the liquid fill. This fluid pressure is transmitted via the capillary to a Bourdon element. The Bourdon element converts the fluid pressure into a corresponding displacement. The converter-amplifier converts the displacement of the Bourdon element into a proportional electric current signal.

5.7 Temperature Measurement

General Considerations. Temperature is a measure of the degree of thermal activity attained by the particles in a body of matter. When two adjacent bodies of matter are at different temperatures, heat is transmitted from the warmer body to the cooler body until the two bodies are at the same temperature (two bodies at the same temperature are said to be in *thermal equilibrium*). The standard temperature scales are illustrated in Figure 5.22. The Celsius and Kelvin scales are the common and absolute temperature scales of the SI system of units.

In the act of measuring the temperature of a body, heat is transmitted between the thermometer and the body until the two are in equilibrium. The thermometer actually measures the equilibrium temperature—not the initial temperature of the body. Thus, the measuring process imposes a change in the original temperature of the body. Except when the thermometer and the measured body are at the same temperature before the measurement, 100 per cent accuracy in temperature measurement is impossible to attain. Additional errors are introduced by heat loss from the portion of the thermometer which is not immersed in the measured body. Careful consideration is required to minimize these measurement errors.

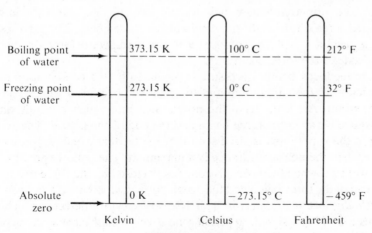

FIGURE 5.22
STANDARD TEMPERATURE SCALES

Differential-Expansion (Bimetallic) Thermostats. A bimetallic element consists of two strips of different metals bonded together to form a leaf, coil, or helix. The two metals must have different coefficients of thermal expansion so that a change in temperature will deform the original shape. A bimetallic thermometer is formed by attaching a scale and indicator to the bimetallic element such that the indicator displacement is proportional to the temperature. A bimetallic thermostat is formed by replacing the dial and indicator with a set of contacts. The bimetallic thermostat is frequently used in on-off temperature control systems. Figure 5.23 is a schematic diagram of a bimetallic thermostat.

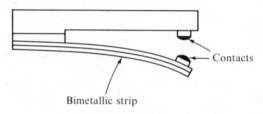

Contacts

Bimetallic strip

FIGURE 5.23
A BIMETALLIC THERMOSTAT

Fluid-Expansion Temperature Transducers. A gas-filled fluid-expansion temperature transducer is illustrated in Figure 5.24. The primary element consists of an inert gas sealed in a bulb which is connected by a capillary to a bellows pressure element. The bulb is immersed in the liquid to be measured until a thermal equilibrium is reached. The inert gas responds to the temperature change with a corresponding change in internal pressure. The primary-element bellows converts the gas pressure into an upward force on the right-hand side of the force beam. The air pressure in the feedback bellows produces a balancing upward force on the left-hand side of the force beam.

The feedback bellows pressure is regulated by the combination of the restriction in the supply line and the relative position of the nozzle and force beam. Air leaks from the nozzle at a rate which depends on the clearance between the force beam and the end of the nozzle. The restriction in the supply line is sized such that the feedback bellows pressure is 3 psi when the nozzle clearance is a maximum. The bellows pressure will rise to 15 psi when the nozzle is completely closed by the force beam. The left end of the force beam and the nozzle form what is called a *flapper-and-nozzle displacement detector.* This is a very sensitive detector which produces a 3- 15-psi signal, depending on a very small displacement of the flapper.

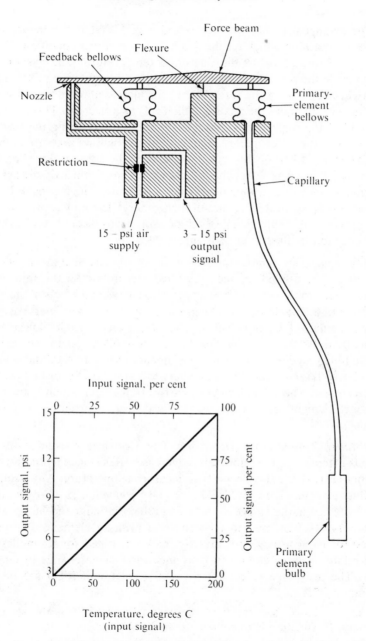

FIGURE 5.24

*A FLUID-EXPANSION TEMPERATURE TRANSDUCER
AND ITS INPUT-OUTPUT GRAPH*

The arrangement of the force beam is such that it automatically assumes the position which results in a balance-of-forces condition. For example, assume a rise in the measured temperature. The gas pressure increases in the primary-element bellows, thereby increasing the force on the right-hand side of the force beam. The increased force tends to rotate the force beam about the flexure, which acts as a fulcrum. This moves the left end of the force beam closer to the nozzle, increasing the feedback bellows pressure. The increase in the feedback bellows pressure increases the balancing force on the left-hand side of the force beam. The force beam is balanced by a feedback bellows pressure which is always proportional to the primary-element bellows pressure. The feedback bellows pressure is an accurate measure of the primary-element gas pressure, and it is used as the output signal of the pressure transducer. A typical input-output graph is illustrated in Figure 5.24.

Fluid-expansion temperature transducers are classified as vapor-filled, gas-filled, liquid-filled, or mercury-filled, depending on the fluid in the sealed measuring system. The vapor-filled type operates on the vapor pressure of a volatile liquid. The gas-filled system uses an inert gas as the filling medium. The liquid-filled system uses an organic liquid as the filling fluid. As the name implies, the mercury-filled system uses mercury as the filling medium. The primary element consists of the fluid sealed in a bulb which is connected by a capillary to a spiral, helix, or bellows pressure element. The secondary element is a transmitter which converts the pressure-element deflection into a standard electric current or air-pressure signal.

Resistance Temperature Detectors. The electrical resistance of most metals increases as the temperature increases. Resistance temperature detectors (RTD) use this property to measure temperature. A typical RTD is illustrated in Figure 5.25. The primary element is a wire-wound resistor located in the end of a protecting tube. Platinum and nickel are the metals most often used to construct the primary element. Platinum is noted for its accuracy and linearity; nickel is noted for its modest cost and relatively large change in resistance for a given change in temperature. The resistance values for a 0-100°C platinum RTD are included in Figure 5.25.

The resistance-to-current transmitter in Figure 5.25 is the secondary element. Its function is to convert the primary-element resistance into an electric current signal suitable for use by the controller. Typical output signals of a resistance-to-current transmitter are included in Figure 5.25.

Another type of secondary element is illustrated in Figure 5.26. This is the self-balancing Wheatstone bridge. The Wheatstone bridge in Figure

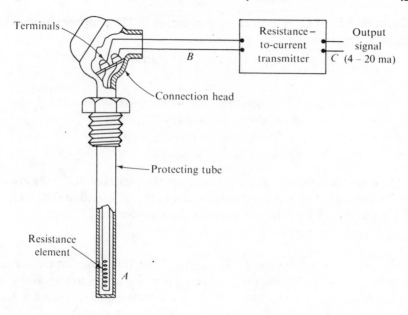

Typical Values for a Platinum RTD

Temperature (A)	Resistance (B)	Output signal (C)
0° C	100 ohms	4 milliamps
25° C	109.9 ohms	8 milliamps
50° C	119.8 ohms	12 milliamps
75° C	129.6 ohms	16 milliamps
100° C	139.3 ohms	20 milliamps

FIGURE 5.25

*A TYPICAL RESISTANCE TEMPERATURE DETECTOR
AND TRANSMITTER*

5.26-a is balanced when voltage points b and b[1] are equal to zero. This occurs when R_1/R_2 is equal to R_3/R_4. If R_1/R_2 is greater than R_3/R_4, then point b[1] is at a higher voltage than point b. If R_1/R_2 is less than R_3/R_4, then point b is at a higher voltage than point b[1]. The self-balancing Wheatstone bridge in Figure 5.26-b uses the voltage between points b and b[1] to balance the bridge. The dc voltage between b and b[1] is first converted to ac, and then amplified. The amplified ac voltage drives a servo-motor connected to the bridge slide wire in a direction that will tend to balance the bridge.

Example 5.9 The resistance of a platinum RTD is approximated by the following equation.

$$R = R_o[1 + a_1T + a_2T^2]$$

where:

R = the resistance at $T°C$, ohms

R_o = the resistance at $0°C$, ohms

T = the temperature, $°C$

a_1 and a_2 are constants

Determine the values of R_o, a_1, and a_2 for the platinum RTD illustrated in Figure 5.25. Use the resistance values at $0°$, $50°$, and $100°C$ to find R_o, a_1, and a_2. Check the accuracy of the equation at $25°C$.

Solution:

There are three unknowns: R_o, a_1, and a_2. Therefore, three equations are required to determine the three values. These equations are obtained by substituting the following three sets of temperature and resistance values from Figure 5.25: ($0°$, 100 ohms), ($50°$, 119.8 ohms), and ($100°$, 139.3 ohms).

$$100 = R_o(1 + a_1 \times 0 + a_2 \times 0^2) \tag{a}$$
$$119.8 = R_o(1 + a_150 + a_250^2) \tag{b}$$
$$139.3 = R_o(1 + a_1100 + a_2100^2) \tag{c}$$

Equation (a) is easily solved for R_o.

$$R_o = 100 \text{ ohms}$$

Substitute 100 for R_o into Equations (b) and (c), and simplify.

$$119.8 = 100(1 + 50a_1 + 2500a_2) \tag{b}$$
$$50a_1 + 2500a_2 = 0.198 \tag{b}$$
$$100a_1 + 10000a_2 = 0.393 \tag{c}$$

Eliminate a_1 from Equation (c) by subtracting Equation (b) twice.

$$5000a_2 = -0.003$$
$$a_2 = -6 \times 10^{-7}$$

Substitute -6×10^{-7} for a_2 in Equation (c) to obtain a_1.

$$100a_1 + 10^4 \times (-6 \times 10^{-7}) = 0.393$$
$$100a_1 = 0.393 + 0.006$$
$$a_1 = 3.99 \times 10^{-3}$$

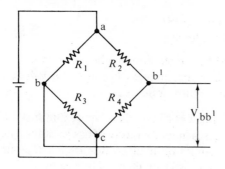

a) A Wheatstone bridge

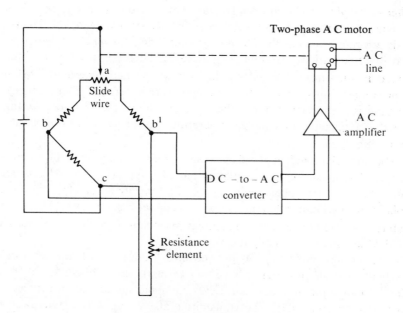

b) A self-balancing Wheatstone bridge

FIGURE 5.26

A SELF-BALANCING WHEATSTONE BRIDGE

The equation for R is

$$R = 100[1 + 3.99 \times 10^{-3}T - 6 \times 10^{-7}T^2]$$

Check the equation at 25°C

$$R = 100[1 + 3.99 \times 10^{-3} \times 25 - 6 \times 10^{-7} \times 25^2]$$
$$R = 100[1 + 0.09975 - 0.000375]$$

$$R = 100[1.099375]$$
$$R = 109.9 \text{ ohms}$$

The value predicted by the equation is the same as the value in the table. The 0.04 difference is meaningless because the values in the table are given to the nearest tenth of a degree.

Thermisters. A thermister is a semiconductor that has a large decrease in resistance with increases in temperature. That is, it has a negative temperature coefficient. A number of metal oxides are used, alone or in combinations, to form the thermister element. The secondary elements used for thermisters are similar to those discussed in the section on resistance thermometers.

Thermocouples. A thermocouple is a temperature sensor that consists of two dissimilar wires each of which is connected only at each end. The two ends are referred to as the *hot junction* and the *cold junction*. When the two junctions are at different temperatures, an EMF (voltage) is generated in the circuit formed by the two wires. The generated EMF is approximately proportional to the temperature difference. Thus, a thermocouple is a primary element that converts temperature into a voltage signal. A typical thermocouple temperature measuring means is illustrated in Figure 5.27.

The material used in the two wires determines the type of thermocouple. The most common combinations are: type E (chromel-constantan), type J (iron-constantan), type K (chromel-alumel), type R-S (platinum-platinum rhodium), type T (copper-constantan). The EMF values generated by an iron-constantan thermocouple with a cold junction at 0°C are included in Figure 5.27.

The thermocouple (T/C) transmitter in Figure 5.27 is the secondary element. It is an electronic device that converts the millivolt signal from the thermocouple into an electric current signal suitable for transmission to a remote controller. The transmitter also compensates for ambient changes in the cold-junction temperature. Thermocouple burnout protection is a safety feature required of most T/C transmitters. This feature provides maximum output from the transmitter whenever the thermocouple is open-circuited. In a control system, the burnout protection will turn off the heat source if the thermocouple is open-circuited for any reason. The current output signal of a typical T/C transmitter is included in Figure 5.27.

A self-balancing potentiometer similar to the one shown in Figure 5.26 is also used as the secondary element in a thermocouple measuring means. The resistance element in Figure 5.26 is replaced by a thermocouple and

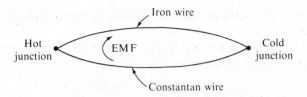

a) An iron-constantan thermocouple

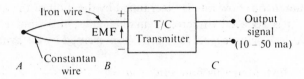

b) A typical thermocouple and T/C transmitter

Typical Values for an Iron-Constantan
Thermocouple

Temperature (A)	EMF (B)	Output signal (C)
0° C	0 millivolts	10 milliamps
50° C	2.61 millivolts	19.7 milliamps
100° C	5.28 millivolts	29.6 milliamps
150° C	8.01 millivolts	39.8 milliamps
200° C	10.77 millivolts	50.0 milliamps

FIGURE 5.27

*A TYPICAL IRON CONSTANTAN THERMOCOUPLE
AND TRANSMITTER*

the Wheatstone bridge is replaced by a slide-wire potentiometer. The servomotor automatically adjusts the slide wire until the potentiometer voltage balances the thermocouple voltage.

5.8 Flow Rate Measurement

General Considerations. The flow rate of liquids and gases is an important variable in industrial processes. The measurement of the flow rate indicates how much fluid is used or distributed in a process. Flow rate is frequently used as a controlled variable to help maintain the economy and efficiency of a given process.

The average flow rate is usually expressed in terms of the volume of liquid transferred in one second or one minute.

$$\text{Flow rate} = Q = \frac{\text{Change in volume}}{\text{Change in time}} = \frac{\triangle V}{\triangle t}$$

The instantaneous flow rate is determined by the limit of the average flow rate as $\triangle t$ is reduced to zero. In mathematics, this limit is called the *derivative of V with respect to t,* and is represented by the symbol dV/dt.

$$\text{Instantaneous flow rate} = q = \frac{\text{limit}}{\triangle t \to 0} \frac{\triangle V}{\triangle t} = \frac{dV}{dt}$$

The flow rate in a pipe can also be expressed in terms of the fluid velocity x and the cross-sectional area (A).

$$Q = Ax$$

The SI unit of flow rate is cubic meters/second.

Sometimes it is preferable to express the flow rate in terms of the mass of fluid transferred per unit time. This is usually referred to as the *mass flow rate.* The mass flow rate (M) is obtained by multiplying the flow rate (Q) by the fluid density (ρ).

$$\text{Average mass flow rate} = M = \rho Q$$

$$\text{Instantaneous mass flow rate} = m = \rho q$$

The SI unit of mass flow rate is kilograms/second.

Differential Pressure Flow Meters. Differential pressure flow meters operate on the principle that a restriction placed in a flow line produces a pressure drop proportional to the flow rate squared. A differential pressure transmitter is used to measure the pressure drop (H) produced by the restriction. The flow rate (Q) is proportional to the square root of the measured pressure drop.

$$Q = K\sqrt{H} \tag{5.25}$$

The restriction most often used for flow measurement is the orifice plate—a plate with a small hole, which is illustrated in Figure 5.28–a. The orifice is installed in the flow line in such a way that all the flowing fluid must pass through the small hole (see Figure 5.28–b).

Special passages transfer the fluid pressure on each side of the orifice to opposite sides of the diaphragm unit in a differential pressure transmitter.

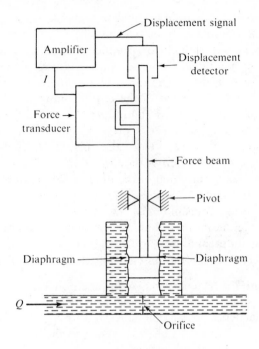

b) A typical differential pressure
flow transducer

a) A typical orifice plate

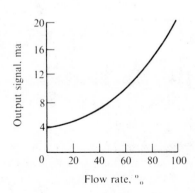

c) A typical calibration curve

FIGURE 5.28

A DIFFERENTIAL PRESSURE FLOW TRANSDUCER

The diaphragm arrangement converts the pressure difference across the orifice into a force on one end of a force beam. A force transducer on the other end of the beam produces an exact counterbalancing force. A displacement detector senses any motion resulting from an imbalance of the forces on the force arm. The amplifier converts this displacement signal into an adjustment of the current input to the force transducer that restores the balanced condition. The counterbalancing force produced by the force transducer is proportional to both the pressure drop and the input current (I). Thus, the current (I) is directly proportional to the pressure drop across the orifice (H). This same electric current is used as the output signal of the differential pressure transducer.

In Figure 5.28, the orifice is the primary element, and the differential pressure transmitter is the secondary element. The orifice converts the flow rate into a differential pressure signal, and the transmitter converts the differential pressure signal into a proportional electric current signal. A typical calibration curve is illustrated in Figure 5.28–c.

Turbine Flow Meters. A turbine flow meter is illustrated in Figure 5.29. A small permanent magnet is imbedded in one of the turbine blades. The magnetic sensing coil generates a pulse each time the magnet passes by. The number of pulses is related to the volume of liquid passing through the meter by the following equation: $V = KN$, where V is the total volume of liquid, K is the volume of liquid per pulse, and N is the number of pulses. The flow rate Q is equal to the total volume V divided by the time interval t.

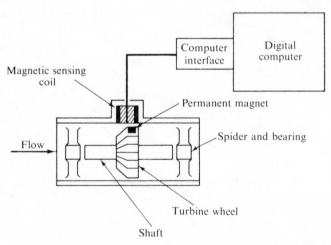

FIGURE 5.29

A TURBINE FLOW METER

$$Q = \frac{V}{t} = K\frac{N}{t}$$

But N/t is the number of pulses per unit time; i.e., the pulse frequency f. Thus,

$$Q = Kf \tag{5.26}$$

The pulse output of the turbine flow meter is ideally suited for digital counting and control techniques. Digital blending control systems make use of turbine flow meters to provide accurate control of the blending of two or more liquids. Turbine flow meters are also used to provide flow rate measurements for input to a digital computer, as shown in Figure 5.29.

Example 5.10   A turbine flow meter has a K value of 12.2 cubic centimeters per pulse. Determine the volume of liquid transferred for each of these pulse counts: 220, 1200, 470. Also determine the flow rate, if each of the above pulse counts occurs during a period of 140 seconds.

Solution:

$$V = KN$$

a.   For 220 pulses in 140 seconds,

$$V = KN = 12.2 \times 220 = 2684 \text{ cubic centimeters}$$

$$Q = \frac{V}{t} = \frac{2684}{140} = 19.2 \text{ cubic centimeters/second}$$

b.   For 1200 pulses in 140 seconds,

$$V = 12.2 \times 1200 = 14{,}640 \text{ cubic centimeters}$$

$$Q = \frac{14{,}640}{140} = 104.6 \text{ cubic centimeters/second}$$

c.   For 470 pulses in 140 seconds,

$$V = 12.2 \times 470 = 5734 \text{ cubic centimeters}$$

$$Q = \frac{5734}{140} = 41 \text{ cubic centimeters/second}$$

Magnetic Flow Meters. The magnetic flow meter has no moving parts and offers no obstructions to the flowing liquid. It operates on the principle that a voltage is induced in a conductor moving in a magnetic field. A magnetic flow meter is illustrated in Figure 5.30. The saddle-shaped coils placed around the flow tube produce a magnetic field at right angles

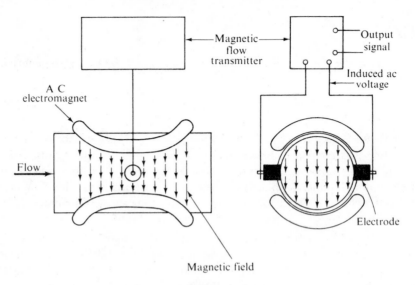

FIGURE 5.30
A MAGNETIC FLOW METER

to the direction of flow. The flowing fluid is the conductor, and the flow of
the fluid provides the movement of the conductor. The induced voltage
is perpendicular to both the magnetic field and the direction of motion of
the conductor. Two electrodes are used to detect the induced voltage,
which is directly proportional to the liquid flow rate. The magnetic flow
transmitter converts the induced ac voltage into a dc electric current signal
suitable for use by an electronic controller.

5.9 Pressure Measurement

General Considerations. Pressure is defined as the force per unit area, with
reference to the force exerted by a liquid or gas. Liquid pressure is the
source of the bouyant force that supports a floating object, such as a boat
or a swimmer. Pressure is also the motivating force that causes liquids and
gases to flow through a pipe.

    An extremely wide range of pressures is measured and controlled in in-
dustrial processes—all the way from 10^{-1} newton/square meter (about
0.001 millimeter of mercury) to above 10^8 newtons/square meter (about
10,000 psi). A great variety of primary elements has been developed to
measure pressure over various parts of this extreme range. The great ma-
jority of these primary elements convert the measured pressure into a
displacement or force. A secondary element then converts the force or displacement into a voltage, current, or air-pressure signal suitable for use by
a controller.

Pressure measurements are always made with respect to some reference pressure. Atmospheric pressure is the most common reference. The difference between the measured pressure and atmospheric pressure is called the *gage pressure*. Weather conditions and altitude both cause variations in the atmospheric pressure, and the gage pressure will vary accordingly. The standard atmospheric pressure is 1.013×10^5 newtons/square meter, or 14.7 pounds/square inch. A pressure less than atmospheric is called a vacuum. A pressure of zero is called a *perfect vacuum*. A perfect vacuum is sometimes used as the pressure reference. The difference between a measured pressure and a perfect vacuum is called the *absolute pressure*. An arbitrary pressure is also used as the reference pressure. The difference between a measured pressure and an arbitrary reference pressure is called the *differential pressure*. A differential pressure measuring means was discussed in the section on flow rate measurement.

Strain-Gage Pressure Transducers. A strain gage is based on the fact that stretching a metal wire changes its resistance. The change in resistance bears an almost linear relationship to the change in length. The strain gage uses the change in resistance to measure extremely small changes in dis-

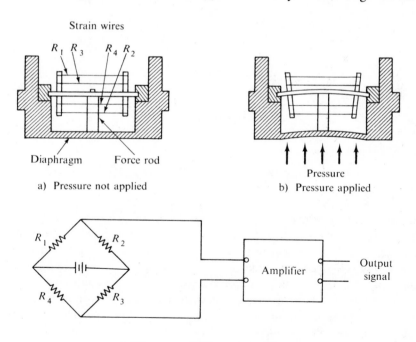

a) Pressure not applied

b) Pressure applied

c) Wheatstone bridge and amplifier

FIGURE 5.31

A STRAIN-GAGE PRESSURE TRANSDUCER

placement. Figure 5.31 is a schematic diagram of a strain-gage pressure transducer. The body acts as a mechanical transducer that converts the pressure into a displacement which increases the length of upper strain wires R_1 and R_3 and decreases the length of lower strain wires R_2 and R_4. A Wheatstone bridge converts the change in resistance of the upper and lower strain wires into an electric voltage proportional to the pressure. The voltage output of the Wheatstone bridge is then amplified to produce a useable signal. The primary element is the body and strain wire assembly. The secondary element consists of the Wheatstone bridge and the amplifier. Strain-gage pressure transducers are used in the high-pressure range (from 10^2 to 10^4 pounds per square inch).

Strain gages are divided into two types: bonded and unbonded. Figure 5.31 is an example of an unbonded type of strain gage. The displacement is transferred to the strain wire by a mechanical linkage. Bonded strain gages are cemented directly onto the body of the transducer, as shown in Figure 5.32.

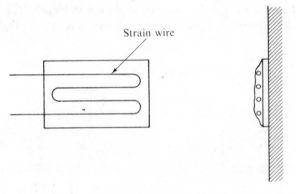

Strain wire

FIGURE 5.32

A BONDED STRAIN GAGE

Deflection Type Pressure Transducers. Deflection-type pressure transducers consist of a primary element and a secondary element. The primary element converts the measured pressure signal into a proportional displacement. The secondary element converts the primary-element displacement into an electric current signal suitable for use by a controller.

Figure 5.33 illustrates the common primary elements used in deflection-type pressure transducers. The Bourdon element is a flattened tube which is shaped into an incomplete circle, spiral, or helix. The tube tends to straighten out as the internal pressure increases, providing a displacement proportional to the pressure. Bourdon elements of bronze, steel, or

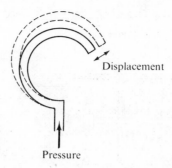

a) A circular Bourdon pressure element

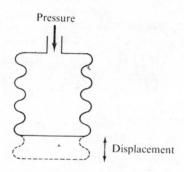

b) A bellows pressure element

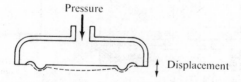

c) A diaphragm pressure element

FIGURE 5.33
DEFLECTION-TYPE PRESSURE ELEMENTS

stainless steel are available to cover the range from 0 to 10,000 psi or greater.

The Bellows element is a thin-walled metal cylinder with corrugated sides. The shape allows the element to extend as the internal pressure is balanced against a calibrated spring. The bellows displacement is proportional to the measured pressure. Bellows elements are widely used to mea-

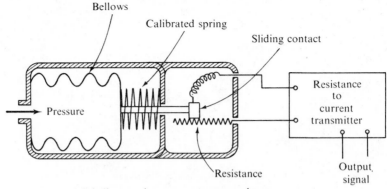

a) A bellows-resistance pressure transducer

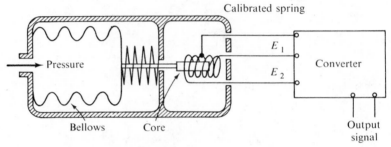

b) A bellows-inductance pressure transducer

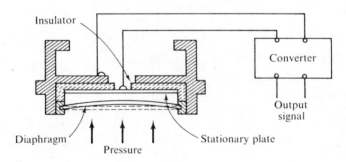

c) A diaphragm-capacitance pressure transducer

FIGURE 5.34

EXAMPLES OF DEFLECTION-TYPE PRESSURE TRANSDUCERS

sure pressures up to 100 psi. They are also used to make vacuum and absolute pressure measurements.

The diaphragm element may be flat or corrugated. The diaphragm allows sufficient movement to balance the pressure against a calibrated spring or force transducer.

The secondary element must convert the primary-element displacement into an electric signal suitable for use in the controller. The conversion is accomplished by using the displacement to adjust one of the three electric circuit elements: resistance, capacitance, or inductance. An electronic circuit then produces an electric signal based on the value of the variable element. Examples of the three types of secondary elements are illustrated in Figure 5.34.

A variable-resistance pressure transducer is illustrated in Figure 5.34–a. The calibrated spring is displaced by an amount proportional to the pressure in the bellows. The sliding contact causes a change in the resistance between the two leads connected to the transmitter. The transmitter, in turn, produces an electric signal based on the resistance value.

A variable-inductance pressure transducer is illustrated in Figure 5.34–b. The inductance value of the coil is determined by the position of the ferromagnetic core within the coil. The center tap arrangement provides two coils. As the core is moved, the inductance of one coil increases while the inductance of the other decreases. This effectively doubles the signal produced by a single coil. The converter is an electronic device that converts the inductance change into an electrical voltage or current signal.

A variable-capacitance pressure transducer is illustrated in Figure 5.34–c. The diaphragm and the stationary plate form the two plates of the capacitor. The displacement of the diaphragm reduces the distance between the two plates, thereby increasing the capacitance. The converter produces an electrical signal based on the capacitance value of the primary element.

Example 5.11   A bellows pressure element similar to Figure 5.34–b has the following values.

> Effective area of bellows = 12.9 square centimeters
> Spring rate of the spring = 80 newtons/centimeter
> Spring rate of the bellows = 6 newtons/centimeter

What is the pressure range of the transducer if the motion of the bellows is limited to 1.5 centimeter?

Solution:

The total spring rate is $80 + 6 = 86$ newtons/centimeter. The force required to deflect the spring 1.5 centimeters is $(1.5 \text{ cm}) \times (86 \text{ newtons/cm}) = 129$ newtons. The pressure required to produce this force is $(129 \text{ newtons}) \div (12.9 \text{ cm}^2) = 10$ newtons/square centimeter.

The range is 0 to 10 newtons/square centimeter, or 0 to 10^5 newtons/ square meter.

5.10 Liquid Level Measurement

General Considerations. The measurement of the level or weight of material stored in a vessel is frequently encountered in industrial processes. Liquid level measurement may be accomplished directly by following the liquid surface, or indirectly by measuring some variable related to the liquid level. The direct methods include sight glasses and various floats with external indicators. Although simple and reliable, the direct method is not easily modified to provide a control signal. Consequently, indirect methods provide most level control signals.

Many indirect methods employ some means of measuring the static pressure at some point in the liquid. These methods are based on the fact that the static pressure is proportional to the liquid density times the height of liquid above the point of measurement.

$$P = \rho g h \qquad\qquad (5.27)$$

where:

P = static pressure, newtons/square meter

ρ = liquid density, kilograms/cubic meter

h = height of liquid above the measurement
point, meters

g = 9.81 meters/second^2

Thus, any static pressure measurement can be calibrated as a liquid level measurement. If the vessel is closed at the top, the differential pressure between the bottom and the top of the vessel must be used as the level measurement.

The following are examples of some of the other indirect methods used to measure liquid level.

1. The *displacement-float method* is based on the fact that the bouyant force on a stationary float is proportional to the liquid level around the float.
2. The *capacitance-probe method* is based on the fact that the capacitance between a stationary probe and the vessel wall depends on the liquid level around the probe.
3. The *gamma-ray system* is based on the fact that the number of gamma rays that penetrate a layer of liquid depends on the thickness of the layer.

Displacement-Float Level Transducer. A displacement-float level transducer is illustrated in Figure 5.35. The float applies a downward force on the force beam equal to the weight of the float minus the buoyant force of the liquid around the float. The force on the beam is given by the following equation.

$$F = Mg - \rho g A H \qquad (5.28)$$

where:

F = net force, newtons

M = mass of the float, kilograms

g = 9.81 meters/second^2

ρ = liquid density, kilograms/cubic meter

A = horizontal cross-sectional area of the float, square meters

h = length of the float below the liquid surface, meters

Equation (5.28) points out that the force, F, bears a linear relationship to the liquid level.

The load cell applies a balancing force on the force beam which is proportional to F and, consequently, bears a linear relationship with the liquid level. The load cell is a strain-gage force transducer that varies its resistance in proportion to the applied force. This resistance is converted to an electrical signal suitable for use by an electronic controller.

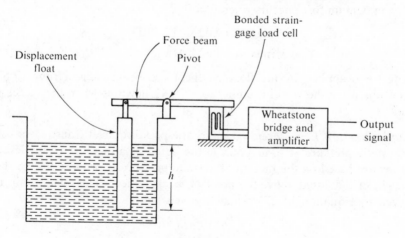

FIGURE 5.35

A DISPLACEMENT-FLOAT LEVEL TRANSDUCER

Example 5.12   The displacement-float level transducer in Figure 5.35 has the following data.

> Mass of the float, $M = 2.0$ kilograms
> Cross-sectional area of float, $A = 20$ square centimeters
> Length of the float, $L = 2.5$ meters
> Liquid in the vessel, kerosene

Determine the minimum and maximum values of the force, F, applied to the force beam by the float.

Solution:

From Table 3 in Appendix A, the density (ρ) of kerosene is 800 kilograms per cubic meter. The force, F, is given by Equation (5.28).

$$F = Mg - \rho g A h$$
$$F = (2)(9.81) - (800)(9.81)\left(\frac{20}{10^4}\right)h$$
$$F = 19.62 - 15.7h \text{ newtons}$$

The minimum force occurs when $h = L$

$$F \min = 19.62 - 15.7(2.5)$$
$$F \min = 19.62 - 39.25$$
$$F \min = -19.43 \text{ newtons}$$

The maximum force occurs when $h = 0$

$$F \max = 19.62 - 15.7(0)$$
$$F \max = 19.62 \text{ newtons}$$

The force applied by the float on the force beam ranges from 19.62 newtons when the vessel is empty to -19.43 newtons when the vessel is full.

Static-Pressure Level Transducers. Static-pressure level transducers use the static pressure at some point in the liquid as a measure of the level. They are based on the fact that the static pressure is proportional to the height of the liquid above the point of measurement. The relationship is given by Equation (5.27) which is given below.

$$P = \rho g h$$

where:

> P = static pressure, newtons/square meter
> ρ = liquid density, kilograms/cubic meter

h = height of liquid above the measurement
  point, meters

g = 9.81 meters/second^2

If the top of the tank is open to atmospheric pressure, then an ordinary
pressure gauge may be used to measure the pressure at some point in the
liquid. A variety of methods is used to measure the static pressure. One
method is illustrated in Figure 5.36. A bellows-resistance pressure transducer and transmitter is used as the measuring means. The output of
the transmitter is a 4- 20-milliamp current signal corresponding to a level
range from 0 to 100 per cent.

If the top of the tank is not vented to the atmosphere, then the static
pressure reflects the pressure in the tank at the liquid surface. The height
of the liquid above the point of measurement is proportional to the difference between the static pressure and the pressure at the top of the tank.
A differential pressure measurement is required. Figure 5.37 illustrates
the use of a differential pressure transducer to measure the level in a
closed tank.

Example 5.13   Water is to be stored in an open vessel. The static-
pressure measurement point is located 2 meters below the top of the tank.

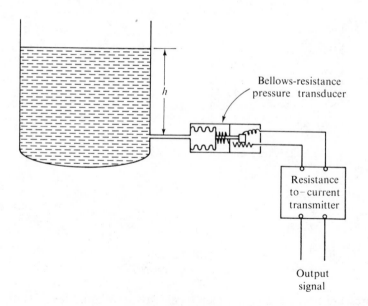

FIGURE 5.36
AN OPEN VESSEL—STATIC PRESSURE—LEVEL TRANSDUCER

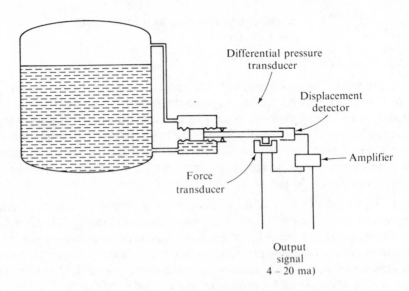

FIGURE 5.37

A CLOSED VESSEL—STATIC PRESSURE—LEVEL TRANSDUCER

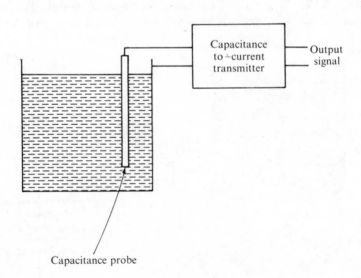

FIGURE 5.38

A CAPACITANCE PROBE LEVEL TRANSDUCER

A pressure transducer is selected to measure the liquid level. Determine the range of the pressure transducer for 100 per cent output when the tank is full.

Solution:

Equation (5.27) gives the desired relationship. The density of water is obtained from Table 3 in Appendix A.

$$\rho = 1000 \text{ kilograms/cubic meter}$$
$$P = \rho g h$$
$$P \text{ max} = (1000)(9.81)(2)$$
$$P \text{ max} = 1.962 \times 10^4 \text{ newtons/square meter}$$

or

$$P \text{ max} = 1.962 \text{ newtons/square centimeter}$$

Capacitance Probe Level Transducers. A capacitance probe level transducer is illustrated in Figure 5.38. The insulated metal probe is one side of the capacitor and the tank wall is the other side. The capacitance varies as the liquid level around the probe varies. The transmitter uses a capacitance bridge to measure the change in capacitance. Solid and liquid levels up to 70 meters can be measured.

EXERCISES

5–1. Determine the number of turns required to produce a potentiometer with a resolution of 0.2 per cent.

5–2. The potentiometer in Figure 5.9 has a resistance of 100,000 ohms. Determine the loading error caused by the following values of R_L and a.

    a. $R_L = 1000$ ohms, $a = 0.25, 0.5, 0.75$
    b. $R_L = 10,000$ ohms, $a = 0.25, 0.5, 0.75$
    c. $R_L = 100,000$ ohms, $a = 0.25, 0.5, 0.75$

5–3. The synchro system in Figure 5.11 operates at a frequency of 60 Hz. The maximum amplitude of the transformer rotor voltage is 6.2 volts. Determine the ac error signal produced by each of the following angular displacements.

    a. $\theta = 75°$ b. $\theta = 45°$
    c. $\theta = 150°$ d. $\theta = 110°$

5–4. For the synchro system in Figure 5.11, determine the angular displacement that will produce each of the following ac error signals.

    a. $3.1 \sin 377t$ b. $-4.8 \sin 377t$
    c. $5.5 \sin 377t$ d. $2.7 \sin (377t + 180°)$

5–5. The synchro system in Figure 5-14 operates at a frequency of 400 Hz. The maximum amplitude of the transformer rotor voltage is 22.5 volts. Determine the ac error signal produced by each of the following pairs of angular displacements.

a. $\theta = 60°, \theta_d = -60°$ b. $\theta = -30°, \theta_d = -20°$
c. $\theta = 45°, \theta_d = 20°$ d. $\theta = -18°, \theta_d = -17°$

5–6. The pulse generator displacement transducer in Figure 5.15 has the following specifications.

$$N_R = 360 \text{ pulses/revolution}$$
$$D = 8.2 \text{ centimeters}$$

Determine the linear displacement per pulse and the displacement measured by each of the following total pulse counts.

a. 85   b. 450
c. 3750   d. 8960

5–7. A dc tachometer has the following specifications.

$$R = 0.025 \text{ meter}$$
$$B = 0.22 \text{ weber/square meter}$$
$$N = 120$$
$$L = 0.25 \text{ meter}$$

Determine K_E and construct a calibration curve for a velocity range of 0 to 5000 revolutions/minute.

5–8. A magnetic pickup velocity transducer uses an 80-tooth gear. Determine the pulse frequency for each of the following gear speeds.

a. 30 rpm   b. 110 rpm
c. 75 rpm   d. 220 rpm

5–9. The magnetic pickup velocity transducer in Figure 5.17 is measuring a shaft rotating at 600 rpm and the pulse frequency is 1000 Hz. How many teeth are there on the gear?

5–10. The accelerometer in Figure 5.18 has the following specifications.

$$M = 0.012 \text{ kilogram}$$
$$K = 320 \text{ newtons/meter}$$
$$X_{max} = \pm 0.25 \text{ centimeter}$$

Determine the following.
a. The maximum acceleration that can be measured.
b. The resonant frequency, f_o.
c. The damping constant, b, required to produce a damping ratio of 0.6.
d. The maximum frequency for which Equation (5.15) can be used with less than 0.5 per cent error.

5–11. The strain-gage force transducer in Figure 5.19 has the following specifications:

cantilever beam
     Material: Steel
     $E = 2 \times 10^{11}$ newtons/square meter
     Max. allowable stress $= 5.0 \times 10^{8}$ newtons/square meter
     $b = 1.25$ centimeter
     $h = 0.25$ centimeter
     $L = 6$ centimeters
Strain gage
     Gage factor $= 2$
     Nominal resistance $= 2$ ohms

Determine the maximum force which can be measured and the change in resistance produced by the maximum force.

5–12. The pneumatic force transducer in Figure 5.20 is to have an input of 0–50 pounds force and an output signal range of 3–15 psi. Determine the required effective area.

5–13. Name the four types of fluids used in fluid expansion temperature transducers.

5–14. The following data were obtained in a calibration test of a fluid expansion temperature transducer, similar to Figure 5.24.

Temperature, °C	0	38	81	120	162	199
Output signal, psi	3.01	5.34	7.86	10.14	12.78	15.01

Construct a calibration graph by plotting the data points. Use judgment to draw a straight line through the data points, and estimate the accuracy of the measuring means in degrees C and percentage of full scale.

5–15. Check the equation developed in Example 5.9 at 75° C.

5–16. The resistance of nickel wire at 20° C is given by the following equation.

$$R = \rho L/A$$

where:

    $R =$ resistance at 20°C, ohms
    $\rho =$ the resistivity of nickel $= 47.0$
    $A =$ the area of the wire in circular
        mils $=$ (diameter in mils)^2
    $L =$ the length of the wire in feet

A nickel resistance thermometer element is to have a resistance of 100 ohms at 20° C. Determine the length of wire required if the diameter of the wire is 0.004 inch (4 mils).

5–17. The EMF produced by a thermocouple may be approximated by the following equation.

$$E = E_o + a_1T + a_2T^2$$

where:

E = the thermocouple EMF at T°C, volts
E_o = the thermocouple EMF at 0°C, volts
T = the temperature, °C
a_1 and a_2 are constants

Determine the values of E_o, a_1 and a_2 for the iron-constantan thermocouple illustrated in Figure 5.27. Use the EMF values at 0°, 100°, and 200° C to find E_o, a_1, and a_2. Check the accuracy of the equation at 50° C (hint: see Example 5.9).

5–18. Construct a graph of the temperature versus the output signal for the iron-constantan thermocouple in Figure 5.27. Determine the nonlinearity of the graph (i.e., the maximum difference between the actual curve and a straight line connecting the two end points).

5–19. A differential pressure flow meter is used as the measuring means in a liquid flow control system. A technician obtained the following data in a series of calibration tests. Each test consisted of measuring the time required to fill a 0.5-gallon container. The tests were conducted at controller settings of 25, 50, 75, and 100 per cent.

Controller Setting	Time to fill a 0.5-gal container
25%	6.63 min
50%	4.78 min
75%	3.95 min
100%	3.45 min

Determine the average flow rate in gal/min for each controller setting by dividing 0.5 gallon by the time in minutes. Then construct a calibration curve from your calculated results.

5–20. The flow control system in Exercise 5–7 is described by the following equation.

$$Q = K\sqrt{SP}$$

where:

Q = the flow rate, gal/min
K = a constant
SP = the controller setting, per cent

Calculate the value of K for each of the four calibration test results in Exercise 5–7. Find the average of the four K-values and use this value to express the relationship between the flow rate Q and the square root of the controller setting SP.

5–21. A turbine flow meter has a K-value of 34.1 cubic centimeters per pulse. Determine the volume of liquid transferred for each of the following pulse counts.

a.   8,200
b.   32,060
c.   680

5–22. Determine the flow rate for each pulse count in Exercise 5-9 if the pulse counts occur during a 210-second period.

5–23. A bellows pressure element similar to Figure 5.34–a has the following values.

Effective area of the bellows = 21 square centimeters
Spring rate of the spring = 200 newtons/centimeter
Spring rate of the bellows = 10 newtons/centimeter

What is the pressure range of the transducer if the motion of the bellows is limited to 2 centimeters?

5–24. A bellows pressure element similar to Figure 5.34–b has the following values.

Effective area of the bellows = 4 square centimeters
Spring rate of the spring = 400 newtons/centimeters
Spring rate of the bellows = 25 newtons/centimeters

What stroke of the bellows is required for a range of 0–50 newtons/square centimeter?

5–25. A displacement-float level transducer has the following data.

Mass of the float, M = 6.5 kilograms
Cross-section area of the float, A = 50 square centimeters
Length of the float, L = 4.0 meters
Liquid in the vessel, water

Determine the minimum and maximum values of the force, F, applied to the force beam by the float.

5–26. Gasoline is to be stored in a vented vessel. The static pressure measurement point is 8 meters below the top of the vessel. A pressure transducer is to be used as the level measuring means. Determine the range of the pressure transducer for 100 per cent output when the tank is filled.

6

Measuring Means
Characteristics

6.1 Introduction

A measuring means has two basic characteristics—*accuracy* and *speed of response*. These characteristics are not completely independent. If the measured variable is changing, the accuracy of the measuring means is influenced by its speed of response. For this reason, the speed-of-response characteristic will be considered first.

6.2 Speed of Response

Many sensors do not give an immediate, complete response to a change in the measured variable. The response takes time to complete its change. The step response of a typical temperature sensor is illustrated in Figure 6.1. Notice that the response is similar to the response of a first-order lag. Many measuring means have dynamic characteristics that very closely approximate a first order lag. Others have the dynamic characteristics of an overdamped or critically damped second-order system. All these measuring means have a step response similar to the one shown in Figure 6.1.

The response of many measuring means is expressed in terms of the time required for the response to reach 63.2 per cent of its final change. This time is called the *lag coefficient* of the measuring means. In Figure

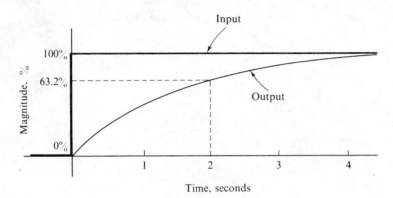

FIGURE 6.1

THE STEP RESPONSE OF A TYPICAL MEASURING MEANS

6.1, the lag coefficient is 2 seconds. If the measuring means is a true first-order lag system, then the lag coefficient is the time constant (τ). The lag coefficient is relatively independent of the size of the step change. In Figure 6.2, for example, the step change of 75°C and the step change of 150°C result in the same value for the lag coefficient.

The speed of response of a measuring means is determined by some or all of the following four primary factors.

1. The resistance of the sensing element.
2. The capacitance of the sensing element.

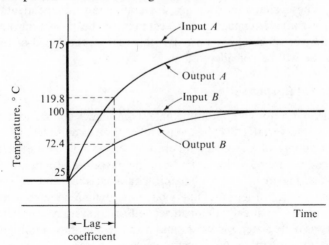

FIGURE 6.2

THE SPEED RESPONSE OF A TEMPERATURE MEASURING MEANS

3.   The mass or inertance of the sensing element.
4.   The transmission lag.

As an example, consider a thermocouple temperature transmitter used to measure the temperature of a fluid. Several factors may affect the thermal resistance. The type of fluid and the fluid velocity have a pronounced effect on the thermal resistance and, hence, on the speed of response. If a protective well is used, the resistance and capacitance of the well also affect the speed of response. The air space between the thermocouple and its protecting well will cause additional lag in the response. Some examples of lag coefficients of temperature measuring means are given below.

Typical Lag Coefficients

Bare thermocouple (in air)	35 sec.
Thermocouple in glass well (in air)	66 sec.
Thermocouple in porcelain well (in air)	100 sec.
Thermocouple in iron well (in air)	120 sec.
Bare thermocouple (in still liquid)	10.0 sec.
Thermometer bulb (water flowing at 2 ft/min)	6.0 sec.
Thermometer bulb (water flowing at 60 ft/min)	2.4 sec.

Example 6.1   The following data were obtained from a thermometer with a protective well which was plunged into a moving liquid at time $t = 0$ seconds. The thermometer and well were maintained at 50°C before the test. The liquid temperature was 150°C. Plot the response curve and determine the speed of response.

Time, seconds	Temperature, °C
0	50
20	64
40	79
60	92
80	104
100	112
120	120
140	125
160	130
180	133
500	150°C

Solution:

The data are plotted in Figure 6.3. The complete response is a change from 50° to 150°. The lag coefficient is the time required for 63.2 per cent of this change to occur. The total change is $150 - 50 = 100°C$, and 63.2 per cent of this change is $63.2°C$. Therefore, the lag coefficient is the time required for the temperature to reach $50 + 63.2 = 113.2°C$. From the graph in Figure 6.3, the lag coefficient is determined as 101 seconds.

6.3   Accuracy

The accuracy of a measuring means is usually expressed in terms of the static error and as a percentage of the full-scale range. The static error is the deviation of the measured value from the true value when the measured variable is maintained at a constant value. For example, a temperature measuring means has a range from 200° to 1200°C and an accuracy of ± 0.25 per cent. The full-scale range is $1200 - 200 = 1000°C$. The maximum error is $\pm 0.0025 (1000) = \pm 2.5°C$. This means that for any temperature between 200°C and 1200°C, the measured value will be within 2.5°C of the true value.

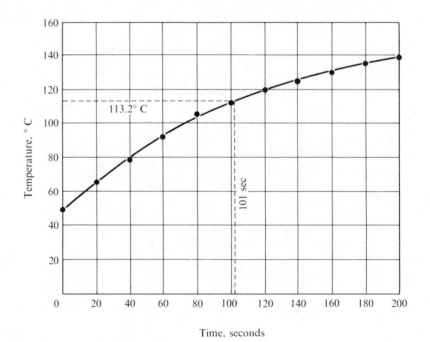

FIGURE 6.3

THE RESPONSE CURVE OF A THERMOMETER AND WELL

The accuracy of a measuring means is usually expressed as a percentage of the full-scale range and not as a percentage of the actual reading.

The reproducibility (or repeatability) of a measuring means is the maximum deviation from the average of repeated measurements of the same static variable. It is the degree of closeness between repeated measurements of the same variable. An automatic controller that is reproducible but not accurate may still be very useful. Figure 6.4 uses the pattern of bullet holes on a target to illustrate reproducibility and accuracy. Notice that a rifle which is reproducible but not accurate produces a tight pattern—but that the pattern is not centered on the bull's-eye. If the rifleman is aware of the offset, he may adjust accordingly and produce an accurate and reproducible pattern during the next try. An experienced operator will make a similar adjustment to compensate for an offset in a controller.

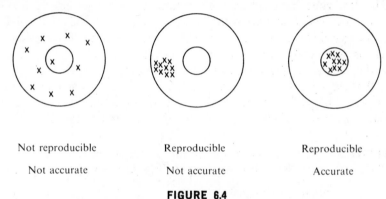

Not reproducible	Reproducible	Reproducible
Not accurate	Not accurate	Accurate

FIGURE 6.4

RIFLE TARGET PATTERNS ILLUSTRATING ACCURACY AND REPRODUCIBILITY

When the measured variable is changing, the lag in the response of the measuring means produces an additional error called the *dynamic error*. The dynamic error depends on the speed of response of the measuring means. It is completely independent of—and additional to—the static error. Figure 6.5 shows a graph of the true value and the measured value of a variable as it changes from one value to another. The lag in the measuring means produces a lag between the measured value and the true value. This lag is referred to as a *dynamic lag* or, simply, as a *lag*. The dynamic error is the difference between the measured value and the true value at any given time. Notice that the dynamic error is not constant and that it reduces to zero when the measured variable stops changing.

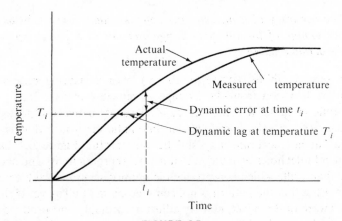

FIGURE 6.5

THE DYNAMIC ERROR AND LAG OF A MEASURING MEANS

The dead zone and dead time of a measuring means are also related to its accuracy. The *dead zone* is the largest change of a measured variable that produces no response in the measuring-means output. The measuring means is not capable of sensing or responding to any change that is less than the dead zone. The *dead time* of a measuring means is the length of time that elapses between a change in the measured variable and the first noticeable change in the output of the measuring means. This is exactly the same as the dead-time characteristic discussed in Chapter Eleven.

Example 6.2   A temperature measuring means has a range of 50° to 250°C. Several calibration tests produced the following results.

> Actual or true value: 212°C
> Measured values: 217°C, 214°C, 215°C. 216°C

a.   What is the worst-case error of the sensor in degrees and in per cent?
b.   What is the reproducibility in degrees and in per cent?

Solution:

a.   The worst-case error is $217 - 212 = 5°C$.
The range of the measuring means is $250 - 50 = 200°C$. The worst-case error is $\dfrac{5}{200} \times 100 = 2.5$ per cent of the full-scale range. Worst case error is 5°C or 2.5 per cent.

b.   The average measured value is

$$(217 + 214 + 215 + 216)/4 = 215.5°C$$

$$\text{Maximum deviation} = 217 - 215.5 = 1.5°C$$

$$\text{Per cent deviation} \quad = \frac{1.5}{200} \times 100 = 0.75$$

Reproducibility is $\pm$ 1.5°C or $\pm$ 0.75 per cent.

Example 6.3 A tachometer-generator is a device used to measure the speed of rotation of gasoline engines, electric motors, speed control systems, etc. The "tach" produces a voltage proportional to its rotational speed. Consider a tachometer-generator that has a nominal rating of 5.0 volts per 1000 rpm, a range of 0 to 5000 rpm, and an accuracy of $\pm$ 0.5 per cent. If the output of the "tach" is 21 volts, what is the nominal value of the actual speed? What are the minimum and maximum possible values of the actual speed?

Solution:

The range of the tachometer is 0 to 25 volts, corresponding to a speed range of 0 to 5000 rpm. The nominal value of the speed is equal to 200 times the output voltage.

$$\text{Nominal speed} = 21 \times 200 = 4,200 \text{ rpm}$$

The accuracy is $\pm$ 0.5 per cent of full scale or $\pm$ 0.005 $\times$ 5000 = $\pm$ 25 rpm. Thus, the actual speed could be anywhere between 4200 $-$ 25 and 4200 $+$ 25 rpm.

$$\text{nominal speed} = 4200 \text{ rpm}$$
$$4175 \leq \text{true speed} \leq 4225 \text{ rpm}$$

Example 6.4 A temperature sensor is used to measure the temperature of an oil bath. The actual temperature is increasing at a constant rate of 1.2°C per minute, and the dynamic lag is 2.3 minutes. What is the dynamic error in degrees Celsius? If the range of the temperature sensor is 75°C to 125°C, what is the dynamic error in per cent?

Solution:

a.   Let θ_1 represent the temperature at any time t_1.
At time $t_1 + 2.3$ minutes, the temperature will be $\theta_1 + (1.2)(2.3) = \theta_1 + 2.76°C$. Thus, at time $t_1 + 2.3$, the actual temperature will be $\theta_1 + 2.76°C$ and the measured value will be θ_1. The dynamic error is the difference between the true value and the measured value.

$$\text{dynamic error} = \theta_1 + 2.76 - \theta_1 = 2.76°C.$$

b.   The range of the sensor is $125 - 75 = 50°C$. The dynamic error in percent is $= \dfrac{2.76}{50} \times 100 = 5.52$ per cent.

6.4 Dynamic Characteristics

The dynamic characteristics of a measuring means determine its speed of response. The simplest dynamic characteristic is one in which there is no appreciable lag between the true value and the measured value. A potentiometer used as a position sensor is an example of a transducer which has negligible lag. The potentiometer produces an output voltage proportional to the position of the shaft, relative to the zero position. Usually, the shaft is mechanically connected to the device whose position is to be measured. For most applications, the lag between a change in shaft position and the corresponding change in voltage is negligible. Thus, there is neither dynamic lag nor dynamic error in this type of measuring means.

A very common dynamic characteristic in measuring means is the first-order lag. This is discussed in Chapter Twelve, and the equations and terminology presented there apply to measuring means as well. The common, liquid-filled thermometer illustrated in Figure 6.6 has a first-order lag dynamic characteristic.

The amount of heat transferred from the fluid surrounding the bulb to the liquid inside the bulb depends on the thermal resistance (R_t) between the two fluids, the difference in temperature ($\theta - \theta_m$), and the time interval $\triangle t$.

$$\triangle Q = \frac{1}{R_t}(\theta_a - \theta_m)\triangle t$$

The change in temperature of the liquid in the bulb ($\triangle\theta_m$) is equal to the amount of heat added ($\triangle Q$) divided by the thermal capacitance (C_t) of the liquid inside the bulb.

$$\triangle\theta_m = \triangle Q/C_t = \frac{1}{R_tC_t}(\theta_a - \theta_m)\triangle t$$

Dividing both sides by $\triangle t$,

$$\frac{\triangle\theta_m}{\triangle t} = \frac{1}{R_tC_t}(\theta_a - \theta_m)$$

or, as $\triangle t \rightarrow 0$

$$\tau\frac{d\theta_m}{dt} + \theta_m = \theta_a \tag{6.1}$$

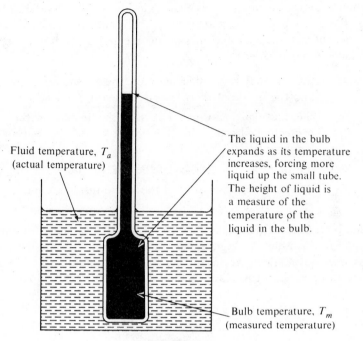

FIGURE 6.6
A LIQUID-FILLED THERMOMETER

The frequency-domain transfer function is:

$$\frac{\theta_m}{\theta_a} = \frac{1}{\tau S + 1}$$

where:

$\tau = R_t C_t =$ the time constant

$R_t =$ thermal resistance between the measured fluid and the liquid inside the bulb, Kelvin/watt

$C_t =$ the thermal capacitance of the liquid in the bulb, joule/Kelvin

$\theta_a =$ the temperature of the measured variable, °C or K

$\theta_m =$ the temperature of the liquid in the bulb, °C or K

Notice the similarity between Equation (6.1) and Equations (12.3) and (12.9) in Chapter Twelve. These are all equations of a first-order lag system. Thus, the liquid-filled thermometer has a first-order lag dynamic characteristic with a time constant (τ) equal to $R_t C_t$.

When a temperature sensor is placed inside a protective sheath (or well), the resistance and capacitance of the sheath become part of the dynamic characteristic of the measuring means. Thus, there are two resistances and two capacitances.

R_1 = the thermal resistance between the measured fluid and the inside of the protective sheath, Kelvin/watt

R_2 = the thermal resistance between the inside of the protective sheath and the liquid inside the bulb, Kelvin/watt

C_1 = the thermal capacitance of the sheath, joule/Kelvin

C_2 = the thermal capacitance of the liquid inside the bulb, Kelvin/watt

The above four elements combine to form a two-capacity, interacting system, as illustrated in Figure 6.7. The equation of the measuring means may be obtained by setting $R_L = \infty$ in Equation (12.21) (i.e., assume an open circuit in place of the load resistor R_L).

$$\tau_1\tau_2\frac{d^2\theta_m}{dt^2} + \left(\tau_1 + \tau_2 + \tau_2\frac{R_1}{R_2}\right)\frac{d\theta_m}{dt} + \theta_m = \theta_a \qquad \textbf{(6.2)}$$

where:

$$\tau_1 = R_1 C_1, \text{ seconds}$$

$$\tau_2 = R_2 C_2, \text{ seconds}$$

The frequency-domain transfer function is:

$$\frac{\theta_m}{\theta_a} = \frac{1}{\tau_1\tau_2 S^2 + \left(\tau_1 + \tau_2 + \tau_2\dfrac{R_1}{R_2}\right)S + 1}$$

Some measuring means employ a spring-mass-damping arrangement to measure a variable. In general, the measured variable is caused to exert a force proportional to its magnitude. This force is applied to a spring or elastic member, causing a deflection proportional to the applied force and, hence, proportional to the measured variable. Electrical voltmeters and accelerometers are examples of this type of measuring means. The dynamic characteristic is that of a second-order system [described by Equation (12.14) in Chapter Twelve and included here for convenience].

Time-domain equation:

$$\frac{1}{\omega_o^2}\frac{d^2h}{dt^2} + 2\left(\frac{\alpha}{\omega_o^2}\right)\frac{dh}{dt} + h = Gm$$

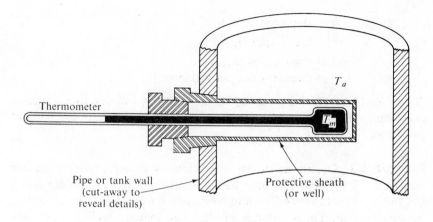

a) A cut-away view of a thermometer and
protective sheath

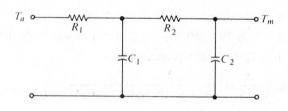

b) The equivalent electrical circuit for the
thermometer and protective sheath.

FIGURE 6.7

*A TEMPERATURE SENSOR IN A PROTECTIVE SHEATH
AND THE ELECTRICAL EQUIVALENT*

The frequency-domain transfer function:

$$\frac{H}{M} = \frac{1}{\dfrac{1}{\omega_o{}^2}S^2 + \dfrac{2\alpha}{\omega_o{}^2}S + 1}$$

Where:

ω_o = the resonant frequency, radian/second

α = the damping coefficient, 1/second

$h = $ the input signal

$m = $ the output signal

$G = $ the steady-state gain

(steady-state output/input)

Second-order measuring means are usually adjusted so that they are critically damped. This provides the most rapid transient response without overshoot.

Example 6.5 Determine the lag coefficient of a mercury thermometer similar to Figure 6.6. The bulb has a diameter of 0.5 centimeter and a length of 1.5 centimeters. The film coefficient is estimated to be 30 watts/meter^2—Kelvin. The resistance of the thermometer wall is negligible compared with the film coefficient. The thickness of the wall may also be neglected.

Solution:

a. Bulb surface $= A = \dfrac{\pi}{4}D^2 + \pi DL.$

$$A = \pi[(0.005)^2/4 + (0.005)(0.015)]$$

$$A = \pi[6.25 \times 10^{-6} + 75 \times 10^{-6}] = \pi 81.25 \times 10^{-6}$$

$$A = 2.55 \times 10^{-4} \text{ square meter}$$

$$R_t = \frac{1}{Ah} = \frac{1}{(2.55 \times 10^{-4})(30)} = \frac{1}{7.65 \times 10^{-3}}$$

$$R_t = 131 \text{ Kelvin/watt}$$

b. Bulb volume $= \dfrac{\pi}{4}D^2L.$

$$\frac{\pi}{4}D^3L = \pi(0.005)^2(0.015)/4$$

$$= 2.94 \times 10^{-7} \text{ cubic meter}$$

From Table 3, Appendix A, the density and specific heat of mercury are:

$$\rho = 13,600 \text{ kilograms/cubic meter}$$

$$S_h = 140 \text{ joule/kilogram—Kelvin}$$

$$C_t = WS_h = \rho V S_h$$

$$C_t = (1.36 \times 10^4)(2.94 \times 10^{-7})(1.4 \times 10^2)$$

$$C_t = 0.56 \text{ joule/kilogram—Kelvin}$$

c. The lag coefficient is equal to the time constant, $(\tau = R_t C_t)$.

$$R_t C_t = (131)(0.56) = 73.4 \text{ seconds}$$
$$\text{lag coefficient} = 73.4 \text{ seconds}$$

6.5 Summary of Terminology

Accuracy: The accuracy of a measuring means is the maximum possible deviation between the true value and the measured value expressed as a percentage of the full-scale range.

Calibration: A procedure in which known values of the measured variable are applied to the measuring means, and the measured value is recorded and plotted on a graph. The calibration provides data on the accuracy, linearity, and hysteresis of the measuring means.

Calibration curve: The graph of the data obtained during the calibration of a measuring means.

Conformity: The measure of the deviation between the true value and the measured value in a non-linear measuring means (see *linearity*).

Critical damping: The value of damping in a second-order system which provides the most rapid transient response without overshoot.

Dead time: The length of time that elapses between a change in the measured variable and the first noticeable change in the output of the measuring means.

Dead zone: The largest change of a measured variable that produces no response in the measuring-means output. The measuring means cannot respond to any change which is less than the dead zone.

Dynamic error: The deviation between the true value and the measured value which is caused by the dynamic lag of the measuring means. The dynamic error is independent of and in addition to the static error. The dynamic error depends on the rate of change of the measured variable, and reduces to zero when the measured variable is not changing (see Figure 6.5).

Dynamic Lag: The time it takes for the measured value to reach a certain level after the true value has reached the same level (see Figure 6.5).

Error: The difference between the true value and the measured value of a variable.

Full scale range: The interval over which a measuring means is calibrated. It is equal to the maximum value of the measured variable minus the minimum value (see *span*). Also, the output over the same interval is called the *full-scale output.*

Hysteresis: The difference in the measured value of the same variable when approached first with an increasing output and then with a decreasing output. Hysteresis is usually expressed as a percentage of the full scale-output.

Lag coefficient: The time required for the response to reach 63.2 per cent of its final change.

Linearity: The degree with which the calibration curve approaches a straight line. Linearity is usually measured in terms of the maximum deviation between the calibration curve and a straight line.

Measurand: Another name for the measured variable.

Reliability: A measure of the probability that a measuring means will continue to perform within specified conditions of accuracy, time, and environment.

Reproducibility: The maximum deviation from the average of repeated measurements of the same static variable.

Resolution: The smallest change in the measured variable that will produce a detectable change in the output of the measuring means.

Sensitivity: The change in output divided by the corresponding change in input.

Span: The interval over which a measuring means is calibrated. It is equal to the maximum value of the measured variable minus the minimum value (see *full-scale range*).

Static error: The deviation of the measured value from the true value when the measured value is maintained at a constant value.

EXERCISES

6–1.   Write a sentence that explains the term, "lag coefficient."

6–2.   The following data were obtained from a temperature measuring means which was plunged from a liquid bath maintained at 50° C into a second bath maintained at 100° C. Plot the response curve and determine the speed of response.

Time, seconds	Temperature, °C
0	50
5	57
10	60
15	71
20	77
25	81
30	85
35	88
40	90
45	91
50	93
55	94
120	100

6–3. A pressure measuring means has a range of 0 to 2×10^5 newtons per square meter. Several calibration tests produced the following results.

     Actual or true value: 0.8×10^5 newtons/meter^2

     Measured values: 0.81×10^5, 0.83×10^5, 0.80×10^5, 0.81×10^5, 0.80×10^5 newtons/square meter

Determine the worst-case error and the reproducibility in newtons/square meter and in per cent.

6–4. A potentiometer is used to measure the position of a shaft. The input range of shaft positions is from $-160°$ to $+160°$. The corresponding output range of the potentiometer is from -20 volts to $+20$ volts. The accuracy is $\pm$ 1 per cent. If the output is $+10$ volts, what is the nominal position? What are the minimum and maximum possible positions?

6–5. A speed sensor is used to measure the speed of an electric motor. The actual speed is increasing at a constant rate of 10 rpm per millisecond. The dynamic lag is 4 milliseconds. What is the dynamic error in rpm? If the range of the speed sensor is 0 to 5000 rpm, what is the dynamic error, in per cent?

6–6. Determine the lag coefficient of an alcohol-filled thermometer. The bulb has a diameter of 0.25 centimeters and a length of 0.85 centimeter. The film coefficient is estimated to be 12 watts/meter^2—Kelvin. Ignore the resistance and thickness of the bulb wall.

6–7. The following results were obtained from the calibration of a pressure measuring means.

True Value (psi)	Measured Value (increasing)	Measured Value (decreasing)
3	2.98	3.05
5	4.96	5.05
7	6.95	7.05
9	8.95	9.05
11	10.95	11.05
13	12.95	13.04
15	14.95	15.02

Determine the hysteresis in psi and in per cent.

6–8. A potentiometer has 1000 turns of wire and a full-scale output of 100 volts. The output changes in steps of 0.1 volt as the wiper moves from wire to wire. Determine the resolution as a percentage of the range.

6–9. The following results were obtained from the calibration of a force transducer.

Input Force, Newtons	Output Voltage
0	0.06
2	0.63
4	1.20
6	1.77
8	2.35
10	2.94
12	3.55
14	4.17
16	4.80
18	5.43
20	6.06

Plot the data on a graph. Draw a straight line through the two points at 20 per cent and 80 per cent of the full scale (i.e., 4 and 16 newtons). Determine the non-linearity in volts and per cent.

6–10. Determine the sensitivity of the position sensor of Exercise 4 in volts per degree.

6–11. Determine the sensitivity of the force transducer of Exercise 9 in volts per newton at 20 per cent and 80 per cent of full scale. Hint: divide the change in output voltage between 2 and 6 newtons by the change in force of $6 - 2 = 4$ newtons to determine the sensitivity at 20 per cent.

7

Microcomputers and Solid-State Components

7.1 Introduction

The evolution of electronic components has come so rapidly that some have called it the microelectronic revolution. The microprocessor, in particular, has completely changed the way that digital computers are used in control systems. In the past, a digital computer was too expensive to devote to a single control loop. The computer had to be shared among many control loops to compete economically with conventional controllers. The low cost and reliability of the microprocessor have changed all that. Not only is it possible to use a microcomputer as a devoted controller for a single loop, but microprocessors are being distributed among almost all parts of the control loop. There are microprocessors in control valves, electric drives, signal conditioners, and soon even in measuring means. New and unforeseen applications of microcomputers will develop as the costs of these systems continue to drop throughout the 1980s.

7.2 Digital Fundamentals

The purpose of this section is to summarize the fundamental concepts and terminology of digital computers. A complete development of these topics is included in books devoted to digital technology.

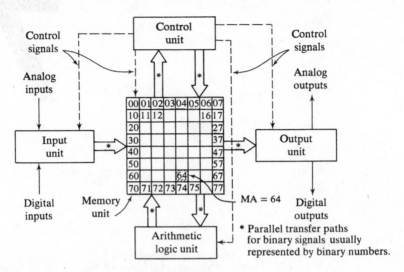

FIGURE 7.1
THE BASIC ELEMENTS OF A COMPUTER

A *digital computer* is a program-controlled device made up of five basic elements as shown in Figure 7.1.

The *input unit* receives both analog and digital input signals and conditions the signals for storage in the memory unit. Analog and digital signals are explained in Sections 3.7 and 3.8. The analog signals are converted into a binary number by a device called an *analog-to-digital converter* (A/D). This is necessary because the computer works only with binary numbers. The digital inputs are already in binary form, so the conditioning is simply a matter of producing the correct voltage levels for use in the computer.

The *output unit* receives binary numbers from the memory unit and conditions them for use in external devices such as a digital indicator, a printer, a control valve, or some other final control element. If the external device requires an analog signal, the binary number from memory is converted by a *digital-to-analog converter* (D/A). If the external device requires a digital signal, then the conditioning is a matter of producing the correct voltage and power for use by the external device.

The *memory unit* stores binary numbers. It can be visualized as a unit consisting of a large number of storage cells, each capable of storing one binary number. Each square in the memory unit in Figure 7.1 represents one storage cell in an eight by eight matrix with a total of 64 cells. Each cell can be identified by the number of its row and column. For example, the cross-hatched cell is in the 6th row and the 4th column, so it would be identified by the number 64. In this manner, each cell has a unique number which is called its *memory address* (MA). A primary characteristic of a memory unit is the number of binary digits (bits) that can be stored in each memory cell. The number of bits each memory cell can store is called the *word length* of the memory unit.

As its name implies, the *arithmetic logic unit* performs the arithmetic and logic operations. All operations are performed with binary numbers. However, some microcomputers use a binary code to represent decimal digits and are capable of performing decimal arithmetic using the binary coded decimal digits (BCD).

The *control unit* coordinates and sequences all operations. It fetches the program instructions one at a time from the memory unit, interprets each instruction, and then executes the operation specified by the instruction. The control unit is capable of making decisions based upon the outcome of an operation and then modifying its operation based on the result of the decision. Since the memory unit can store only binary numbers, the program instructions are in the form of binary numbers.

Digital computers use binary numbers to represent numerical values and various binary codes to represent other information such as decimal digits, alphabetic characters, punctuation marks, etc. The binary system has only two symbols, 1 and 0, which can be implemented by any ON–OFF signal. A signal that is ON is given the value one (1), and a signal that is OFF is given the value zero (0). The ON–OFF signals can be produced by mechanical switches, electromechanical relays, transistor switches, or integrated circuits which may contain thousands of switching elements.

The binary numbering system is a positional numbering system similar to the more familiar decimal system. A positional numbering system is one in which the position of a symbol affects the value that the symbol represents. In the decimal system, each digit position represents a different power of ten, i.e., $10^0 = 1$, $10^1 = 10$, $10^2 = 100$, $10^3 = 1,000$, etc. For example, the decimal number 345 has the following representation:

$$345$$

$$5 \times 10^0 = \quad 5 \times 1 = \quad 5$$
$$4 \times 10^1 = \quad 4 \times 10 = \quad 40$$
$$3 \times 10^2 = 3 \times 100 = 300$$

In decimal 345, the numeral 5 represents five ones, the numeral 4 represents four tens, and the numeral 3 represents three hundreds. In a similar way, each digit position in a binary number represents a power of two. The digits in a binary number are usually referred to as bits, a contraction of binary digit. Starting on the right and moving to the left, the bit positions represent ascending powers of two, i.e., $2^0 = 1$, $2^1 = 2$, $2^2 = 4$, $2^3 = 8$, etc. For example, the binary number 1011 has the following representation:

$$
\begin{array}{l}
1011 \\
\quad\;\; 1 \times 2^0 = 1 \times 1 = 1 \\
\quad\;\; 1 \times 2^1 = 1 \times 2 = 2 \\
\quad\;\; 0 \times 2^2 = 0 \times 4 = 0 \\
\quad\;\; 1 \times 2^3 = 1 \times 8 = 8
\end{array}
$$

It is easy to convert a binary number into the equivalent decimal number by simply multiplying each bit by the power of two that it represents.

Example 7.1   Convert 1011 (binary) and 1100 (binary) to the decimal equivalent.

Solution:

$$1011 = 1 \times 2^3 + 0 \times 2^2 + 1 \times 2^1 + 1 \times 2^0 = 11 \quad \text{(decimal)}$$
$$1100 = 1 \times 2^3 + 1 \times 2^2 + 0 \times 2^1 + 0 \times 2^0 = 12 \quad \text{(decimal)}$$

The value of the first-order position is called the *base* or *radix* of the numbering system. The decimal system is base 10, and the binary system is base 2. Two other bases are used to work with digital computers. These are the octal system, base 8, and the hexadecimal system, base 16. These two numbering systems are used as a convenient way for people to handle binary numbers.

The octal system uses the symbols 0–7. Each digit position in the octal system represents a power of 8, e.g., $8^0 = 1$, $8^1 = 8$, $8^2 = 64$, etc. Each octal digit is equivalent to three binary digits, and the octal system is used as a convenient way of handling binary numbers. For example, the octal number 371 is easier to handle than its binary equivalent 011111001 (or 011 111 001). Table 7.1 gives the binary equivalent of each octal digit. Examples 7.2 and 7.3 illustrate the methods for converting from octal to binary and from binary to octal.

The hexadecimal system uses the symbols 0–9 and A–F. Each digit position in the hexadecimal system represents a power of 16, e.g., $16^0 = 1$, $16^1 = 16$, $16^2 = 256$, etc. Each hexadecimal digit is equivalent to four binary digits, and the hexadecimal system is also used as a convenient way of handling binary numbers. For example, the hexadecimal number 9B is easier to handle than its binary equivalent 10011011 (or 1001 1011). Table 7.2 gives the binary equivalent of each hexadecimal digit.

TABLE 7.1

BINARY EQUIVALENT OF THE OCTAL DIGITS

Octal digit	Binary equivalent
0	000
1	001
2	010
3	011
4	100
5	101
6	110
7	111

TABLE 7.2

BINARY EQUIVALENT OF THE HEXADECIMAL DIGITS

Hexadecimal digit	Binary equivalent	Hexadecimal digit	Binary equivalent
0	0000	8	1000
1	0001	9	1001
2	0010	A	1010
3	0011	B	1011
4	0100	C	1100
5	0101	D	1101
6	0110	E	1110
7	0111	F	1111

Examples 7.4 and 7.5 illustrate the methods for converting from binary to hexadecimal and from hexadecimal to binary.

Example 7.2   Convert 101 100 111 010 (binary) to octal digits.

Solution:

Start at the right end of the number and form three bit groups. Then use Table 7.1 to convert each group to an octal digit.

$$101 \quad 100 \quad 111 \quad 010$$
$$5 \quad\; 4 \quad\; 7 \quad\; 2$$
101 100 111 010 (binary) = 5472 (octal).

Example 7.3   Convert 4621 (octal) to binary digits.

Solution:

Use Table 7.1 to convert each octal digit to the equivalent three binary bits.

$$
\begin{array}{cccc}
4 & 6 & 2 & 1 \\
100 & 110 & 010 & 001
\end{array}
$$
$$
4621 \ \text{(octal)} = 100\,110\,010\,011 \ \text{(binary)}
$$

Example 7.4   Convert 1011 0011 0010 (binary) to hexadecimal digits.

Solution:

Start at the right end of the number and form four bit groups. Then use Table 7.2 to convert each group of bits to its hexadecimal equivalent.

$$
\begin{array}{ccc}
1011 & 0011 & 0010 \\
B & 3 & 2
\end{array}
$$
$$
1011\,0011\,0010 \ \text{(binary)} = \text{B32} \ \text{(hexadecimal)}
$$

Example 7.5   Convert E2C (hexadecimal) to binary digits.

Solution:

Use Table 7.2 to convert each hexadecimal digit to the equivalent four binary bits.

$$
\begin{array}{ccc}
E & 2 & C \\
1110 & 0010 & 1100
\end{array}
$$
$$
\text{E2C} \ \text{(hexadecimal)} = 1110\ 0010\ 1100 \ \text{(binary)}
$$

Binary numbers are fine for digital computers, but we live in a world of decimal numbers. Consequently, we would like to use computers to process decimal numbers and give us decimal results. There are two ways that decimal numbers can be processed in a computer. The first is to convert the decimal numbers to their binary equivalent. The computer then processes the binary equivalents and converts the results back to decimals for presentation to the user. There are several methods for converting numbers from decimal to binary. One method is illustrated in Example 7.6.

A second method of handling decimal numbers is to use a code to convert each digit of the decimal number to a four-bit binary character. The codes used for this purpose are called BCD (*Binary Coded Decimal*) codes. There are almost 30 billion possible BCD codes, but only a few are used to any extent. Table 7.3 shows two of the most useful BCD codes. Special provisions are required to enable a computer to process numbers represented by a BCD code. Some microprocessors include a decimal mode of operation that allows arithmetic operations to be performed directly on the BCD representation. Other microprocessors include a DECIMAL ADJUST instruction that adjusts the result of a normal binary arithmetic operation to give the correct result when BCD representations are used. Other conversion methods can be found in books devoted to digital technology.

TABLE 7.3

BINARY CODED DECIMAL (BCD) CODES

Decimal	8421 code	Excess 3 code
0	0000	0011
1	0001	0100
2	0010	0101
3	0011	0110
4	0100	0111
5	0101	1000
6	0110	1001
7	0111	1010
8	1000	1011
9	1001	1100

Example 7.6   Convert 217 (decimal) to its binary equivalent.

Method:

Begin the conversion by finding the greatest power of 2 that is less than or equal to the number to be converted. Subtract this power of 2 from the number to be converted. Then find the greatest power of 2 that is less than or equal to the remainder and subtract this second power of 2 from the remainder. Repeat this procedure, always using the most recent remainder, until a remainder of 0 is obtained. The binary equivalent will have a 1 in the bit position of each power of 2 that was subtracted and 0 in all other bit positions. Some powers of 2 are given below.

x	0	1	2	3	4	5	6	7	8
2^x	1	2	4	8	16	32	64	128	256

Solution:

The greatest power of 2 contained in decimal 217 is $2^7 = 128$

$$217 - 128 = 89$$

The greatest power of 2 contained in decimal 89 is $2^6 = 64$

$$89 - 64 = 25$$

The greatest power of 2 contained in decimal 25 is $2^4 = 16$

$$25 - 16 = 9$$

The greatest power of 2 contained in decimal 9 is $2^3 = 8$

$$9 - 8 = 1$$

The greatest power of 2 contained in decimal 1 is $2^0 = 1$

$$1-1 = 0$$

The binary equivalent of decimal 217 is 11011001

Example 7.7 Use the 8421 code to convert the decimal number 49 to an 8-bit BCD representation with the lower four bits used for the lower-order decimal digit and the upper four bits used for the higher-order decimal digit.

Solution:

The BCD representation of 4 is 0100 and the BCD representation of 9 is 1001.

$$49 \text{ (decimal)} = 01001001 \text{ (BCD)}$$

A number of codes are used for communication between computers and various peripheral devices such as teletypes, CRT terminals, line printers, etc. The two most common codes are the ASCII (American Standard Code for Information Interchange) and EBCDIC (Extended Binary-Coded-Decimal Interchange). The ASCII code is the more popular of the two codes. There are 128 ASCII code words which require seven bits to represent. An eighth bit is provided for use in an error-detecting scheme known as *parity*. The parity bit is used to make the total number of ones in a code word even or odd, depending on the convention used. For example, the seven-bit code for the letter A is 1000001. If even parity is used, then a 0 is placed at the left end of the number to give 01000001 as the code for an A. In a like manner, the seven-bit code for C is 1000011, so a 1 is placed at the left end to give 11000011 as the code for a C. Both codes now have an even number of ones. If this is done for all the coded characters, then the receiver can determine if one bit has been changed during transmission by checking the number of ones to see if the sum is even. If the sum is not even, then there was an error during the transmission of the coded word. Some of the ASCII charatcer codes are shown in Table 7.4.

7.3 Microprocessors and Microcomputers

A *microprocessor* is a single electronic component that contains the entire arithmetic and control section of a computer. It is a large-scale integrated circuit (LSI), a term applied to an integrated circuit that contains more than 1,000 transistors. Included in the microprocessor are several storage registers for holding binary numbers, an instruction

TABLE 7.4

SOME ASCII CHARACTER CODES*

Code	Character	Code	Character	Code	Character
x010 0000	space	x011 1000	8	x101 0000	P
x010 0001	!	x011 1001	9	x101 0001	Q
x010 0010	"	x011 1010	:	x101 0010	R
x010 0011	#	x011 1011	;	x101 0011	S
x010 0100	$	x011 1100	<	x101 0100	T
x010 0101	%	x011 1101	=	x101 0101	U
x010 0110	&	x011 1110	>	x101 0110	V
x010 0111	'	x011 1111	?	x101 0111	W
x010 1000	(	x100 0000	@	x101 1000	X
x010 1001	)	x100 0001	A	x101 1001	Y
x010 1010	*	x100 0010	B	x101 1010	Z
x010 1011	+	x100 0011	C	x101 1011	[
x010 1100	,	x100 0100	D	x101 1100	\
x010 1101	−	x100 0101	E	x101 1101	]
x010 1110	.	x100 0110	F	x101 1110	↑
x010 1111	/	x100 0111	G	x101 1111	←
x011 0000	0	x100 1000	H		
x011 0001	1	x100 1001	I	x000 1011	Line feed
x011 0010	2	x100 1010	J	x000 1101	Carriage
x011 0011	3	x100 1011	K		return
x011 0100	4	x100 1100	L	x000 0111	Bell
x011 0101	5	x100 1101	M	x111 1111	Rubout
x011 0110	6	x100 1110	N	* The x is the parity bit.	
x011 0111	7	x100 1111	O		

decoder to interpret instructions, an arithmetic logic unit, and a control unit. The microprocessor controls the fetching and executing of the instructions in the program. A microprocessor is an electronic component whose function is determined by the sequence of instructions that make up the program, i.e., it is a programmable electronic component.

A microcomputer consists of a microprocessor, a section of read-only memory (ROM, PROM, or EPROM), a section of random-access memory (RAM), an input/output (I/O) interface, an address bus, and a data bus. A simplified schematic of a microprocessor is shown in Figure 7.2.

The type of memory currently used in microcomputers is volatile memory, i.e., the contents of memory are lost when the power supply is interrupted or turned off. For this reason, a large portion of the memory in the microprocessor is permanently set during the last stage of its manufacture. This type of memory is called read-only memory, or simply ROM. Once set, the contents of ROM cannot be changed,

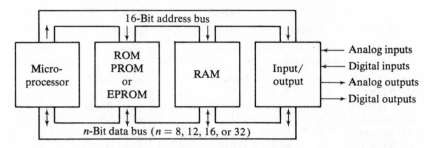

FIGURE 7.2
A MICROCOMPUTER

so it is used to store the permanent programs and data that will be used by the microcomputer. A major disadvantage of ROM is that the user cannot alter the contents to make changes or correct errors in the permanently stored program.

Programmable read-only memory (PROM) is a type of memory that contains all ones or all zeros as manufactured. The user can create a permanent pattern of ones and zeros by applying a specified voltage that destroys a fusible link. Once the PROM is burned in, it also is permanent, but it does give the user more flexibility than ROM. Erasable PROM (EPROM) is also programmable electrically, but not permanently. The user can use an ultraviolet light source to restore the EPROM so that a new program can be set into the memory. In this way, the EPROM can be reprogrammed many times.

Random-access memory can be written in or read out at any time. The operator can enter data or instructions into RAM from an input keyboard or terminal. The microprocessor can also write in RAM on command and then read out as required. The term "random access" refers to the ability of the microprocessor to go directly to any location in the memory unit for a READ or WRITE operation. This is in contrast to a nonrandom-access memory, such as magnetic tape, in which the contents are read out in sequence as they appear on the tape.

A *bus* is a set of conductors intended to provide a transfer path between various components in the microprocessor. The number of conductors determines how many binary signals can be transferred at the same time (in parallel). Figure 7.2 shows two such transfer paths; the data bus and the address bus. As the name implies, the data bus is intended to transfer data (or instructions) from one component to another. The number of lines in the data bus coincides with the number of bits that can be transferred in parallel, and it is called the *word length* of the microcomputer. Standard word lengths of microcomputers are 4, 8, 12, 16, and 32 bits. The term *byte* is used to describe the 8-bit word length. The address bus is also a set of conductors, usually 16, used to transfer the binary address of one storage cell in the computer memory. The memory unit decodes the number on the address bus and selects the specified storage cell for a READ or WRITE operation.

The input/output interface connects the microcomputer to the outside world. The most common means of input or output is through a parallel I/O port. There are as many lines in the I/O port as there are bits in the microcomputers word length. This means that the input or output data can be handled in the same way as all other data inside the microcomputer. In some microprocessors, each I/O port is identified by a unique memory address, just as if it were a storage cell in the memory unit. Obviously, the I/O port address cannot be used by a memory cell, but it does mean that the microprocessor can treat the I/O port as if it was a storage cell in the memory. This feature is called *memory-mapped I/O*. In this way, the input and output data can be handled as easily as data stored in memory. Further details of I/O interfaces are included in the next section.

7.4 Communications Interfaces

A major concern in the utilization of microcomputers in control systems is the need for communication among the various parts of the system. In its simplest form, this communication is typically a current signal on a pair of wires from a measuring means transmitter to an analog-to-digital converter (A/D) in the microprocessor I/O interface. Beyond this point, communications may connect a microprocessor to a large supervisory computer or to a wide variety of terminals, printers, and other peripheral devices. An understanding of the basics of communications is becoming increasingly important with the increased distribution of microcomputers throughout the control system. Figure 7.3 illustrates the variety of communications interfaces in a distributed control system.

Several characteristics distinguish digital communications interfaces. Digital data may be transmitted in serial (one bit at a time) or in parallel (all bits in a word at the same time). In either serial or parallel, the data may be transmitted synchronously or asynchronously. Synchronous transmission means that a timing signal is used to make sure the transmitter and receiver act together. Asynchronous transmission requires the use of control signals between the transmitter and receiver to make sure the receiver accepts the data before the transmitter terminates the transmission. This procedure is sometimes referred to as "handshaking" or the "please-and-thank-you" routine. Protocol is an-

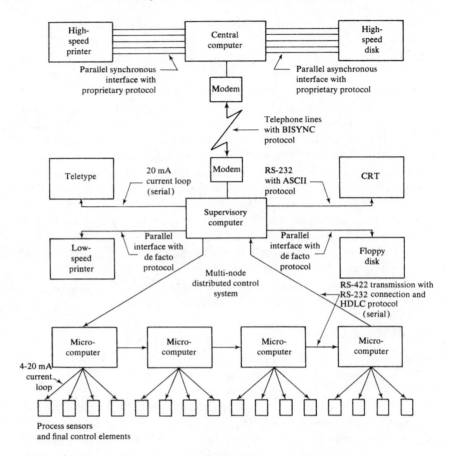

FIGURE 7.3 [1]

A DISTRIBUTED CONTROL SYSTEM

other characteristic of a communications interface. It defines the format of the data, the meaning of the control signals, the order in which messages are transmitted, and the method used to check for errors in transmission. The final characteristic is the mechanical and electrical configuration of the interface. This includes such items as the voltage levels (or current levels) of the signals, the number of conductors, and the mechanical configuration of the interface.

Four interfaces designed for use in the process industry will be examined in more detail. These are the CAMAC serial ASCII, IEEE-488, and SDLC/HDLC. All four interfaces have been standardized by at

---

[1] J. Washburn, "Communications Interface Primer—Part I" Instruments and Control Systems, March 1978, p. 43.

least one of several national and international standards-making organizations.

The CAMAC (or IEEE-583) interface is the most completely specified interface system. It was originally designed for nuclear instrumentation labs where instruments were constantly swapped among systems. The IEEE-583 standard completely defines the mechanical configuration, the electrical connectors, the data transmission paths, and the protocol. The mechanical configuration consists of a "crate" containing slots for 25 modules. The two slots on the right are occupied by a controller that manages all CAMAC communications. Five types of controllers are available: one for direct access to a computer, one for parallel transmission, two for different types of serial transmission, and one for stand-alone operation. Figure 7.4 illustrates these different possibilities with a CAMAC interface.

By far the most universal communication interface is serial ASCII. Some ASCII character codes are presented in Table 7.4. The ASCII code was originally designed for teletype communication. It is usually used with a 20-milliamp current loop for long distances or with an

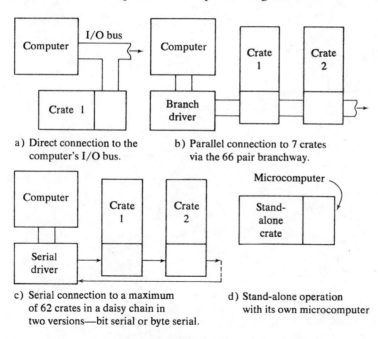

a) Direct connection to the computer's I/O bus.

b) Parallel connection to 7 crates via the 66 pair branchway.

c) Serial connection to a maximum of 62 crates in a daisy chain in two versions—bit serial or byte serial.

d) Stand-alone operation with its own microcomputer

FIGURE 7.4 [2]

CAMAC'S DIFFERENT CONFIGURATIONS

[2] J. Washburn, "Communications Interface Primer—Part II," Instruments and Control Systems, April, 1978, p. 61.

RS232-C interface for short distances (less than 50 feet). The RS232-C is a very popular interface that is available as a standard feature or an

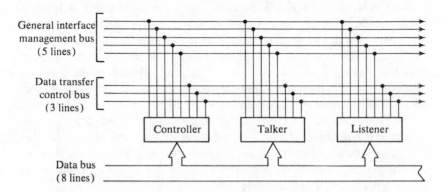

The bus controller uses the management and transfer control lines to handle bus traffic. The open-ended bus connects up to 15 devices, but total cable distance cannot exceed 66 ft. Device addresses are set by five-bit jumpers or switches on the back panel of each instrument.

FIGURE 7.5 [3]

THE IEEE-488 INTERFACE

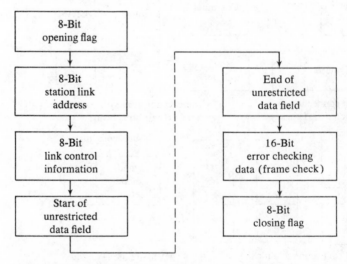

An SDLC message frame contains a number of information fields that serve to synchronize messages, identify the destination station and check the transmission for errors. The data field itself is unrestricted.

FIGURE 7.6 [3]

AN SDLC MESSAGE FRAME

---

[3] Ibid, pp. 62–63.

option on most microcomputers. Most of the operator terminals (tele-type, CRT displays, etc.) use an ASCII interface. The RS232 standard specifies a 25-pin connector. Two lines are used for data transmission, one for signal ground, and seven for handshaking signals.

The IEEE-488 data bus was originally developed by Hewlett–Packard for use with its own minicomputers and instruments. The interface was well received and both IEEE and ANSI (American National Standards Institute) made it a standard. Often referred to as the general-purpose interface bus (GPIB), the IEEE-488 interface has been built in to a wide range of instruments, peripherals, and calculators. The IEEE-488 interface is described in Figure 7.5.

The synchronous data link control (SDLC) and the high-level data link control (HDLC) are primarily designed to control the logical flow of information between two or more computers. The protocols define the method by which the remote computer stations are addressed and how the stations communicate with each other. The two systems are very similar in the way they work. An SDLC message frame is illustrated in Figure 7.6.

7.5 D/A and A/D Converters

Most measuring means in control systems produce an electrical ana-log signal such as the 4- or 20-milliamp output from a temperature transmitter (see Figure 1.7). These analog signals are unsuitable for use in a microcomputer because they are not in a binary form. An analog-to-digital (A/D) converter is used to convert the analog signal into a binary signal that represents the approximate value of the analog signal. Ofter a device, called a *sample/hold,* is included with the A/D converter. As the name implies, the sample/hold samples the analog signal and holds (or stores) the sampled value until the AD conversion is completed. This guarantees that the analog signal does not change during the conversion process.

Many of the final control elements in a control system require an analog signal. Since a microcomputer is not capable of producing an analog output signal directly, a digital-to-analog (D/A) converter is used to convert the binary signal from the microcomputer into an analog signal. A D/A converter is much easier to build than an A/D converter. In fact, many A/D converters include a D/A converter as part of the system. Two resistive network D/A converters are shown in Figure 7.7. The binary ladder D/A converter is simpler to build because only two resistor sizes are required.

Example 7.8 The logic levels of the input to a binary ladder D/A converter are 4 volts for a logic 1 and 0 volts for a logic 0. Determine the analog output voltage when the binary input is 1011.

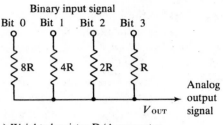

a) Weighted resistor D/A converter

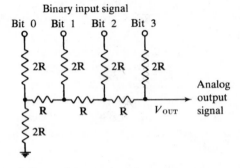

b) Binary ladder D/A converter

Binary input	Analog output Weighted resistor	Analog output Binary ladder
0000	0	0
0001	V/15	V/16
0010	2V/15	V/8
0011	3V/15	3V/16
0100	4V/15	V/4
0101	5V/15	5V/16
0110	6V/15	3V/8
0111	7V/15	7V/16
1000	8V/15	V/2
1001	9V/15	9V/16
1010	10V/15	5V/8
1011	11V/15	11V/16
1100	12V/15	3V/4
1101	13V/15	13V/16
1110	14V/15	7V/8
1111	V	15V/16

FIGURE 7.7

RESISTIVE NETWORK D/A CONVERTERS

Solution:

$$\text{Analog output} = 11V/16 = (11)(4)/16 = 11/4 = 2.75 \text{ volts}$$

A counter-type A/D converter is illustrated in Figure 7.8. This is a relatively easy circuit to construct using the components specified in the schematic. The timing diagram is quite useful to help explain the conversion operation. The 555 clock waveform is shown at the top of the timing diagram. The output of the clock is fed into the binary counter so that each clock pulse will cause the counter to increase its count by one bit. The output of the counter will sequence from 0000 to 1111 in the same sequence as the binary inputs in Figure 7.7. When the counter reaches 1111, the next clock pulse restores the 0000 count and the sequence starts over. The counter output is shown in binary form on the second section of the timing diagram.

The binary output of the counter is fed into a binary ladder D/A converter producing the stairstep output shown in the third section of the timing diagram. The stairstep levels are determined from the table

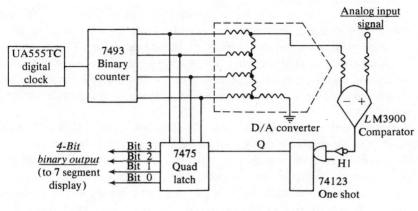

a) Schematic diagram

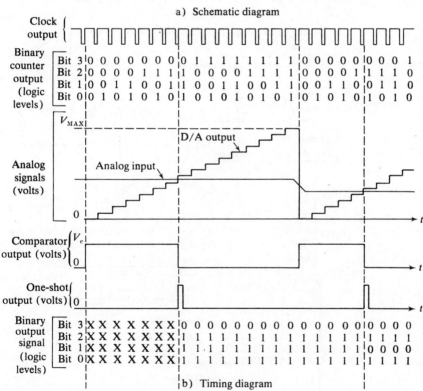

b) Timing diagram

FIGURE 7.8

*SCHEMATIC DIAGRAM AND WAVEFORMS OF A
COUNTER-TYPE A/D CONVERTER*

in Figure 7.7. The analog input signal (the signal that is to be converted to digital) is superimposed on the stairstep waveform to show the relative size of the two inputs to the comparator. The output of the comparator is positive when the analog input signal is greater than the analog signal from the D/A converter. When the relative magnitude of the two analog signals reverses, the output of the comparator reverses also.

Notice that the comparator output switches from $+V_c$ to 0 at the point that the analog input signal crosses the stairstep output from the D/A converter. The one-shot circuit converts the $+V_c$ to 0 change in the comparator output into a pulse that is used to trigger the quad latch. Each time the quad latch is triggered, the binary signal from the counter is transferred to the output of the latch and held there until

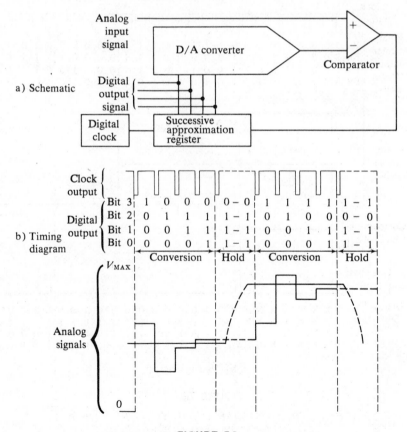

FIGURE 7.9

*SCHEMATIC AND WAVEFORMS OF A SUCCESSIVE
APPROXIMATION A/D CONVERTER*

the next triggering pulse. The result is that the output of the quad latch reflects the count for which the analog input signal was approximately equal to the stairstep voltage. The binary output signal from the A/D converter is shown on the bottom line of the timing diagram.

The successive approximation A/D converter shown in Figure 7.9 is the type most frequently used for control systems. It is much faster than the counter type discussed previously or the dual slope type that will be discussed next. The successive approximation starts at the first clock pulse by setting the most significant bit (bit 3) of the counter to a 1 and resetting the remaining bits to a 0 as shown on the second line of the timing diagram in Figure 7.9. This puts the D/A output at the middle step of the stair. The analog input is less than the D/A output, so Bit 3 is reset to a 0 and Bit 2 is set to a 1 at the second clock pulse. This time the analog output is greater than the D/A output, so at the next clock pulse, Bit 2 remains a 1 and Bit 1 is set to a 1. Again the analog input is smaller, so at the next clock pulse, Bit 1 remains a 1 and Bit 0 is set to a 1. This completes the A/D conversion. Notice that it took only 4 clock pulses instead of the 16 required by the counter-type A/D converter. The difference increases as the number of bits in the binary number is increased. A 12-bit A/D converter requires only 12 clock pulses to complete the conversion while the counter-type requires 2^{12} or 4,096 clock pulses to complete the cycle.

The dual slope A/D converter is illustrated in Figure 7.10. When V_{in} is connected to the integrator, the output of the integrator increases at a rate proportional to V_{in}. When $-V_{ref}$ is connected to the input of the integrator, the output of the integrator decreases at a rate proportional to $-V_{ref}$. Voltage V_{in} is the analog input signal to be converted. Voltage $-V_{ref}$ is a reference voltage. The converter operates as follows.

The conversion begins with the counter set to a count of 0 and the integrator output also at 0. Switch 1 is closed and the counter is started at the same time. The integrator ramps up until the counter reaches its full count and resets to 0. When this happens, switch 1 opens and switch 2 closes. The counter begins its count at 0 as the integrator output begins to decrease. The integrator decreases in a ramp until the voltage crosses 0. When this happens, the counter stops and the conversion is complete. The count in the counter is a measure of V_{in}.

The dual slope A/D converter has a high degree of accuracy with relatively little affect due to time or temperature changes. This is due to the inherent self-calibrating characteristics of the dual slope method. However, the dual slope converter is slow, so it is used primarily where accuracy is important and speed is not a primary consideration.

The number of bits in the binary signal is called the *resolution* of the A/D converter. The resolution determines how close the binary

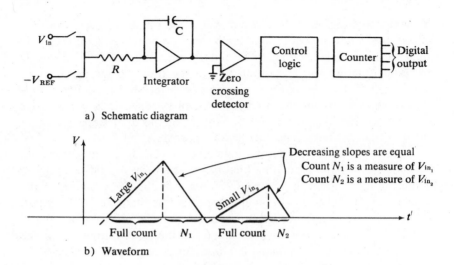

a) Schematic diagram

b) Waveform

FIGURE 7.10

A DUAL SLOPE A/D CONVERTER

output comes to an exact representation of the analog input signal. Figure 7.8 shows that it is a rare instant when the digital output is the exact representation of the analog input signal. The difference between the exact value of the analog signal and the value represented by the binary output of the A/D converter is called the *quantization error*. The quantization error can be expressed as a per cent by dividing the quantization error by the largest value that can be represented by the binary number.

Example 7.9 An A/D converter has a resolution of 4 bits and a quantization error of $\pm\frac{1}{2}$ bit. The output voltage range is from 0 to 5 volts. Determine the quantization error in volts and per cent of the full-scale range.

Solution:

The binary number 0000 represents 0 volts and the number 1111 represents 5 volts. There will be 2^4-1 or 15 steps between 0000 and 1111. A 1-bit change in the binary number is equal to a $5/15 = 0.333$ volt change in the voltage it represents. The quantization error is $\pm\frac{1}{2}$ bit which is $\pm\frac{1}{2}$ $(0.333) = \pm0.1\overline{66}$ volts. The per cent accuracy is given in the following equation:

$$\text{Per cent accuracy} = \pm(0.1\overline{66}/5)(100) = \pm3.33\%$$

7.6   Programming a Microcomputer

The study of microcomputers is incomplete without a discussion of programming. A program is a sequence of instructions that defines the operations to be performed by the microcomputer. Programming is not a trivial task. It often represents a major portion of the development cost of a computer system. The purpose of this section is to illustrate the basic procedure for programming a microcomputer without getting excessively complicated. A hypothetical microcomputer called the ICST-1 will be used for this purpose.

The ICST-1 is an 8-bit microcomputer that uses one or two bytes (8 or 16 bits) for each instruction. The first byte in each instruction specifies the operation to be performed. If the second byte is used, it specifies an address in memory used in the operation to be performed. The 8-bit operation byte is enough to specify 256 different operations, and the 8-bit address byte is enough to specify 256 addresses in memory. Most microcomputers have considerably more memory than the ICST-1, but 256 locations is enough to illustrate the programming procedure.

The arithmetic unit of the ICST-1 includes an 8-bit register called the *accumulator* (A) and two additional 8-bit registers called the *X index register* (X) and the *Y index register* (Y). The instructions used in the example are listed in Table 7.5. These instructions will be explained as the program is developed. Figure 7.11 is a simplified diagram of ICST-1 registers and transfer paths.

Writing a program for a microcomputer proceeds in five steps:

1. Define the problem in a word statement that describes the desired result.

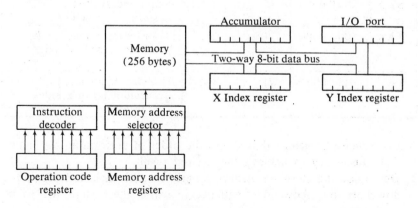

FIGURE 7.11
THE ICST-1 REGISTERS AND TRANSFER PATHS

TABLE 7.5

A PARTIAL LIST OF ICST-1 INSTRUCTIONS

Name	Binary code	Hex code	Number of bytes	Operation performed
DEX	0001 0000	10	1	Decrement X by 1
DEY	0001 0001	11	1	Decrement Y by 1
JMP	0010 0000	20	2	Jump to the memory location given in Byte 2
JMS	0011 0000	30	2	Jump to the subroutine at the address given in Byte 2
LDA	0100 0000	40	2	Load accumulator with the number stored in the memory location given in Byte 2
RTN	1111 0000	F0	1	Return from subroutine to the instruction immediately after the JMS instruction that called the subroutine
SAP	0110 0011	63	1	Skip if the accumulator is equal to or greater than 0
SIR	0110 0000	60	2	Skip the next instruction if the I/O device given in Byte 2 is ready
STA	1001 0000	90	2	Transfer the number in the accumulator to the address given in Byte 2
STP	1111 1111	FF	1	Stops the computer
SXO	0110 0001	61	1	Skip if index register X is storing the number 0
SYO	0110 0010	62	1	Skip if index register Y is storing the number 0
TAX	0111 0000	70	1	Transfer the number in the accumulator to X index register
TAY	0111 0001	71	1	Transfer the number in the accumulator to Y index register

2. Construct a flowchart that shows the order of performing the major tasks necessary to achieve the desired result.
3. Determine the steps necessary to perform each task and prepare a flowchart that shows the sequence in which these steps are to be performed.

4. Use the flowcharts to help code the instructions using names for operation codes and addresses. This is called an *assembly language program*.

5. Use the assembly language program to prepare a machine language program. The machine language program is the only program the computer understands—the operation codes and memory addresses must all be given in a binary form (i.e., binary, octal, or hexa-decimal numbers).

Example 7.10 Write a program for the ICST-1 microcomputer that will cause the teletype terminal to print the pattern shown below. The pattern should be placed in the center of the page with 35 spaces on the left of the top X. The teletype is identified by the address code 1111 0000 (binary) or FO (hexadecimal). Use a 0 for the ASCII parity bit (designated by an x in Table 7.4).

X

XXX

XXXXX

Solution:

Step 1.   Problem definition: In order to print the above pattern, the teletype must be given a series of commands using the ASCII code (see Table 7.4). The first task should be to send a carriage return and line feed command to the teletype to make sure the carriage is ready to print a new line. The next commands should be to print the first line of spaces and an X followed by a second carriage return and line feed. The remaining tasks should complete each line of the pattern with each line followed by a carriage return and line feed. Two sub-routines will be used, one to print a given number of spaces or x's and one to do the carriage return and line feed. A subroutine is a program within the program that may be used at any place in the remainder of the program by using the JMS instruction to go to the subroutine and the RTN instruction to go back when the subroutine is completed.

Step 2.   The flowchart for Step 2 is given in Figure 7.12. Although at first glance there appear to be three tasks, the operations necessary to print spaces and x's are almost identical. Therefore two tasks are defined. One task is to generate the carriage return and line feed. The second task is to print a specified number of spaces or x's.

Step 3.   The flowcharts for Step 3 are given in Figure 7.12. Task 1 is completely contained in a subroutine. The first substep is to load

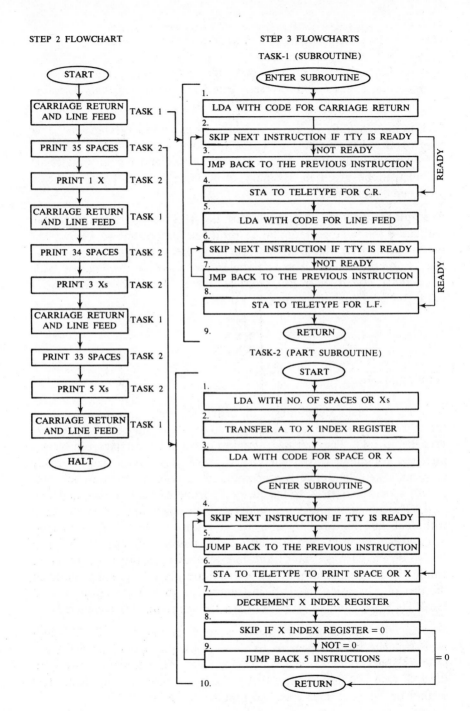

FIGURE 7.12
FLOWCHARTS FOR STEP 2 AND STEP 3

the accumulator with the ASCII code for a carriage return (see Table 7.4). Note the ASCII codes for a carriage return and a line feed must be stored in the computer memory for use when the program is run. Substeps 2 and 3 cause the computer to check if the teletype is ready to receive a command. If the teletype is not ready, the computer is directed to go back and check again. This loop back will be repeated until the teletype is ready to receive a command. This check is necessary because the computer is about 100,000 times faster than the teletype. The computer could run through the entire program before the teletype could complete the first carriage return. When the teletype is ready at substep 2, the computer is directed to skip the instruction at substep 3 and go to the instruction at substep 4. The instruction at substep 4 directs the computer to transfer the number in the accumulator to the teletype I/O port which will cause the teletype to execute a carriage return. The teletype takes about 0.1 seconds to complete the carriage return. The process of waiting until the teletype is ready and then sending out an ASCII code is repeated in substeps 5, 6, 7, and 8. This time the code is for a line feed. Substep 9 directs the computer to go back to the instruction immediately after the instruction that directed it to go to the subroutine. This return back to the source is a feature of all subroutines.

Task 2 is partly in a subroutine and partly not. The first three substeps determine whether spaces or x's will be printed and how many will be printed. The number of spaces and x's for each line must be stored in the memory unit. For example, the binary equivalent of 35 (decimal) must be stored in memory so the computer can use it to direct the teletype to print 35 spaces. Also, the ASCII codes for a space and an x must be stored in memory so they can be loaded into the accumulator in substep 3. Substeps 4 and 5 are the same as substeps 2 and 3 in Task 1. These are the instructions that make the computer cycle until the teletype is done printing the last character and is ready to receive the next character. The sixth substep directs the computer to transfer the ASCII code from the accumulator to the I/O port so the teletype will print the space or x. The seventh substep decrements the number in the x register (i.e., subtracts one from the number in the X register). If the X index register is not 0 after substep 7, the jump at substep 9 directs the computer to go back to substep 4 to print again. If the X index register is 0 at substep 8, then substep 9 is skipped and the computer returns to the instruction after the JMS instruction that directed it to the subroutine. In completing Task 2, Substeps 1, 2, 3, and a JMS instruction will be in the main program. Substeps 4 through 10 will be in the second subroutine.

TABLE 7.6

NAMES OF DATA AND ADDRESSES USED IN THE PROGRAM

Name	Data value	Data address	Description
CODECR	0000 1101		ASCII code for carriage return
CODELF	0000 1011		ASCII code for line feed
CODESP	0010 0000		ASCII code for space
CODEX	0101 1000		ASCII code for X
SPACE1	0010 0011		No. of spaces in line 1
SPACE2	0010 0010		No. of spaces in line 2
SPACE3	0010 0001		No. of spaces in line 3
X1	0000 0001		No. of x's in line 1
X2	0000 0011		No. of x's in line 2
X3	0000 0101		No. of x's in line 3
TTY		F0	Teletype address code
SUB1			Address of Subroutine 1
SUB2			Address of Subroutine 2

Step 4. Code the program using names for the operations and addresses in memory. The first step is to assign names to the data that must be stored in memory. Table 7.6 is the list of assigned names. The columns for data value and data address will be completed in the process of completing Step 5.

The assembly language program is the left half of Table 7.7. The corresponding machine language program in the right half is developed in Step 5. The flowcharts in Figure 7.12 are used as a guide in preparing the assembly language program. The first instruction in the program is a JMS to SUB1 which directs the computer to Subroutine 1 for the carriage return and line feed. When the computer completes Subroutine 1, it returns to the second instruction which loads the accumulator with the number of spaces on line 1 of the pattern. The third instruction transfers the number of spaces from the accumulator to the X index register for use in counting the spaces as they are printed in Subroutine 2. The fifth instruction directs the computer to Subroutine 2 to complete the printing of 35 spaces. The computer returns from subroutine 2 to the sixth instruction which begins the sequence to print an x followed by a carriage return and line feed. The above sequences are repeated until the pattern is completed and the STP instruction stops the computer run.

The first instruction in Subroutine 1 has an address name (SUB1) which is used in the address byte of the JMS instructions. The second instruction and the sixth instruction also have address names (LOC1 and LOC2) which are used in the address byte of the JMP instructions.

TABLE 7.7

THE ASSEMBLY AND MACHINE LANGUAGE PROGRAMS

Assembly-coded program				Machine-coded program		
Name of address	Opera-tion byte	Address byte	Task	Address	Opera-tion byte	Address byte
START	JMS	SUB1	Task 1	00	30	40
	LDA	SPACE1		02	40	84
	TAX		Task 2	04	70	
	LDA	CODESP		05	40	82
	JMS	SUB2		07	30	60
	LDA	X1		09	40	87
	TAX		Task 2	0B	70	
	LDA	CODEX		0C	40	83
	JMS	SUB2		0E	30	60
	JMS	SUB1	Task 1	10	30	40
	LDA	SPACE2		12	40	85
	TAX		Task 2	14	70	
	LDA	CODESP		15	40	82
	JMS	SUB2		17	30	60
	LDA	X2		19	40	88
	TAX		Task 2	1B	70	
	LDA	CODEX		1C	40	83
	JMS	SUB2		1E	30	60
	JMS	SUB1	Task 1	20	30	40
	LDA	SPACE3		22	40	86
	TAX		Task 2	24	70	
	LDA	CODESP		25	40	82
	JMS	SUB2		27	30	60
	LDA	X3		29	40	89
	TAX		Task 2	2B	70	
	LDA	CODEX		2C	40	83
	JMS	SUB2		2E	30	60
	JMS	SUB1	Task 1	30	30	40
	STP			32	FF	
SUB1	LDA	CODECR		40	40	80
LOC1	SIR			42	60	
	JMP	LOC1		43	20	42
	STA	TTY		45	90	F0
	LDA	CODELF		47	40	81
LOC2	SIR			49	60	
	JMP	LOC2		4A	20	49
	STA	TTY		4C	90	F0
	RTN			4E	F0	
SUB2	SIR			60	60	
	JMP	SUB2		61	20	60
	STA	TTY		63	90	F0
	DEX			65	10	
	SXO			66	61	
	JMP	SUB2		67	20	60
	RTN			69	F0	

These address names will be assigned actual binary addresses when the machine language program is coded.

Step 5.   The assembly language program is used as the guide in coding the machine language program. The first step is to assign the binary address of the operation byte of each instruction. The address byte of all 2 byte instructions is stored in the memory cell with the next higher binary address. The programmer selects the address of the first instruction in the main program and each subroutine. The programmer must be certain that there are enough unused memory cells following the starting address to accommodate the entire program or subroutine. Address 00 is the starting address of the main program. The remaining addresses are filled in allowing two memory locations for each 2-byte instruction and one location for the 1-byte instructions (see Table 7.5 for the number of bytes in each instruction).

The programmer also selects the addresses of all the data and enters the values from Table 7.6. The complete list of names and the assigned address is given below. Notice that several names have been added since the original list in Table 7.6. This is normal because it is not possible to anticipate all the names that will be required. The names are added to the list as the program is developed.

The last step is to write the hexadecimal value of each address byte. The machine language program is now ready to be entered into the microcomputer for final debugging and operation.

Name	Address	Name	Address	Name	Address
CODECR	80(hex)	SPACE1	84	TTY	F0
CODELF	81	SPACE2	85	SUB1	40
CODESP	82	SPACE3	86	SUB2	60
CODEX	83	X1	87	START	00
LOC1	42	X2	88		
LOC2	49	X3	89		

Example 7.11   The ICST microcomputer is to be used as a sequential controller for an industrial process. The process consists of a blending tank with provisions to add water, syrup, and solids to the tank. A ready light indicates when the process has completed the sequence of operations and is ready to begin a new sequence. A switch closure initiates the following sequence of operations.

1.   Turn the tank drain valve OFF, turn the ready light ON, and wait for a switch closure to start sequence of operations.
2.   Turn the ready light OFF, turn the water valve ON, and delay for 30 seconds with only the water valve ON.

3. Turn the syrup valve ON and delay for 85 seconds with both the water and syrup valves ON.
4. Turn the syrup valve OFF, turn the solids feeder ON, and delay for 120 seconds with both the water and solids feeder ON.
5. Turn the water valve OFF, turn the solids feeder OFF, and delay for 175 seconds with everything OFF.
6. Turn the tank drain valve ON and delay 115 seconds to drain the tank.

Solution:

Step 1. Problem definition: First we must define the input and output connections. Connect the switch to the I/O port ready line, so the SIR instruction can be used to test the condition of the switch. Use the output port to control the inputs to the tank such that bit 0 controls the water valve, bit 1 controls the syrup valve, bit 2 controls the solids feeder, bit 3 controls the tank drain valve, and bit 4 controls the ready light. All signals are arranged such that a 1 turns the device ON and a 0 turns the device OFF.

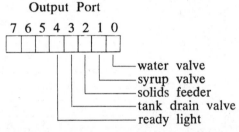

The delays can be obtained by the use of a subroutine that takes 1 second each time it is used. The 30-second delay can be obtained by repeating the delay subroutine 30 times, the 85-second delay by repeating the subroutine 85 times, etc.

Step 2. The flowchart for Step 2 is given in Figure 7.13. There are three tasks identified in the Step 2 flowchart. Task 1 involves the operation of the various output devices (output with reference to the computer). Task 2 checks the status of the switch repeatedly until the switch is closed and the program advances to the next step. Task 3 accomplishes the various delays.

Step 3. The flowchart for Step 3 is also given in Figure 7.13. Task 1 involves the loading of an 8-bit binary number into the I/O port to turn devices ON or OFF. The I/O port assignment in Step 1 is used to determine the binary number required to accomplish the desired

STEP 2 FLOWCHART

STEP 3 FLOWCHART

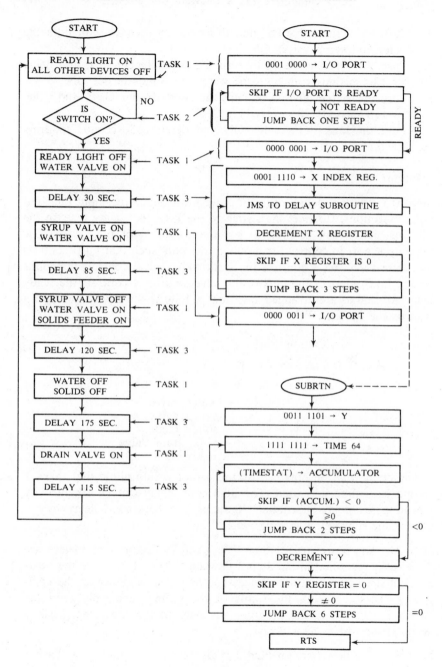

FIGURE 7.13
FLOWCHARTS FOR EXAMPLE 7.11

194

results. For example, the number 0000 0101 would turn ON the solids feeder and the water valve.

Task 2 is accomplished by a SKIP if the I/O port is ready, followed by a jump back to the skip instruction. The computer will cycle back and forth between these two instructions until the switch is closed and the skip passes over the jump instruction.

Task 3 uses the X index register as a counter to count the desired number of seconds of delay. The skip instruction checks the counter after each decrement. If the counter has not reached zero, the program loops back and delays for an additional second. When the counter reaches zero, the skip moves to the next section of the program. The initial value in X is the binary equivalent of the number of seconds of delay that will result. Thus 0001 1110 (binary) $=$ 30 (decimal) will result in a delay of 30 seconds.

The delay subroutine in Task 3 makes use of an internal timer in the ICST-1 microcomputer. The timer is started by storing a number at the memory address with the label TIME64. The number in TIME64 is decremented once each 64 microseconds until it passes zero (i.e., when the content of TIME64 is decremented from 0 to -1). When the number in TIME64 passes zero, the number in the memory address labeled TIMESAT is changed from a positive value to a negative value. If the number 1111 1111 (binary) $= 255$ (decimal) is loaded into TIME64, the timer will require 256 counts to pass zero. Each count takes 64 microseconds so the total time required for the timer to run down is $64 \times 256 = 16,384$ microseconds. The program repeatedly checks the content of TIMESTAT until it becomes negative, indicating that the timer has run down.

Each time the timer runs down, the Y index register is decremented. The program loops back to run the timer again until the content of Y is decremented to zero. The Y register is used as a counter of the number of times the timer is used. If the number 0011 1101 (binary) $=$ 61 (decimal) is loaded into the Y register, the timer will be used 61 times and the total delay will be $61 \times 16,384 = 999,424$ microseconds or 0.999424 seconds.

Step 4.   Code the program using names for the operations and addresses in memory. Table 7.8 is the list of names of data and addresses used in the program. The numbers that are underlined were added as the machine language program was developed. The assembly language program is in the left half of Table 7.9.

Step 5.   The assembly language program is used as a guide in coding the machine language program. The first step is to assign the address

TABLE 7.8

NAMES OF DATA AND ADDRESSES USED IN THE PROGRAM

Name	Data value	Data address	Name	Address
OUT1	0001 0000	60	SUBRTN	4B
OUT2	0000 0001	61	OUTPUT	F1
OUT3	0000 0011	62	TIME64	F2
OUT4	0000 0101	63	TIMESTAT	F3
OUT5	0000 0000	64	LOC1	04
OUT6	0000 1000	65	LOC2	0F
DELAY1	0001 1110	66	LOC3	1C
DELAY2	0101 0101	67	LOC4	29
DELAY3	0111 1000	68	LOC5	36
DELAY4	1010 1111	69	LOC6	43
DELAY5	0111 0011	6A	LOC7	4E
COUNT	0011 1101	6B	LOC8	52
TIME	1111 1111	6C	START	00

of the operation byte of each instruction. The next step is to assign the data addresses in Table 7.8. The final step is to assign the hex values of the operation byte and the address byte of each instruction.

TABLE 7.9

THE ASSEMBLY AND MACHINE LANGUAGE PROGRAMS

Assembly-coded program			Machine-coded program		
Name of address	Operation byte	Address byte	Hex address	Opera-tion byte	Address byte
START	LDA	OUT1	00	40	60
	STA	OUTPUT	02	90	F1
LOC1	SIR	OUTPUT	04	60	F1
	JMP	LOC1	06	20	04
	LDA	OUT2	08	40	61
	STA	OUTPUT	0A	90	F1
	LDA	DELAY1	0C	40	66
	TAX		0E	70	
LOC2	JMS	SUBRTN	0F	30	4B
	DEX		11	10	
	SXO		12	61	
	JMP	LOC2	13	20	0F
	LDA	OUT3	15	40	62
	STA	OUTPUT	17	90	F1
	LDA	DELAY2	19	40	67
	TAX		1B	70	
LOC3	JMS	SUBRTN	1C	30	4B
	DEX		1E	10	
	SXO		1F	61	
	JMP	LOC3	20	20	1C

Name of address	Operation byte	Address byte	Address	Operation byte	Address byte
	LDA	OUT4	22	40	63
	STA	OUTPUT	24	90	F1
	LDA	DELAY3	26	40	68
	TAX		28	70	
LOC4	JMS	SUBRTN	29	30	4B
	DEX		2B	10	
	SXO		2C	61	
	JMP	LOC4	2D	20	29
	LDA	OUT5	2F	40	64
	STA	OUTPUT	31	90	F1
	LDA	DELAY4	33	40	69
	TAX		35	70	
LOC5	JMS	SUBRTN	36	30	4B
	DEX		38	10	
	SXO		39	61	
	JMP	LOC5	3A	20	36
	LDA	OUT6	3C	40	65
	STA	OUTPUT	3E	90	F1
	LDA	DELAY5	40	40	6A
	TAX		42	70	
LOC6	JMS	SUBRTN	43	30	4B
	DEX		45	10	
	SXO		46	61	
	JMP	LOC6	47	20	43
	JMP	START	49	20	00
SUBRTN	LDA	COUNT	4B	40	6B
	TAY		4D	71	
LOC7	LDA	TIME	4E	40	6C
	STA	TIME64	50	90	F2
LOC8	LDA	TIMESTAT	52	40	F3
	SAP		54	63	
	JMP	LOC8	55	20	52
	DEY		57	11	
	SYO		58	62	
	JMP	LOC7	59	20	4E
	RTN		5B	F0	

7.7 Solid-State Components

The number and variety of solid-state components used in control is sufficient to fill several books the size of this one. Solid-state components are found in relays, timers, contactors, motor starters, temperature controllers, a variety of switching circuits, ac to dc converters, dc to ac inverters, battery chargers, variable speed drives, cycloconverters, and rectifiers—just to name a few. The purpose of this section is to present a few typical examples that will illustrate how solid-state components are used in control components. Table 7.10 gives the symbols and V–I characteristics of the solid-state components most frequently used in control.

One of the major uses of solid-state components is in switching circuits of all sizes. Although transistors are used in some switching circuits, power switching is the domain of the *silicon controlled rectifier* (SRC) and the *triac*. The SCR has a number of useful characteristics. First, it is a rectifier—an SCR will conduct current only in one direction. Second, the SCR is a latching switch—the SCR can be turned on by a short pulse of control current into the gate and it remains on when the control current is removed. Third, the SCR has a very high gain—the anode current is about 3000 times as large as the control current. This means that a small amount of power in the turn-on circuit is used to control a large amount of power in the main circuit. Figure 7.14 shows the volt–ampere graph of an SCR.

When the anode-to-cathode voltage, V_{AC}, is negative, the SCR is said to be reverse biased. With a reverse bias, the anode current is negligible until the breakdown occurs—typically 50 to 2000 volts. When the anode-to-cathode voltage is made positive, the anode current is still negligible. However, when V_A is positive and a small positive voltage is applied between the gate and cathode, the SCR switches on. When the SCR is on, it remains on even if the triggering voltage is removed. The SCR has three operating regions.

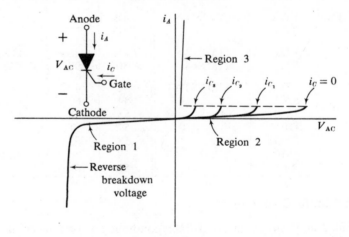

FIGURE 7.14

THE VOLT–AMPERE CHARACTERISTICS OF AN SCR

TABLE 7.10

A PARTIAL LIST OF SOLID-STATE COMPONENTS

Name of Solid-state Component	Graphical Symbol	Volt-ampere Characteristic	Typical Applications
Diode			Rectifier block bypass
N–P–N transistor			Amplifiers, switches, oscillators
P–N–P transistor			Amplifiers, switches, oscillators
Unijunction transistor (UJT)			Timers, oscillators, SCR trigger
Programmable unijunction transistor (PUT)			Timers, oscillators, SCR trigger
Silicon-controlled rectifier (SCR)			Power switches, inverters, frequency converters
Triac			Switches, relays
Diac			Triac and SCR trigger

Region 1: V_{AC} is negative and I_A is negligible.
Region 2: V_{AC} is positive and I_A is negligible.
Region 3: V_{AC} is positive and I_A is large; the SCR is ON.

There are four ways that an SCR can make the transition from Region 2 to Region 3. The first way is to increase the anode voltage until the forward breakdown voltage is reached and the device turns on. The second way is by applying a positive voltage pulse to the gate input to "trigger" the SCR on. A third method is by applying light to the gate/cathode junction. The light energy is sufficient to turn the SCR on. The fourth method is to rapidly increase the anode-to-cathode voltage, V_A. The rapid increase in voltage will also turn the SCR on. The *triggering pulse method* is usually used in control applications.

Once the SCR is turned on, it will remain on. There are three ways than an SCR can be turned off.

1. The anode current can be reduced below a minimum value called the holding current. The holding current is typically about 1 per cent of the rated current. This causes a transition from Region 3 to Region 2. The SCR is still forward biased, but it is off instead of on.
2. If the anode voltage is reversed (i.e., $V_A < O$), the SCR will go from Region 3 to Region 1. This method is used in ac circuits where the voltage reverses each half-cycle. It is also used in inverters and frequency converters where a charged capacitor is switched into the SCR circuit to force a reversal of the anode-to-cathode voltage.
3. In small current applications, the SCR can also be turned off by supplying a negative gate current to increase the holding current. When the increased holding current exceeds the load current, the SCR switches into Region 2.

The second method, reversing V_A, is the method most often used in control circuits.

One application of the SCR is the inverter circuit illustrated in Figure 7.15. An inverter is a device that converts the voltage from a dc source

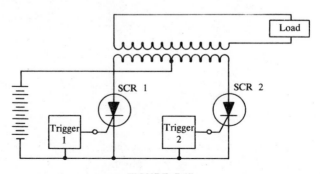

FIGURE 7.15
CENTER-TAPPED, PRIMARY, SCR-CONTROLLED INVERTER

into an ac voltage. The dc source might be a bank of storage batteries used to store electrical energy collected by solar cells or a wind generator. The ac voltage is needed to operate ac equipment such as fluorescent lights, appliances, electric power tools, etc. The circuit operation consists of the SCR's alternately switching the dc current from one half of the primary winding to the other half. The effect is equivalent to an alternating current in a single primary winding so the secondary winding delivers an alternating current to the load. The term *commutation* is used to name the transfer of current from one SCR to the other SCR. Reliable commutation is one of the major problems in inverters. In Figure 7.15, additional commutation circuitry is required to turn off the SCR's by reverse biasing (method 2) each time the other SCR is turned on by a triggering pulse.

The *triac* was developed as a means of providing improved controls for ac power. The major difference between the triac and the SCR is that the triac can conduct in both directions whereas the SCR can only conduct in one direction. The V–I characteristic of the triac is shown in Table 7.10. A positive or negative gate current of sufficient amplitude will trigger the triac on when V_{21} is either positive or negative. A typical triac circuit is shown in Figure 7.16. The trigger circuit produces a pulse that turns the triac on. The triac remains on until the ac voltage reverses and turns the triac off. The next triggering pulse turns the triac on to conduct the remainder of the half wave. The trigger circuit determines when the trigger pulse will turn the triac on. This in turn determines how much current is delivered to the load.

A number of devices are used to produce the trigger pulse for SCR's and triacs. These include the *unijunction transistor* (UJT) and a number of other devices. A basic unijunction trigger circuit and its waveforms are shown in Figure 7.17. In this circuit, the capacitor is charged

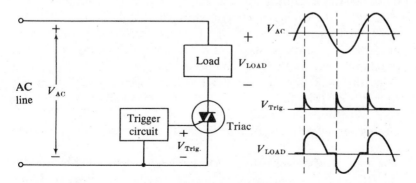

FIGURE 7.16

THE BASIC TRIAC CIRCUIT FOR CONTROL OF AN AC LOAD

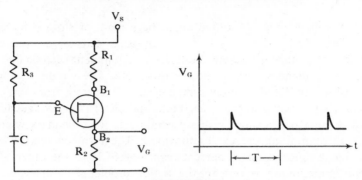

FIGURE 7.17
A UJT TRIGGER CIRCUIT

by the current through R_3 until the emitter voltage reaches V_p (see Table 7.10) and the UJT turns on. This discharges the capacitor through R_2 producing the triggering spike in V_G. When the emitter voltage reaches 2 volts, the UJT turns off and the cycle is repeated. The values of R_3 and C determine the time between the triggering pulses.

Figure 7.18 shows an example of a UJT used as an SCR trigger circuit. The circuit operates as follows:

1. The ac input is applied to the load in series with the SCR. The load voltage remains at zero until the SCR is switched on by a gate pulse.
2. e_{in} is also applied to zener diode clipper circuits (R_s and V_z). The zener diode breaks down at V_z volts and thus limits or clips the positive peaks of e_{in} (see v_{xy} waveform in the figure). During the negative half-cycle of e_{in} the zener is forward biased and maintains $\simeq 0$ V between x and y.
3. This clipped sine wave is present as v_{xy} and is the supply voltage for the UJT circuit. During the positive half-cycle of e_{in} while v_{xy} is at $+V_z$ volts, the capacitor charges through R until it reaches V_P of the UJT. When it does, the UJT turns "on" and discharges C, producing a positive pulse across $R1$ (see v_{B1} waveform). This pulse is fed to the gate of the SCR and turns "on" the SCR.
4. Once the SCR is "on," the load voltage becomes approximately equal to e_{in} for the duration of the positive half-cycle. (See the v_{load} waveform.)
5. During the negative half-cycle the SCR stays "off" and v_{load} stays at zero.
6. The amount of power delivered to the load is controlled by varying the RC time constant, which causes C to charge slower or faster, thereby triggering the UJT and SCR later or earlier in the e_{in} half-cycle. In other words, the RC time constant controls the SCR trigger

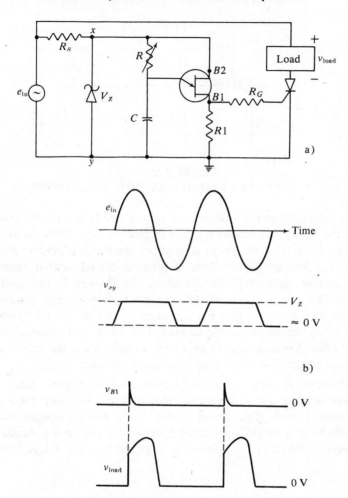

FIGURE 7.18 [4]

A UJT AS AN SCR TRIGGER IN AN AC POWER CONTROL CIRCUIT

angle and therefore the load power. As RC is increased, the trigger angle increases and the load power decreases.[5]

Diodes are used as rectifiers in dc power supplies and as one-way "valves" to block or bypass undesired electric currents. The volt–ampere characteristic of the diode is shown in Figure 7.19. The diode

---

[4] R. Tocci, *Fundamentals of Electronic Devices* (Charles E. Merrill Publishing Co., 1975), p. 371.

[5] Ibid., p. 370.

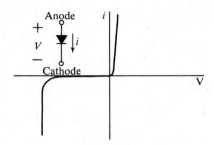

FIGURE 7.19
THE VOLT–AMPERE CHARACTERISTIC OF A DIODE

has three operating regions depending on the anode-to-cathode voltage (V). If the anode-to-cathode voltage is positive, the diode is said to be *forward biased;* if the voltage is negative, the diode is *reverse biased.* In operating Region 1, the diode is forward biased, and it conducts electric current with very low resistance. In Region 2, the diode is reverse biased and it blocks electric current with very high resistance. In Region 3, the reverse bias has increased to the point of breakdown and the diode conducts electric current. For most applications, operation in Region 3 is avoided. An exception to this rule is the zener diode which is operated in Region 3 as a voltage regulator.

A full-wave dc power supply is illustrated in Figure 7.20. The diodes are used as one-way switches that pass current only when they are forward biased. The center-tapped transformer configuration is such that the two diodes are forward biased by opposite polarities of the ac input voltage (V_{in}). When V_{in} is positive, diode $D1$ is forward

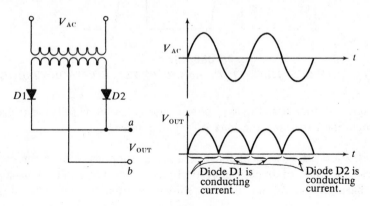

FIGURE 7.20
A FULL-WAVE DC POWER SUPPLY

biased and when V_{in} is negative, diode $D2$ is forward biased. There-fore, diode $D1$ conducts current during each positive half-cycle of the input voltage, and diode $D2$ conducts current during each negative half-cycle. The output of each diode is connected to output terminal a producing the full-wave rectified output voltage shown in Figure 7.20. A capacitor is usually placed across terminals a and b to produce a more constant output voltage.

Transistors are used in control systems in switches, amplifiers, and oscillators. There are two types of transistors, the N–P–N type and the P–N–P type. The graphic symbols and volt–ampere characteristics of both types are included in Table 7.10. The operating characteristics of N–P–N and P–N–P transistors are the same except that the directions of the voltages and currents are opposite. The following discussion is confined to the N–P–N transistor.

An N–P–N transistor has three terminals called the collector, base, and emitter, as shown in Figure 7.21. Transistors are essentially con-trolled current amplifiers. A relatively small current entering the base is used to control a much larger current entering through the collector. Both currents leave through the emitter. The ratio of the collector current (i_C) over the base current (i_B) is called the β (beta) of the tran-sistor ($\beta = i_C/i_B$). The β of a transistor ranges from 20 to 200, and it varies from one transistor to another, even if they are of the same type. Aging and changes in temperature may also cause the β to change.

A transistor may be used in one of three different amplifier con-figurations, depending on which terminal is common to the input and the output. The three configurations and a summary of their character-istics are shown in Figure 7.22.

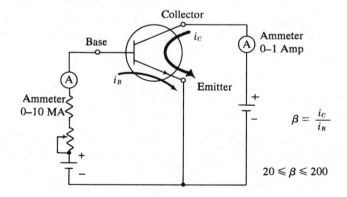

FIGURE 7.21
AN N–P–N TRANSISTOR TEST CIRCUIT

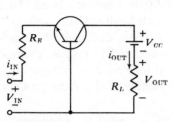

a) Common base amplifier

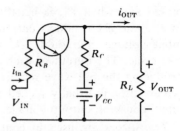

b) Common emitter amplifier

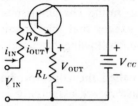

c) Common collector amplifier

Transistor circuit	CB	CE	CC
Voltage gain	Yes	Yes	No
Current gain	No	Yes	Yes
Signal inversion	No	Yes	No

FIGURE 7.22

TRANSISTOR AMPLIFIER CONFIGURATIONS AND CHARACTERISTICS

The input/output graphs of a typical common emitter amplifier are shown in Figure 7.23. Notice the signal inversion in both graphs (i.e., as the input voltage or current increases, the output voltage or current decreases). The voltage gain is determined as follows. Select two points

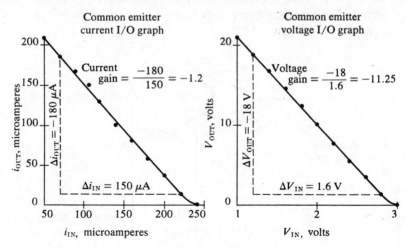

FIGURE 7.23

INPUT/OUTPUT GRAPHS OF A COMMON EMITTER AMPLIFIER

on the straight portion of the graph that are spaced as far apart as possible. Measure the change in the input voltage (ΔV_{in}) and the change in the output voltage (ΔV_{out}) between the two points. The voltage gain is the change in output voltage divided by the change in input voltage. The current gain is determined in a similar manner. In the example illustrated in Figure 7.23, the voltage gain is -11.25 and the current gain is -1.2. The negative sign is due to the signal inversion. This signal inversion is a characteristic of the common emitter amplifier. The inversion is eliminated when two common emitter amplifiers are connected in series because the inverted output of the first stage is inverted again by the second stage. The overall voltage gain of the two-stage amplifier is the product of the voltage gains of the two stages. The same is true for the current gain. For example, if stage 1 has a voltage gain of -12 and stage 2 has a voltage gain of -10, the overall voltage gain is $(-12)(-10) = 120$.

The input/output graphs of a typical common collector amplifier are shown in Figure 7.24. The voltage gain of this amplifier is 0.97, and the current gain is 181. Notice the absence of the signal inversion that characterized the common emitter amplifier. This direct relationship between the input and output signals is a characteristic of the common collector amplifier: so is the absence of voltage gain. The common collector amplifier is a current amplifier.

The common emitter and common collector amplifiers may be combined to form multistage amplifiers with increased voltage and current

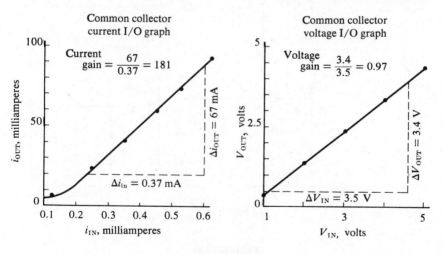

FIGURE 7.24

INPUT/OUTPUT GRAPHS OF A COMMON COLLECTOR AMPLIFIER

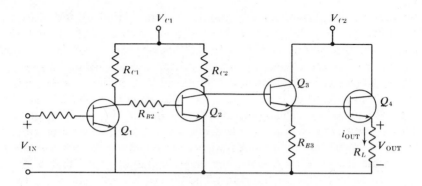

FIGURE 7.25

A FOUR-STAGE TRANSISTOR AMPLIFIER

gains. A four-stage amplifier is shown in Figure 7.25. Transistors $Q1$ and $Q2$ form the common emitter amplifiers for stages 1 and 2. Transistors $Q3$ and $Q4$ form common collector amplifiers for stages 3 and 4.

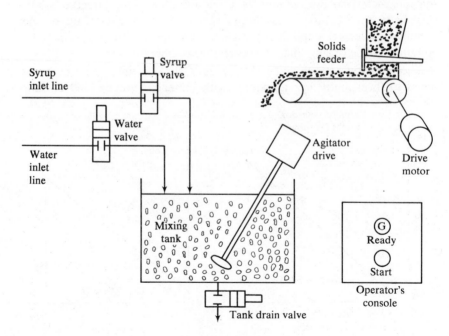

FIGURE 7.26

AN INDUSTRIAL BLENDING PROCESS

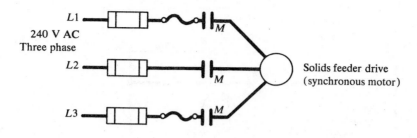

240 V AC
Three phase

FIGURE 7.27
SOLIDS FEEDER DRIVE CIRCUIT

Example 7.12   A schematic diagram of the industrial process described in Example 7.11 is shown in Figure 7.26. In Example 7.11, a
control program was developed for the ICST-1 microcomputer to control the required sequence of operations. Two-position solenoid valves
are used to control the water and syrup flows into the tank. A third
solenoid valve is used to control the flow out of the tank. The solids
feeder is driven by a synchronous motor, and a universal motor is used
for the agitator drive. The operator's console completes the main components of the process.

The main contacts for starting the solids feeder drive motor are
shown in Figure 7.27. The control circuit for energizing the solids
feeder starter coil is shown in Figure 7.28. Triac 4 in Figure 7.28 is
triggered by a dc signal controlled by bit 2 of the output port of the
ICST-1 microcomputer. Triac 4 is ON when bit 2 is a logic 1 and it
is OFF when bit 2 is a logic 0. Transistor $T3$ acts as a solid-state switch
to control the triggering of the triac. Similar triac circuits are used to
control the solenoid valves and the ready light.

The agitator drive motor is controlled by Triac 1 in Figure 7.28.
Variable resistor $R2$, capacitor $C2$, and the diac make up the triggering circuit. Resistor $R2$ and capacitor $C2$ form a variable time constant
circuit that triggers Triac 1 when the voltage across $C2$ reaches the
breakover voltage of the diac. Resistor $R1$ and capacitor $C1$ reduce
the transient voltages developed during switching due to the inductive
energy stored in the motor windings. The diac triggering circuit provides a smooth control of motor speed from some intermediate value
up to full speed. Other triggering techniques are used when the full
control range is required.

The control circuit shown in Figure 7.28 is simpler than the actual
circuits used in an industrial process, but it does illustrate some of the
variety of applications of solid-state components in control systems.

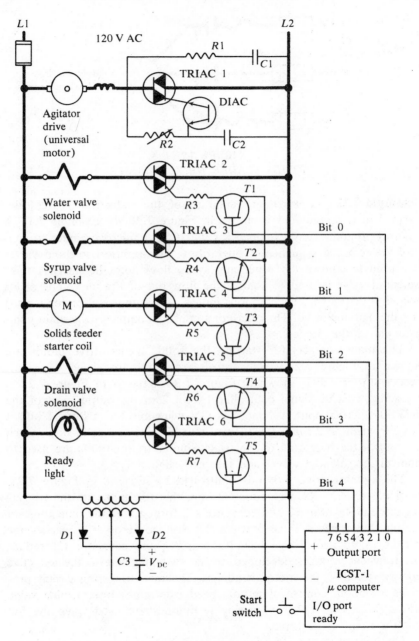

FIGURE 7.28
BLENDING PROCESS CONTROL CIRCUIT

EXERCISES

7–1. Convert the following binary numbers to the decimal equivalent.
   a. 0101   b. 1101
   c. 100110   d. 1100 1010

7–2. Prepare a table of the six-bit binary equivalents of the decimal numbers from 0 to 64.

7–3. Convert the following binary numbers to the equivalent octal numbers.
   a. 110   b. 101 111
   c. 010 001 011   d. 011 000 100

7–4. Convert the following octal numbers to the equivalent binary numbers.
   a. 530   b. 4017
   c. 6231   d. 35546

7–5. Convert the following binary numbers to the equivalent hexadecimal numbers.
   a. 0110   b. 1010
   c. 1101 1111   d. 1000 1001 1010

7–6. Convert the following hexadecimal numbers to the equivalent binary numbers.
   a. 26   b. CB
   c. 2AFE   d. FB76

7–7. The following 16-bit binary number represents a 4-digit decimal number encoded in the 8421 code. Determine the decimal number.
   0100 1001 0011 0101

7–8. Use the Excess 3 code to encode the following decimal number into a 16-bit binary number.
   8426

7–9. The following sequence of hexadecimal numbers represents a word encoded with the ASCII code. A 0 is used for all parity bits. Decode each character and determine the coded word.
   4D, 49, 43, 52, 4F, 43, 4F, 4D, 50, 55, 54, 45, 52

7–10. Write the names represented by the following abbreviations and briefly explain each one.
   a. LSI   b. RAM
   c. I/O   d. ROM
   e. PROM   f. EPROM
   g. A/D   h. D/A
   i. ASCII   j. SDLC
   k. HDLC   m. GPIB

7–11. The logic levels of the input to a 4-bit binary ladder D–A converter are 5 volts for a logic 1 and 0 volts for a logic 0. Determine the analog output for each of the following binary inputs.

a. 0101   b. 1011

c. 1000   d. 0111

7–12. Determine the quantization error in volts and per cent of the full-scale range for each of the following A–D converters. The output voltage range is from 0 to 10 volts.

a. resolution = 4 bits; error = ±½ bit

b. resolution = 8 bits; error = ±1 bit

c. resolution = 10 bits; error = ±½ bit

d. resolution = 12 bits; error = ±½ bit

7–13. Write the assembly language and the machine language programs for the ICST-1 microcomputer that will cause the teletype terminal to print each of the following patterns in the center of the page.

a.   X   b.   CCC

     XXX   C

     X   C

c. X X   CCC

     X   d.   OOO

     X X   O O

                                         OOO

7–14. Describe the four ways that an SCR can be turned on (i.e., move from Region 2 to Region 3).

7–15. Describe the three ways that an SCR can be turned off and indicate which region the SCR is in when it is off.

7–16. Describe the major difference between the triac and the SCR and explain how the triac is equivalent to two SCR's connected in parallel but in opposite directions.

7–17. A half-wave dc power supply uses one diode instead of the two diodes used in a full-wave power supply. Sketch the output voltage waveform of the power supply in Figure 7.19 if diode D2 is removed from the circuit.

7–18. The following voltage and current measurements were obtained from a common emitter amplifier similar to Figure 7.22 (b). Draw voltage and current graphs similar to Figure 7.23 and determine the voltage gain and the current gain.

a. $V_{in} = 0.70$ V, $V_{out} = 24.2$ V, $i_{in} = 0$ μA, $i_{out} = 0.95$ mA

b. $V_{in} = 0.75$ V, $V_{out} = 20.0$ V, $i_{in} = 10.6$ μA, $i_{out} = 0.78$ mA

c. $V_{in} = 0.80$ V, $V_{out} = 15.8$ V, $i_{in} = 21$ μA, $i_{out} = 0.61$ mA

d. $V_{in} = 0.85$ V, $V_{out} = 11.6$ V, $i_{in} = 31.5$ μA, $i_{out} = 0.45$ mA

  e. $V_{in} = 0.90$ V, $V_{out} = $ 7.4 V, $i_{in} = 42.2$ μA, $i_{out} = 0.27$ mA
  f. $V_{in} = 0.95$ V, $V_{out} = $ 3.4 V, $i_{in} = 52.4$ μA, $i_{out} = 0.12$ mA
  g. $V_{in} = 1.00$ V, $V_{out} = 0,$ $i_{in} = 62.8$ μA, $i_{out} = 0$

7–19. A four-stage transistor amplifier similar to Figure 7.25 has the following voltage and current gains for each stage. Determine the overall voltage gain and current gain.
  a. Stage 1, voltage gain $= -8$, current gain $= -1.4$
  b. Stage 2, voltage gain $= -10$, current gain $= -1.2$
  c. Stage 3, voltage gain $= 0.95$, current gain $= 110$
  d. Stage 4, voltage gain $= 0.96$, current gain $= 150$

8

An Introduction to
Analog Computers

8.1 Introduction

The common thread uniting control systems of all types is the mathematical equations which describe the operations of each system. It doesn't matter whether the controller operates on pneumatic, electric, hydraulic, or mechanical signals; the same mathematical equations are used to describe the control action. Every process is modeled by combinations of the same mathematical elements. Systems which are modeled by the *same* combination of mathematical elements are said to be *analogous*.

Analog computers are used to simulate control systems by utilizing electrical analogs of the real system. The analog computer is programmed so that the mathematical equation which describes the computer operation is identical to the equation which describes the operation of the real system. Each electrical voltage corresponds to some variable in the real system. The response of the real system to some input condition is simulated by applying the same input condition to the analog computer. The response of some variable in the real system is determined by observing the response of the corresponding signal in the analog computer.

Analog computers are particularly well suited to solving the algebraic and differential equations which describe most control systems. An analog computer is a set of mathematical building blocks: each block is able to perform a specific mathematical operation. The blocks are easily interconnected to form an electrical model of a real system. The standard operations of an analog computer are:

1. Inversions.
2. Summation.
3. Multiplication.
4. Integration.
5. Function generation.

The heart of an analog computer is a high-gain dc amplifier called an *operational amplifier*. The gain of most commercial operational amplifiers is between 10^5 and 10^8. Section 8.3 will explain the reason for such a high gain. The common computer symbols are presented in Figure 8.1.

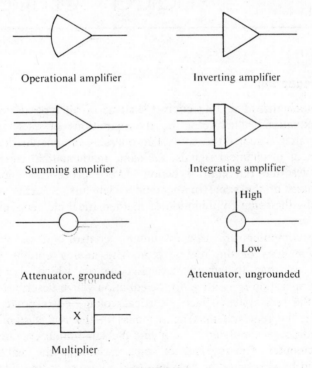

FIGURE 8.1
ANALOG COMPUTER SYMBOLS

8.2 Inverting Amplifiers

The inverting amplifier simply changes the sign of a voltage signal. The output voltage has the same magnitude as the input, but it has the opposite sign. Thus, a 4.6-volt input signal produces a -4.6-volt output signal, and a -3.2-volt input signal produces a 3.2-volt output signal.

The inverter consists of an operational amplifier and two resistors (see Figure 8.2). All three components are connected to a common node called the *summing junction (SJ)*. According to Kirchhoff's current law for electric circuits, the total current leaving a node must equal the total current entering the node. The input resistance of the operational amplifier is very large, so current i_a is negligible. Thus, current i_1 and i_2 must be equal to satisfy Kirchhoff's law.

$$i_1 = i_2 \qquad\qquad (8.1)$$

By Ohm's law for electric circuits, we know that the current through a resistor is equal to the voltage drop across the resistor divided by the resistance value. The voltage drop across R_1 is $(e_{in} - e_{sj})$. The voltage drop across R_2 is $(e_{sj} - e_{out})$. Thus, i_1 is equal to $(e_{in} - e_{sj})/R_1$ and i_2 is equal to $(e_{sj} - e_{out})/R_2$. Substituting these values in Equation (8.1), we have

$$\frac{e_{in} - e_{sj}}{R_1} = \frac{e_{sj} - e_{out}}{R_2} \qquad\qquad (8.2)$$

In the inverting amplifier, R_1 and R_2 are equal. Thus, R_2 may be replaced by R_1 and Equation (8.2) may be solved for e_{out}.

$$e_{out} = -e_{in} + 2e_{sj} \qquad\qquad (8.3)$$

Note that in Equation (8.3) that e_{sj} must be zero if the inverting amplifier is to work as it is supposed to. *The purpose of the operational amplifier is to maintain the summing junction at zero volts.* The high gain of

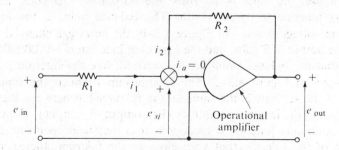

FIGURE 8.2

AN INVERTING AMPLIFIER

the amplifier makes this possible. Let $-G$ represent the gain of the operational amplifier. Normally, G ranges between 10^5 and 10^8. The gain of an amplifier is the ratio of the output voltage over the input voltage.

$$-G = \frac{e_{\text{out}}}{e_{sj}} \tag{8.4}$$

or

$$e_{sj} = -e_{\text{out}}/G \tag{8.5}$$

The summing junction voltage, e_{sj}, may be eliminated in Equation (8.3) by substituting $-e_{\text{out}}/G$ for e_{sj}.

$$e_{\text{out}} = -e_{\text{in}} + 2(-e_{\text{out}}/G)$$
$$e_{\text{out}} + 2e_{\text{out}}/G = -e_{\text{in}}$$
$$e_{\text{out}}(1 + 2/G) = -e_{\text{in}}$$
$$e_{\text{out}}(G + 2)/G = -e_{\text{in}}$$

$$e_{\text{out}} = -\left(\frac{G}{G + 2}\right)e_{\text{in}} \tag{8.6}$$

If G is very large, then $G/(G + 2)$ will be very nearly equal to one. For example, if $G = -10^6$, then $G/(G + 2)$ will be about 1.000002. A difference of 0.000002 is certainly negligible, and Equation (8.6) may be simplified to the following equation.

$$e_{\text{out}} = -e_{\text{in}} \tag{8.7}$$

The operational amplifier maintains a balanced condition between the input voltage and the output voltage, as illustrated in Figure 8.3. Notice the direct analogy between the height at different points on a see-saw and the voltage at different points in an inverting amplifier, as illustrated in Figure 8.3. A value of 10^6 was used for G. In Figure 8.3–a, the input is 10 volts, the output is -10 volts, and the summing junction is 0.00001 volt. The balance point is the location where the voltage is zero. In Figure 8.3–b, the input has changed to -5 volts, the output is 5 volts, and the summing junction is -0.000005 volt. Notice that the balance point has not moved, like the fulcrum point in the see-saw in Figure 8.3–c. The relationship between the input and output of the see-saw is the same as the relationship between the input and output of the inverting amplifier (i.e., output = −input). Notice also that the summing junction is analogous to a measuring point on the input side of the fulcrum, but very close to the fulcrum. Increasing the gain of the operational amplifier is analogous to moving the measuring point closer to the fulcrum and thereby reducing the measuring point height.

> *The inverting amplifier produces an output equal to the input in magnitude, but opposite in sign.*

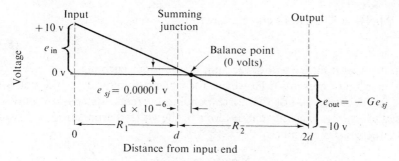

a) Input voltage of 10 volts and output voltage of −10 volts

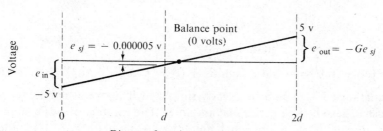

b) Input voltage of −5 volts and output voltage of + 5 volts

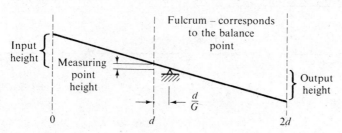

c) The see-saw analogy

FIGURE 8.3

A SEE-SAW ANALOGY OF THE INVERTING AMPLIFIER

8.3 Summation with Summing Amplifiers

The simplest electrical summing device is the summing junction shown in Figure 8.4. Voltages e_1, e_2, and e_3 are the input signals, and voltage e_0 is the output signal. Kirchhoff's current law is the basis for summation: *The total current leaving a node must equal the total current entering the node.*

Applying Kirchhoff's current law to Figure 6.4

$$i_o = i_1 + i_2 + i_3 \qquad (8.8)$$

Thus, the output current is the sum of the input currents (i.e., electrical summation of currents). However, the signals of interest are the voltages e_1, e_2, e_3, and e_0—not the corresponding currents. Ohm's law may be used to express each current in terms of the voltage drop across a resistor and the resistance value.

$$i_i = (e_1 - e_{sj})/R_1$$
$$i_2 = (e_2 - e_{sj})/R_2$$
$$i_3 = (e_3 - e_{sj})/R_3$$
$$i_o = (e_{sj} - e_o)/R_o$$

By substitution in Equation (8.8), we have

$$(e_{sj} - e_o)/R_o = (e_1 - e_{sj})/R_1 + (e_2 - e_{sj})/R_2 + (e_3 - e_{sj})/R_3 \quad (8.9)$$

Equation (8.9) would be greatly simplified if e_{sj} were maintained very close to zero by an operational amplifier. The combination of a summing junction with an operational amplifier to maintain $e_{sj} \approx 0$ is called a

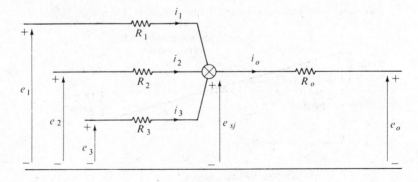

$$i_o = i_1 + i_2 + i_3$$

FIGURE 8.4

AN ELECTRICAL SUMMING JUNCTION

summing amplifier (see Figure 8.5). With $e_{sj} \approx 0$, Equation (8.9) becomes

$$- e_o/R_o = e_1/R_1 + e_2/R_2 + e_3/R_3$$

or

$$e_o = - [(R_o/R_1)e_1 + (R_o/R_2)e_2 + (R_o/R_3)e_3] \qquad \textbf{(8.10)}$$

or

$$e_o = - [k_1e_1 + k_2e_2 + k_3e_3] \qquad k_i = R_o/R_i \qquad \textbf{(8.11)}$$

Equations (8.10) and (8.11) are the defining equations for the summing amplifier. In the process of summation, each input voltage (e_i) is multiplied by constant (k_i) equal to the output resistance (R_o) divided by the corresponding input resistance (R_i). If the output and input resistances are equal, then the constant has the value of unity (i.e., $k_i = 1$ if $R_o = R_i$).

Example 8.1   A summing amplifier similar to Figure 8.5 has the following resistance values.

$$R_1 = 100,000 \text{ ohms}$$
$$R_2 = 10,000 \text{ ohms}$$
$$R_3 = 1,000,000 \text{ ohms}$$
$$R_o = 100,000 \text{ ohms}$$

Determine the output voltage if $e_1 = -5$ volts, $e_2 = 0.1$ volt, and $e_3 = 8$ volts.

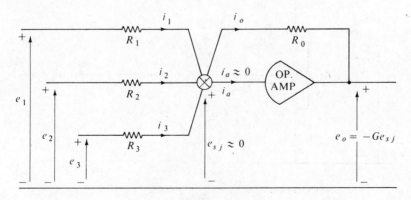

FIGURE 8.5

A SUMMING AMPLIFIER

Solution:

From Equation (8.10),

$$e_o = -\left[\left(\frac{100,000}{100,000}\right)(-5) + \left(\frac{100,000}{10,000}\right)(0.1) + \left(\frac{100,000}{1,000,000}\right)(8)\right]$$

$$e_o = [(1)(-5) + (10)(0.1) + (0.1)(8)]$$

$$e_o = -[-5 + 1 + 0.8]$$

$$e_o = -[-3.2]$$

$$e_o = 3.2 \text{ volts}$$

8.4 Multiplication by a Constant with Attenuators

The attenuator is a voltage divider potentiometer which may be used to multiply a voltage signal by any constant less than or equal to one. The attenuator is illustrated in Figure 8.6. The output voltage of the attenuator is given by the following expression.

$$e_o = (R_a/R_t)e_i$$

or

$$e_o = ae_i$$

where

$$a = R_a/R_t$$
$$0 \le a \le 1$$

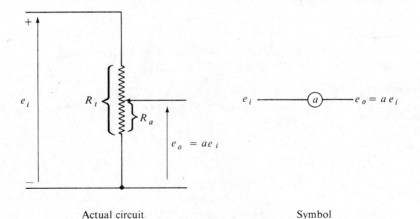

Actual circuit   Symbol

FIGURE 8.6
AN ATTENUATOR

Attenuators are usually used at the input of an inverter, a summing amplifier, or an integrator, as shown in Figure 8.7. Multiplication by a constant greater than one is accomplished by making the k constant greater than one. In Figure 8.7, for example, if $a_1 = 0.3$ and $k_1 = 10$, then $a_1 k_1 = (0.3)(10) = 3$. Voltage e_1 is effectively multiplied by the constant 3 by the combination of the attenuator and the summing amplifier.

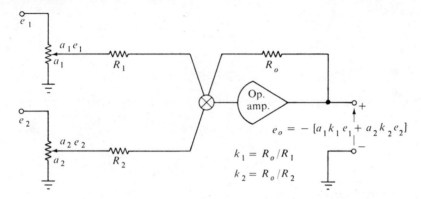

FIGURE 8.7

MULTIPLICATION BY CONSTANTS WITH ATTENUATORS

Example 8.2   The circuit shown in Figure 8.7 has the following resistance and coefficient values.

$$a_1 = 0.3$$
$$a_2 = 0.82$$
$$R_1 = 10,000 \text{ ohms}$$
$$R_2 = 100,000 \text{ ohms}$$
$$R_o = 100,000 \text{ ohms}$$

Determine the output voltage (e_o) if the input signals are:

$$e_1 = -2.1 \text{ volts}$$
$$e_2 = 4 \text{ volts}$$

Solution:

$$k_1 = R_o/R_1 = 100,000/10,000 = 10$$
$$k_2 = R_o/R_2 = 100,000/100,000 = 1$$
$$e_o = -(a_1 k_1 e_1 + a_2 k_2 e_2)$$
$$e_o = -[(0.3)(10)(-2.1) + (0.82)(1)(4)]$$
$$e_o = -(-6.3 + 3.28)$$
$$e_o = 3.02 \text{ volts}$$

8.5 Solving Algebraic Equations

Algebraic equations of the form $y = c_1x_1 + c_2x_2 + c_3x_3$ are easily solved by using attenuators and a summing amplifier, as illustrated in Figure 8.8. The c_i-terms represent constant coefficients, the x_i-terms represent variable input voltage signals, and y represents the output voltage signal. The c_i-coefficients are obtained by the combination of attenuator a_i and the corresponding summing amplifier constant k_i as follows.

$$c_1 = a_1k_1$$
$$c_2 = a_2k_2$$
$$c_3 = a_3k_3$$

The summing amplifier adds the three terms to complete the solution.

$$y = (a_1k_1x_1 + a_2k_2x_2 + a_3k_3x_3) \qquad \textbf{(8.12)}$$

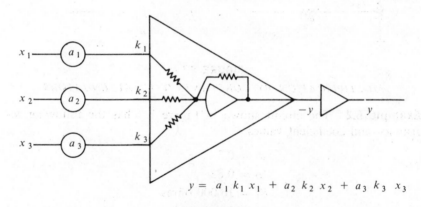

$$y = a_1\,k_1\,x_1 + a_2\,k_2\,x_2 + a_3\,k_3\,x_3$$

FIGURE 8.8
*AN ANALOG COMPUTER PROGRAM FOR SOLVING
AN ALGEBRAIC EQUATION*

Example 8.3   Use the analog computer program in Figure 8.8 to solve the following algebraic equation.

$$y = 0.7x_1 + 3.7x_2 + 8.2x_3$$

Determine the necessary attenuator and summing amplifier coefficients, and the output y corresponding to the following input conditions.

$$x_1 = 2.0 \text{ volts}$$
$$x_2 = 1.0 \text{ volts}$$
$$x_3 = -0.5 \text{ volts}$$

Solution:

$$a_1 k_1 = 0.7 \quad \text{therefore } a_1 = 0.7 \text{ and } k_1 = 1$$
$$a_2 k_2 = 3.7 \quad \text{therefore } a_2 = 0.37 \text{ and } k_2 = 10$$
$$a_3 k_3 = 8.2 \quad \text{therefore } a_3 = 0.82 \text{ and } k_3 = 10$$

By Equation (8.12),

$$y = (a_1 k_1 x_1 + a_2 k_2 x_2 + a_3 k_3 x_3)$$
$$y = [\,(0.7)\,(2) + (3.7)\,(1) + (8.2)\,(-0.5)\,]$$
$$y = (1.4 + 3.7 - 4.1)$$
$$y = (5.1 - 4.1)$$
$$y = 1 \text{ volt}$$

8.6 Integration with Integrating Amplifiers

Integration is accomplished in an analog computer by replacing resistor R_o in an inverter by a capacitor, as shown in Figure 8.9. The result is called an *integrating amplifier* or, simply, an *integrator*.

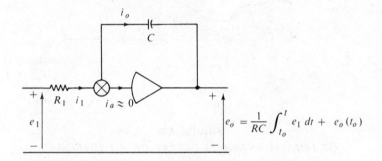

FIGURE 8.9

AN INTEGRATING AMPLIFIER

The voltage current relationship in a capacitor is given by the following equation.

$$e_c(t) = \frac{1}{c} \int_{t_o}^{t} i_c(t)dt + e_c(t_o) \tag{8.13}$$

where:

$$e_c(t) = \text{capacitor voltage at time } t$$
$$e_c(t_o) = \text{capacitor voltage at time } t_o$$

$$c = \text{the capacitance value in farads}$$
$$i_c(t) = \text{the capacitor current at time } t$$
$$\int_{t_o}^{t} i_c(t)dt = \text{the integral of current } i_c(t) \text{ from time } t_o \text{ to time } t$$

The integral of the current has a very simple geometric interpretation. Consider the graph of the current versus time shown in Figure 8.10. The integral of $i(t)$ from time t_o to time t is the area under the curve between t_o and t. In simple terms, Equation (8.13) states that the voltage at time t is equal to $1/c$ times the area under the current curve from t_o to t plus the voltage at time t_o.

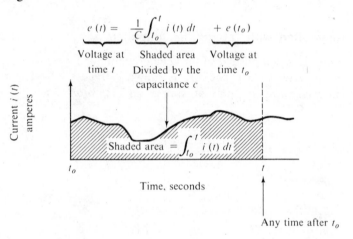

FIGURE 8.10

GRAPHICAL INTERPRETATION OF AN INTEGRAL

According to Kirchhoff's current law, currents i_1 and i_o in Figure 8.9 must be equal; and according to Ohm's law, current i_1 is equal to voltage $e_1(t)$ divided by R_1.

$$i_1 = i_o = e_1(t)/R_1 \qquad\qquad (8.14)$$

Equation (8.13) applies to the capacitor in Figure 8.9, where $(e_{sj} - e_o)$ is the capacitor voltage and i_o is the capacitor current. Since e_{sj} is maintained very close to zero by the operational amplifier, $(e_{sj} - e_o) = -e_o$.

$$-e_o(t) = \frac{1}{c} \int_{t_o}^{t} i_o(t)dt - e_o(t_o) \qquad\qquad (8.15)$$

Substituting Equation (8.14) into (8.15),

$$-e_o(t) = \frac{1}{c} \int_{t_o}^{t} \frac{e_1(t)}{R_1} dt - e_o(t_o)$$

The resistance R_1 is a constant and may be moved outside the integral sign. The result is Equation (8.16), which is the defining equation for an integrating amplifier.

$$e_o(t) = -\frac{1}{R_1 c} \int_{t_o}^{t} e_1(t) dt + e_o(t_o) \qquad \textbf{(8.16)}$$

The voltage at time zero, $e_o(t_o)$, is usually referred to as the *initial condition*—a constant voltage which is applied to the capacitor at the start of an analog computer run. When in *reset position,* a *reset-hold-operate* switch on the computer automatically applies the initial condition; the *hold* position maintains the voltage constant; and the *operate* position causes the integrator to integrate. The integrator will compute the area under the curve during the interval that the switch is in the *operate* position. Time t_o is the time the switch is placed in the *operate* position and time t is any time after t_o when the switch is placed in the *hold* position.

Like the summing amplifier, the integrator may have several inputs, as shown in Figure 8.11. The individual input voltages, e_i, are multiplied by the constant $k_i = 1/(R_i c)$ and added together; then the sum is integrated. All these steps are included in Equation (8.17) below.

$$e_o(t) = -\int_{t_o}^{t} (k_1 e_1 + k_2 e_2 + k_3 e_3) dt + e_o(t_o) \qquad \textbf{(8.17)}$$

where:

$$k_1 = 1/(R_1 c)$$
$$k_2 = 1/(R_2 c)$$
$$k_3 = 1/(R_3 c)$$

Example 8.4 The integrator in Figure 8.9 is used to integrate a constant 0.2-volt signal for 5 seconds (i.e., the operate switch will be closed for 5 seconds). The initial condition is 4 volts, R_1 is 100,000 ohms, and

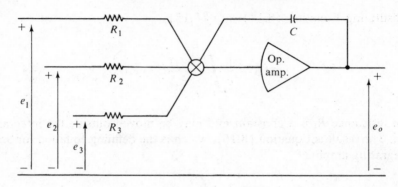

FIGURE 8.11

AN INTEGRATOR WITH THREE INPUTS

c is 10 microfarads. Determine the output voltage at time $t = 5$ seconds (the end of the run).

Solution:

$$k_1 = 1/(R_1 c) = 1/10^5 \cdot 10^{-5}) = 1$$

By Equation (8.16),

$$e_o(5) = -\frac{1}{1} \int_0^5 0.2 dt + e_o(0)$$

$$\int_0^5 0.2 dt = \text{area under the curve from 0 to 5 seconds}$$
$$= 0.2(5 - 0) = 1.0 \text{ volt}$$

$$e_o(5) = -\frac{1}{1}(1) + 4$$

$$e_o(5) = 3 \text{ volts}$$

Example 8.5 Use the integrator in Figure 8.11 to perform the following integration.

$$e_o(t) = -\int_{t_o}^{t} (e_1 + 10e_2 + 0.1e_3) dt + e_o(t_o)$$

The capacitor has a value of 10 microfarads. Determine the value of each input resistor (R_1, R_2, and R_3).

Solution:

$$k_1 = \ \ 1 = 1/(R_1c) \quad \text{therefore, } R_1c = 1$$
$$R_1(10^{-5}) = 1$$
$$R_1 = 10^5 = 100{,}000 \text{ ohms}$$

$$k_2 = \ \ 10 = 1/(R_2c) \quad \text{therefore, } 10R_2c = 1$$
$$10R_2(10^{-5}) = 1$$
$$R_2 = 10^4 = 10{,}000 \text{ ohms}$$

$$k_3 = 0.1 = 1/(R_3c) \quad \text{therefore, } 0.1R_3c = 1$$
$$R_3 = 10^6 = 1{,}000{,}000 \text{ ohms}$$

8.7 Analog Computer Symbols

The analog computer symbols in Figure 8.1 are easier to use than the more complete diagrams presented in Figures 8.2, 8.5, 8.7, and 8.9. In Figure 8.8, the simple triangle is used to represent a summing amplifier. The summing junction, resistors, and operational amplifier are drawn in to indicate that the circuit diagram has been replaced by the triangle. The symbol for an integrator also replaces the summing junction, resistors, capacitor, and operational amplifier. Figure 8.12 shows examples of analog computer symbols, which are used to describe analog computer programs, along with the defining equations.

8.8 Solving Differential Equations

Differential equations are used to define the relationship between the input signal and the output signal of most control system components. In Chapter Five, differential equations were introduced to define the input-output relationship of a liquid tank [Equation (4.6)] and an electrical RS circuit [Equation (4.10)]. Both equations have the following general form.

$$\tau\frac{dx}{dt} + x = y \tag{8.18}$$

where:

τ is a constant called the "time constant"

dx/dt is the rate of change of the output signal

x is the output signal

y is the input signal

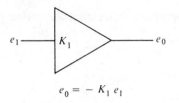

$$e_0 = - K_1 e_1$$

a) Inverting amplifier

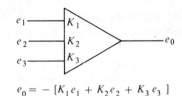

$$e_0 = - [K_1 e_1 + K_2 e_2 + K_3 e_3]$$

b) Summing amplifier

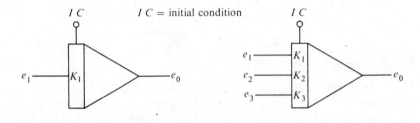

$I C$ = initial condition

$$e_0 = - K \int_{t_0}^{t} e_1 \, dt + I C$$

c) Integrating amplifier – 1 input

$$e_0 = - \int_{t_0}^{t} (K_1 e_1 + K_2 e_2 + K_3 e_3) \, dt + I C$$

d) Integrating amplifier – 3 inputs

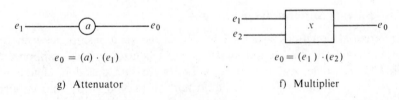

$$e_0 = (a) \cdot (e_1)$$

g) Attenuator

$$e_0 = (e_1) \cdot (e_2)$$

f) Multiplier

FIGURE 8.12

ANALOG COMPUTER SYMBOLS WITH DEFINING EQUATIONS

In words, Equation (8.18) states that the input signal (y) is equal to the output signal (x) plus the product of the time constant (τ) times the rate of change of the output (dx/dt).

The solution of a differential equation is the determination of the output signal x for a given input signal y. Usually, y and x have different values at different times; a solution defines the value of x at each instant of time. Differential equations like Equation (8.18) are easily solved by the analog computer. The solution is based on the fact that the integral is the inverse of the derivative—that is, the integral of dx/dt is x.

$$\int (dx/dt)dt = x \qquad (8.19)$$

For simplicity, the symbol $\dot{x}$ is often used to represent dx/dt. Using the dot notation, Equations (8.18) and (8.19) may be written as follows.

$$\tau\dot{x} + x = y \qquad (8.20)$$
$$\int \dot{x}dt = x \qquad (8.21)$$

In solving differential equations, the integrator is used as an inverse differentiator. The analog computer is patched (programmed) so that the input to the integrator is $\dot{x}$ (see Figure 8.13). The output of the integrator is equal to minus the integral of the input (i.e., $-\int\dot{x}dt$). But from Equation (8.21), the integral of $-\dot{x}$ is $-x$. Thus to solve Equation (8.20) we need only to program the analog computer so that the input is equal to $\dot{x}$. The output of the integrator will be $-x$, and an inverter may be used to obtain x if the minus sign is undesirable.

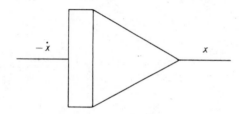

FIGURE 8.13

THE INTEGRATOR AS AN INVERSE DIFFERENTIATOR

The first step in programming the analog computer is to solve the differential equation for $\dot{x}$.

$$\dot{x} = -\left(\frac{1}{\tau}\right)x + \left(\frac{1}{\tau}\right)y \qquad (8.22)$$

The right-hand side of Equation (8.22) gives us the necessary instructions to obtain $\dot{x}$ as the input to the integrator. The output of the integrator is used as a source of $-x$, and an attenuator provides the $1/\tau$-coefficient. The step-by-step development is shown in Figure 8.14.

The method just described may be used to solve more complex differential equations. As an example, consider the spring-mass-damping system shown in Figure 8.15. The input to the system is the external force (F) applied to the weight. This force is equal to the sum of three reactionary forces.

1. The force exerted by the spring (F_s).
2. The force required to overcome resistance in the dashpot (F_r).
3. The force required to overcome the inertia of the mass (F_i).

$$F_i + F_r + F_s = F \qquad\qquad (8.23)$$

$$\dot{x} = -\frac{1}{\tau} x + \frac{1}{\tau} y$$

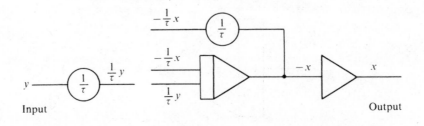

a)   Assign the integrator input signals required
to obtain $\dot{x}$

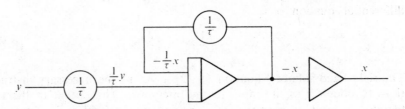

b)   Generate the required integrator input signals

c)   The complete analog computer diagram of the system
described by the equation $\tau \dot{x} + x = y$

FIGURE 8.14

THE DEVELOPMENT OF AN ANALOG COMPUTER DIAGRAM
TO SOLVE A FIRST-ORDER DIFFERENTIAL EQUATION

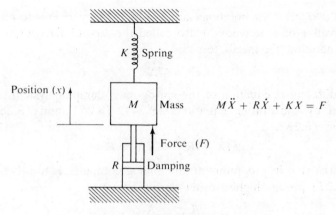

FIGURE 8.15
A SPRING-MASS-DAMPING SYSTEM

The external force (F) is the input to the system, and the position of the mass (X) is the output. The spring force (F_s) depends on the position of the mass (X) and the spring constant (K).

$$F_s = KX \tag{8.24}$$

The resistance force (F_r) depends on the velocity of the mass (V) and the resistance (R).

$$F_r = RV$$

However, the velocity of the mass is equal to the change in position ($\triangle X$) divided by the corresponding change in time ($\triangle t$).

$$V = \triangle X / \triangle t$$

As $\triangle t$ becomes very small, the term $\triangle X / \triangle t$ approaches dX/dt (or $\dot{X}$, if we use the dot notation). Thus, the velocity of the mass is $\dot{X}$ and

$$F_r = R\dot{X} \tag{8.25}$$

The inertia force is equal to the product of the mass (M) times the acceleration of the mass (a). But the acceleration of the mass is equal to the change in velocity ($\triangle V$) divided by the corresponding change in time ($\triangle t$).

$$a = \triangle V / \triangle t$$

As $\triangle t$ becomes very small, the term $\triangle V / \triangle t$ approaches dV/dt. But $V = dX/dt$ and

$$a = dV/dt = \frac{d(dX/dt)}{dt} = \frac{d^2 X}{dt^2} = \ddot{X}$$

The last two terms are notations adopted by mathematicians to represent the derivative of a derivative (also called the *second derivative*). Using the last notation, the inertia force is

$$F_i = M\ddot{X} \qquad (8.26)$$

The differential equation of the spring-mass-damping system may be obtained by substituting equations (8.24), (8.25), and (8.26) into Equation (8.23).

$$M\ddot{X} + R\dot{X} + KX = F \qquad (8.27)$$

The first step in programming the analog computer is to solve Equation (8.27) for the highest-order derivative $(\ddot{X})$.

$$\ddot{X} = -\frac{R}{M}\dot{X} - \frac{K}{M}X + \frac{1}{M}F \qquad (8.28)$$

The right-hand side of Equation (8.28) provides the input to the first integrator (see Figure 8.16). The input to the first integrator is the second derivative, $\ddot{X}$. The output of the first integrator is minus the first derivative $(-\dot{X})$. A second integrator is required to obtain the desired output, X. The development of the analog computer diagram is shown in Figure 8.16.

> *When programming an analog computer to solve a differential equation, the first step is to solve the equation for the highest-order derivative. The right-hand side of the derived equation defines the required inputs to the first integrator. Each integrator reduces the order of the derivative by one (i.e., removes one dot). Include enough integrators to reduce the order of the derivative to zero (i.e., remove all dots). Use attenuators and inverters to generate all inputs to the first integrator. The output of the last integrator is either + or − the desired solution.*

8.9 Multiplication of Two Variables

Several techniques are used to multiply two variables. The operation and limitations of the technique used in a particular computer are usually explained in the instruction manual accompanying the computer. References that deal with analog computers contain a full treatment of multiplication and division techniques.[1] The analog computer symbol for a multiplier is shown in Figure 8.12.

---

[1] See, for example, C. L. Johnson, *Analog Computer Techniques* (New York: McGraw-Hill, 1956).

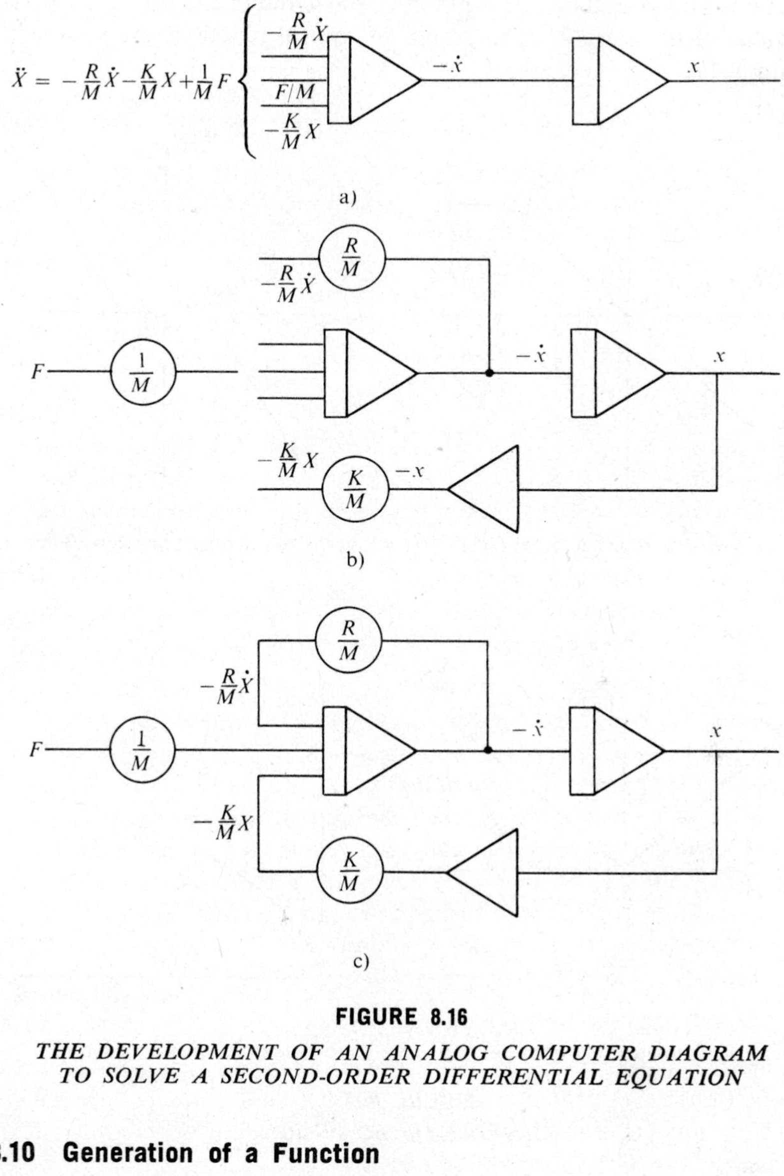

$$\ddot{X} = -\frac{R}{M}\dot{X} - \frac{K}{M}X + \frac{1}{M}F \begin{cases} -\frac{R}{M}\dot{X} \\ F/M \\ -\frac{K}{M}X \end{cases}$$

a)

b)

c)

FIGURE 8.16

*THE DEVELOPMENT OF AN ANALOG COMPUTER DIAGRAM
TO SOLVE A SECOND-ORDER DIFFERENTIAL EQUATION*

8.10   Generation of a Function

A function of a variable has a definite value for each value of the
variable. Functions are designated by a letter followed by the variable
name enclosed in parentheses. For example, $f(x)$ denotes the function f
of the variable x, and $G(t)$ denotes the function G of the variable t.

Functions may be defined by a graph of the value of the function versus the value of the variable. The graphs of several functions are shown in Figure 8.17.

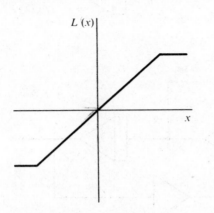

a) The saturation function $L(x)$

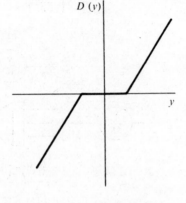

b) The dead-zone function $D(y)$

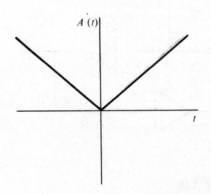

c) The absolute-value function $A(t)$

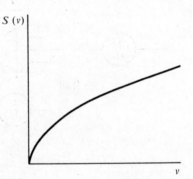

d) The square-root function $S(v)$

FIGURE 8.17

EXAMPLES OF FUNCTIONS

The analog computer module used to generate a function is called a *function generator*. A variety of techniques are used to generate functions, in general, any function defined by a graph may be generated by a function generator.

EXERCISES

8–1. Determine the output voltage (e_{out}) of the inverting amplifier illustrated in Figure 8.2 for each of the following condition sets.

a. $R_1 = 100,000$ ohms
    $R_2 = 100,000$ ohms
    $e_{in} = 8$ volts

b. $R_1 = 100,000$ ohms
    $R_2 = 100,000$ ohms
    $e_{in} = -2.78$ volts

c. $R_1 = 10,000$ ohms
    $R_2 = 100,000$ ohms
    $e_{in} = 0.6$ volt

d. $R_1 = 10,000$ ohms
    $R_2 = 100,000$ ohms
    $e_{in} = -0.468$ volt

e. $R_1 = 27,000$ ohms
    $R_2 = 100,000$ ohms
    $e_{in} = 2.1$ volts

f. $R_1 = 10,000$ ohms
    $R_2 = 34,000$ ohms
    $e_{in} = 7.4$ volts

8–2. A summing amplifier similar to Figure 8.5 has the following resistance values.

$R_o = 100,000$ ohms $R_1 = 10,000$ ohms, $R_2 = 100,000$ ohms,
$R_3 = 100,000$ ohms

Determine the output voltage for each of the following sets of input voltages.

a. $e_1 = 0.23$ volt
    $e_2 = -3.6$ volts
    $e_3 = 7.6$ volts

b. $e_1 = -0.11$ volt
    $e_2 = 8.6$ volts
    $e_3 = 1.2$ volts

c. $e_1 = 0.08$ volt
    $e_2 = 1.12$ volts
    $e_3 = 1.67$ volts

d. $e_1 = 0.78$ volt
    $e_2 = -0.8$ volt
    $e_3 = 0.95$ volt

8–3. The circuit shown in Figure 8.7 has the following resistance and coefficient values.

$a_1 = 0.62$, $a_2 = 0.47$, $R_1 = 100,000$ ohms, $R_2 = 10,000$ ohms,
$R_o = 100,000$ ohms

Determine the output voltage (e_o) for each of the following sets of input voltages.

a. $e_1 = 4.2$ volts
    $e_2 = 0.12$ volt

b. $e_1 = -6.3$ volts
    $e_2 = 2.8$ volts

c. $e_1 = 0.8$ volt
    $e_2 = 0.06$ volt

d. $e_1 = 0.89$ volt
    $e_2 = -6.3$ volts

8–4. The analog computer program for solving algebraic equations is shown in Figure 8.8. Determine the attenuator coefficients (K_1, K_2, and K_3) to solve each of the following algebraic equations. Use only values of 1 or 10 for the K-coefficients.

    a. $y = 8.3\ x_1 + 0.73\ x_2 + 6.4\ x_3$
    b. $y = 0.8\ x_1 + 4.7\ \ x_2 + 0.3\ x_3$
    c. $y = 3.4\ x_1 + 0.27\ x_2 + 1.2\ x_3$

8–5. Determine the output (y) of each algebraic equation in Exercise 4 for the following input conditions.

$$x_1 = 1.3 \text{ volts}, \qquad x_2 = 2.2 \text{ volts}, \qquad x_3 = -3.1 \text{ volts}$$

8–6. Sketch analog computer diagrams similar to Figure 8.8 to solve each of the following algebraic equations. Indicate the necessary attenuator coefficient values (a_1, a_2, etc.) and summing amplifier coefficients (K_1, K_2, etc.). Use only values of 1 or 10 for the K-coefficients.

    a. $y = -(0.7\ x_1 + 3.7\ x_2 + 8.2\ x_3)$
    b. $y = 6.4\ x_1 + 8.2\ x_2 + 0.41\ x_3 + 0.62\ x_4$
    c. $y = 0.37\ x_1 - 0.42\ x_2 + 3.7\ x_3$
    d. $y = 0.77\ x_1 + 2.7\ x_2 - 7\ x_3$

8–7. The integrator in Figure 8.9 is used to integrate a constant 0.32-volt signal for 6 seconds. The initial condition is 0 volts, R_1 is 100,000 ohms, and C is 10 microfarads. Determine the output voltage at $t = 3$ seconds and $t = 6$ seconds.

8–8. Use the integrator in Figure 8.11 to perform the following integration.

$$e_o(t) = -\int_{t_o}^{t} (10e_1 + e_2 + e_3)dt + e_o(t_o)$$

The capacitor has a value of 10 microfarads. Determine the value of each input resistor (R_1, R_2 and R_3).

8–9. Sketch circuit diagrams similar to Figures 8.5 and 8.11 that will solve each of the following mathematical operations. Assume that you have a 10-microfarad capacitor available, and use a value of 100,000 ohms for R_o. Determine the proper size of each input resistor.

    a. $e_o(t) = -e_1(t)$
    b. $e_o(t) = -10e_1(t)$
    c. $e_o(t) = -\int_{t_o}^{t} e_1(t)\, dt + e_o(t_o)$
    d. $e_o(t) = -\int_{t_o}^{t} 10e_1(t)\, dt + e_o(t_o)$
    e. $e_o(t) = -[e_1(t) + e_2(t)]$
    f. $e_o(t) = -[e_1(t) + e_2(t) + 10e_3(t) + 10e_4(t)]$

g. $e_o(t) = -\int_{t_o}^{t} [e_1(t) + e_2(t) + 10e_3(t) + 10e_4(t)]\, dt + e_o(t_o)$

h. $e_o(t) = -e_1(t) + \int_{t_o}^{t} 10e_2(t)\, dt - e_o(t_o)$

8–10. Sketch circuit diagrams similar to Figure 8.5 that will solve the following algebraic equations.

a. $y = 0.8\, x_1 + 8\, x_2$

b. $y = -(0.65\, x_1 + 3.1\, x_2 + 0.2\, x_3)$

8–11. Sketch a circuit diagram similar to Figure 8.5 for the analog computer diagram of the system described by the equation $T\dot{x} + x = y$ (see Figure 8.14–c). Note that the resistor (R_o) must be replaced by a capacitor (C) in the integrator. Determine the attenuator coefficient settings for a value of $T = 10$ seconds.

8–12. Sketch analog computer diagrams similar to Figure 8.16–c to solve each of the following differential equations.

a. A liquid tank system (see Exercises 4–6).

$$2000\frac{dh}{dt} + h = 1000\, m$$

b. A typical temperature-measuring means (see Exercises 4–9).

$$6\frac{di}{dt} + i = 0.4\theta$$

c. A process-control valve and converter (see Exercises 4–10).

$$0.0001\frac{d^2x}{dt^2} + 0.02\frac{dx}{dt} + x = 0.3i$$

d. A tubular heat exchanger (see Exercises 4–11).

$$25\frac{d^2\theta}{dt^2} + 26\frac{d\theta}{dt} + \theta = 125x$$

e. A liquid surge tank (see Exercises 4–12).

$$h(t) = 0.5\int m(t)\, dt$$

f. A spring-mass-damping system (see Exercises 4–13).

$$M\frac{d^2x}{dt^2} + R\frac{dx}{dt} + Kx = F$$

8–13. Sketch a circuit diagram similar to Figure 8.5 for the analog computer diagram of each system in Exercises 8–12.

8–14. Explain the operation of the circuit in Exercise 11 in terms of the currents entering and leaving the summing junction. Assume that the input signal (y) changes from 0 to 10 volts at time $t = 0$ seconds, and that the output signal $(-x)$ is zero volts until time $t = 0$. Recall that the summing junction voltage is always zero, and that the current through each resistor is equal to the voltage across the resistor divided by the resistance value.

The integrator has two inputs—one from the y-signal and the other from the $-x$-signal—and two input resistors $(R_1$ and $R_2)$. Assume that the attenuated $-x$-feedback signal is connected to R_1 and the attenuated y-input signal is connected to R_2. Answering the following questions will help develop your explanation.

a. Input y is attenuated by $1/T$ and then applied to input resistor R_2. What is the voltage drop across R_2? How much current enters the summing junction through input resistor R_2? Does the current in R_2 change after time $t = 0$?

b. Output $-x$ is attenuated by $1/T$ and then applied to input resistor R_1. What is the voltage drop across R_1? How much current leaves the summing junction through R_1 when $-x_1 = 0$? When $-x_1 = -\dfrac{y}{2}$ volts? When $-x_1 = -y$ volts?

c. The current entering the integrator summing junction through R_2 has two paths for leaving the summing junction—through R_1, and through the feedback capacitor. The current through the capacitor will charge the capacitor, causing output voltage $-x$ to become negative. The current through the capacitor determines the rate of change of voltage $-x$. What happens to the current through R_1 as $-x$ becomes more negative? What happens to the current through the capacitor as $-x$ becomes more negative?

d. When the capacitor finally becomes fully charged, the current through the capacitor will be zero. How much current will leave through R_1 when this occurs? What is the value of $-x$ when this occurs?

9

Controllers

9.1 Introduction

The basic characteristic of a controller is the way in which it acts to restore the controlled variable to its desired value. The different control actions are called the modes of control. Common modes of control include: (1) two-position, (2) floating, (3) proportional, (4) integral, and (5) derivative. The integral mode is sometimes called *reset* and the derivative mode is sometimes called *rate* or *pre-act*. The reason for this alternate terminology will be explained within the discussion of each control mode.

The block diagram of a control system is illustrated in Figure 9.1. Control is achieved by performing the following three steps:

1. *Measuring* the controlled variable.
2. *Comparing* the measured value with the desired value (setpoint) and applying the difference to a controller.
3. *Correcting* the error by manipulating some variable with the final control element in response to the output of the controller.

The measuring means is used to accomplish step 1 (measurement). The output of the measuring means is a signal (C_m) which is used as a measure of the controlled variable. The measured value (C_m) may be

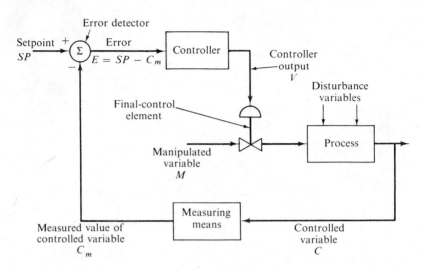

FIGURE 9.1

BLOCK DIAGRAM OF A CLOSED-LOOP CONTROL SYSTEM

an electric current signal, an electric voltage signal, a pneumatic pressure signal, or any other convenient signal.

The error detector accomplishes step 2 (comparison). An example of an electrical error detector is illustrated in Figure 9.2 in which the output of the measuring means is a 4- to 20-milliamp electric current signal. Each value of current represents a unique value of the controlled variable (C). The 4-ma signal represents the minimum value of C, and the 20-ma signal represents the maximum value. The current signal is applied to a 62.5-ohm resistor, resulting in a 0.25- to 1.25-volt signal across the resistor. The setpoint signal is produced by a potentiometer with a 0.25- to 1.25-volt output range. The two voltage signals are connected in opposition so that the voltage between points a and b is equal to the setpoint signal minus the measured value signal.

$$E = SP - C_m$$

The controller accomplishes step 3 (correction). The error signal (E) is the input to the controller. The controller acts on the error signal to produce the controller output signal (V). The control modes determine the size and timing relationship between the controller input (E) and the output (V). The controller output signal is applied to some type of final control element which regulates the manipulated variable. The control action changes the manipulated variable in a manner that will tend to drive the error toward zero.

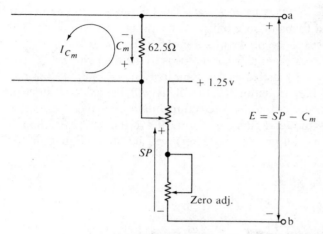

FIGURE 9.2
AN ELECTRICAL ERROR DETECTOR

A brief review of the effect of load changes will be quite helpful in developing an understanding of the function of the controller. The first-
order lag process illustrated in Figure 9.3 will be used to develop the
concept of a process load curve. The liquid process in Figure 9.3–a consists of a tank with liquid level (h), input valve resistance (R_m), and
output resistance (R_L). The input valve is the final control element, and
the input flow rate (m) is the manipulated variable. The liquid level
(h) is the controlled variable. The output resistance (R_L) is the disturbance variable. Load changes are created whenever R_L changes.

The process is maintained in a balanced condition by adjusting the
input flow rate (m) until it balances the output flow rate (q). Assuming
laminar flow, the output flow rate (q) is given by

$$q = \frac{9.81\rho}{R_L}h \qquad (9.1)$$

where

$\rho = $ liquid density kilograms/cubic meter

$R_L = $ liquid resistance newton—seconds/meter

$h = $ liquid level, meters

If R_L is constant, then q is proportional to h. For each level (h), there
will be a specific output flow rate (q) and, hence, a specific input flow
rate (m) necessary to maintain a balanced condition. Each input flow
rate (m) requires a specific position of the final control element (R_m)
which, in turn, requires a specific controller output signal (V). Elimi-

nating the middle variables, each level requires a different controller output signal to maintain a balanced condition in the process. This concept is the basis of the process load curve in Figure 9.3–b.

Now let us examine what happens to the load curve if R_L changes. For example, if R_L increases for some reason, the output flow rate (q) will decrease [see Equation (9.1)]. This will be reflected in a decrease in the controller output required to maintain a balanced condition, as illustrated by the load curve in Figure 9.3–c. On the other hand, if R_L decreases, the output flow rate (q) will increase [Equation (9.1)]. This

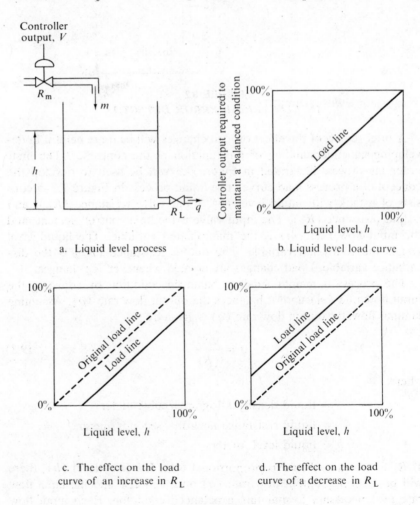

a. Liquid level process

b. Liquid level load curve

c. The effect on the load curve of an increase in R_L

d. The effect on the load curve of a decrease in R_L

FIGURE 9.3
THE LOAD CURVE OF A LIQUID-LEVEL PROCESS

will be reflected in an increase in the controller output, as indicated in Figure 9.3–d. A load change is any condition that changes the location of the load line. After a load change, the controller must adjust its output to the value determined by the new load line. This change is accomplished by the action of the control modes in the controller.

9.2 Two-Position Control

Two-position control is the simplest, most common mode of control. The controller output has only two possible values, depending on the sign of the error. If the two positions are fully open and fully closed, the controller is called an on-off controller. Most two-position controllers have a neutral zone to prevent chattering. The neutral zone is a set of values around zero which the error must pass through before any control action takes place. The input-output curve and block diagram of a two-position controller are given in Figures 9.4 and 9.5.

Two-position control supplies energy in pulses to the process. This causes a cycling of the controlled variable. The amplitude of the cycling depends on three factors: The capacitance of the process, the dead-time lag of the process, and the size of the load changes the process is capable of handling. The amplitude of the oscillation is decreased by either increasing the capacitance, decreasing the dead-time lag, or decreasing the size of the load change that can be accommodated. For these reasons, two-position control is only used on processes that have a capacitance large enough to counteract the combined effect of the dead-time lag and

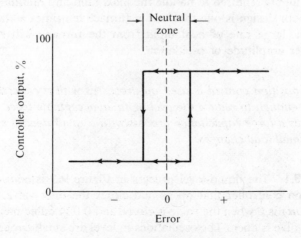

FIGURE 9.4
THE INPUT-OUTPUT CURVE OF A TWO-POSITION CONTROLLER

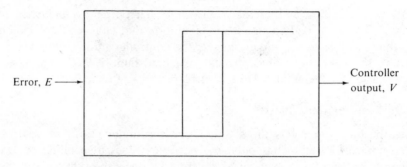

FIGURE 9.5
THE BLOCK DIAGRAM OF A TWO-POSITION CONTROLLER

the load-change capability of the process. Two-position control is simple and inexpensive. Its use is preferred whenever the cycling can be reduced to an acceptable level.

A household heating system is an example of a two-position control system. The air in the house has relatively large thermal capacitance. The dead time is relatively small and the rate of heat input from the furnace is also small, compared with the capacitance of the room. The room temperature cycles with an amplitude which is well within the acceptable limits of human comfort. The rate of heat input from the furnace is just sufficient to heat the house on the most severe winter day. During the summer, no heat is required. This is an example of good two-position design. The two positions provide inputs just slightly above and slightly below the inputs required to handle the maximum and minimum process loads. A poor design is one that uses a furnace ten times as large as required. The large rate of heat input from the furnace will result in a much larger amplitude of oscillation.

> *Two-position control is used on processes with a capacitance large enough to reduce the cycling to an acceptable level. This implies a large capacitance process with a small dead-time lag and small load changes.*

Example 9.1 The liquid-level process in Figure 9.3 is controlled by a two-position controller that opens and closes the inlet valve. The inlet flow rate (m) is 0 when the valve is closed and 0.004 cubic meter/second when the valve is open. The oscillations in level are small enough that the outlet flow rate (q) is essentially constant at 0.002 cubic meter/second. The tank has a cross-sectional area of 2.0 square meters, and the process

dead-time lag is 10 seconds. The neutral zone of the controller is equivalent to a ± 0.05-meter change in level. Determine the amplitude and
period of the oscillation in level (h).

Solution:

1. The rate of accumulation of liquid in the tank (a) is equal to the
inflow minus the outflow.

$$a = m - q, \text{ cubic meters/second}$$

The rate of change of the level $\left(\dfrac{dh}{dt}\right)$ is equal to the rate of accumulation divided by the area of the tank (A).

$$\frac{dh}{dt} = \frac{a}{A} = \frac{m - q}{2.0}, \text{ meters/second}$$

When the valve is open,

$$\frac{dh}{dt} = \frac{0.004 - 0.002}{2.0} = 0.001 \text{ meter/second}$$

When the valve is closed,

$$\frac{dh}{dt} = \frac{0 - 0.002}{2} = -0.001 \text{ meter/second}$$

2. A graph of the oscillation is illustrated in Figure 9.6. At point a, the
valve is closed. The level is changing at a rate of -0.001 meter/second

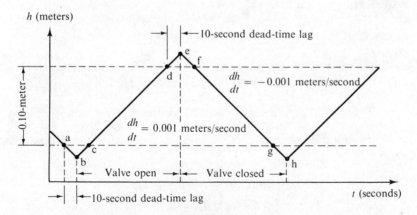

FIGURE 9.6
*THE OSCILLATION OF THE LEVEL OF THE LIQUID TANK
IN EXAMPLE 9.1*

(decreasing), and the level has reached the lower limit of the neutral zone. After the 10-second dead-time lag, the controller opens the valve at point b. The level has decreased an additional amount equal to (10 seconds) (0.001 meter/second) = 0.01 meter.

$$h_{ab} = -0.01 \text{ meter}$$
$$t_{ab} = 10 \text{ seconds}$$

3.   The valve is open from point b to point e, and the level is increasing at a rate of 0.001 meter/second. It takes 10 seconds to reach point c.

$$h_{bc} = 0.01 \text{ meter}$$
$$t_{bc} = 10 \text{ seconds}$$

The time required to move from point c to point d is equal to the change in level (0.1 meter) divided by the rate of change of level (0.001 meter/second).

$$t_{cd} = \frac{0.1}{0.001} = 100 \text{ seconds}$$
$$h_{cd} = 0.1 \text{ meter}$$
$$t_{cd} = 100 \text{ seconds}$$

At point d, the level has reached the upper level of the neutral zone. After the 10-second dead-time lag, the controller closes the valve at point e.

$$h_{de} = h_{bc} = 0.01 \text{ meter}$$
$$t_{cd} = 10 \text{ seconds}$$

4.   Since the rate of increase and the rate of decrease of the level are equal, the time from e to h is the same as the time from b to e.

$$t_{eh} = t_{be} = 120 \text{ seconds}$$

5.   The amplitude of oscillation is equal to h_{be}, and the period is equal to t_{bh}.

$$\text{Amplitude} = h_{bc} + h_{cd} + h_{de}$$
$$= 0.01 + 0.10 + 0.01$$
$$= \underline{0.12 \text{ meter}}$$

$$\text{Period} = t_{be} + t_{eh}$$
$$= 120 + 120$$
$$= \underline{\underline{240 \text{ seconds}}}$$

9.3   Floating Control

Floating control is a special application of two-position control in which the final control element is stationary as long as the error remains within the neutral zone. When the controlled variable is outside the neutral zone, the final control element changes at a constant rate in a direction determined by the sign of the error. The final control element continues to change until the error returns to the neutral zone, or until the final control element reaches one of its extreme positions. The input-output curve and block diagram of a floating controller are illustrated in Figures 9.7 and 9.8.

Floating control has a tendency to produce cycling of the controlled variable. The amplitude of the cycling depends on the dead-time lag of the process, the capacitance of the process, and the speed at which the controller increases and decreases the final control element. The speed of the final control element determines the fastest load change that it can keep pace with, but not the size of the load change. The main advantage of floating control is its ability to handle large load changes by gradually adjusting the final control element. As with two-position control, the amplitude is decreased by increasing the capacitance, decreasing the

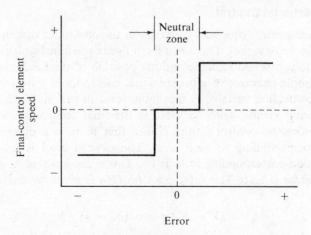

FIGURE 9.7
THE INPUT-OUTPUT CURVE OF A FLOATING CONTROLLER

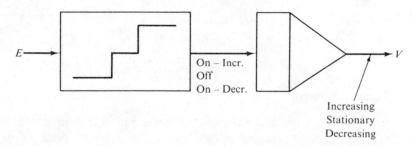

On – Incr.
Off
On – Decr.

Increasing
Stationary
Decreasing

FIGURE 9.8
THE BLOCK DIAGRAM OF A FLOATING CONTROLLER

dead-time lag, or decreasing the speed of the final control element. Floating control is used when large load changes are anticipated, and the capacitance is large enough to counteract the effects of the dead-time lag and the speed of the final control element. Floating control is frequently used because it is inherent in the type of actuator used to drive the final control element (e.g., electric motors and hydraulic cylinders operated by on-off relays, or solenoid valves provide a floating control action).

Floating control is used on processes with large, slow-moving load changes and a capacitance large enough to reduce the cycling to an acceptable level. This implies a large capacitance process with a small dead-time lag. Floating control is inherent in some final control elements.

9.4   Proportional Control

Proportional control produces a change in the controller output proportional to the error signal. There is a fixed linear relationship between the value of the controlled variable and the position of the final control element. A simple example of a proportional controller is shown in Figure 9.9. The controlled variable is the liquid level in the tank. The float is the measuring means, and the valve is the final control element. The lever provides the control action. Notice that there is a different valve position corresponding to each level. The desired level is h_o, and the valve position corresponding to h_o is V_o. This is the position of the valve when the error is zero. The valve position (V) is given by the following equation.

$$(V - V_o) / a = (h_o - h) / b$$

But $h_o - h$ = error signal = e.

$$V = V_o + \frac{a}{b} e \qquad\qquad (9.2)$$

where

$$V = \text{the valve position, meters}$$
$$V_o = \text{the valve position with zero error, meters}$$
$$e = \text{the error signal, meters}$$

Equation (9.2) is one version of the defining equation of the proportional control mode.

The gain (K) of the proportional controller in Figure 9.9 is the change in valve position divided by the corresponding change in level. Both are expressed in percentage of the full-scale range.

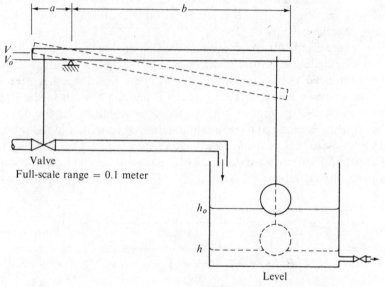

Valve
Full-scale range = 0.1 meter

Level
Full-scale range = 1 meter

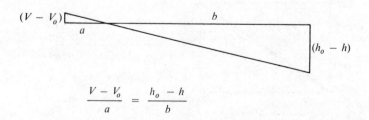

$$\frac{V - V_o}{a} = \frac{h_o - h}{b}$$

FIGURE 9.9
A SIMPLE PROPORTIONAL CONTROLLER

percent change in valve position $= 100(V - V_o)/0.1 = 1000(V - V_o)$

per cent change in level $= 100(h_o - h)/1 = 100(h_o - h)$

$$\text{gain, } K = \frac{1000(V - V_o)}{100(h_o - h)} = 10\frac{(V - V_o)}{(h_o - h)} = 10\frac{a}{b}$$

Figure 9.10 includes schematic diagrams and input-output curves of proportional controllers with gains of 0.5, 1, and 2. In general, an increase in the gain reduces the size of the error required to produce a 100 per cent change in the valve position. In other words, a high gain requires a small error to produce the change in valve position necessary to balance the process. Although this seems to imply that the gain should be as high as possible, unfortunately, increasing the gain increases the tendency for oscillation of the controlled variable. A compromise is necessary in which the gain is made as large as possible without producing unacceptable oscillations.

The main disadvantage of proportional control is that it cannot completely eliminate the error caused by a load change. A residual error is always required to maintain the valve at some position other than V_o. This is obvious in Equation (9.2), and it is equally obvious in the simple system illustrated in Figure 9.9. A load change means that a different valve position is required to maintain a balanced condition in the process. With a proportional control mode, a change in level is the only way that the valve position can be changed. This change or residual error is called the proportional offset. The size of the offset is directly proportional to

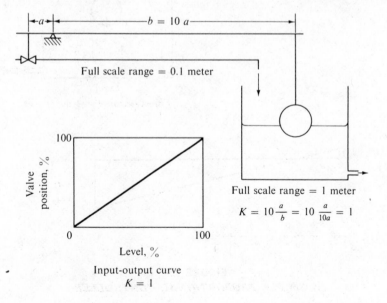

Input-output curve
$K = 1$

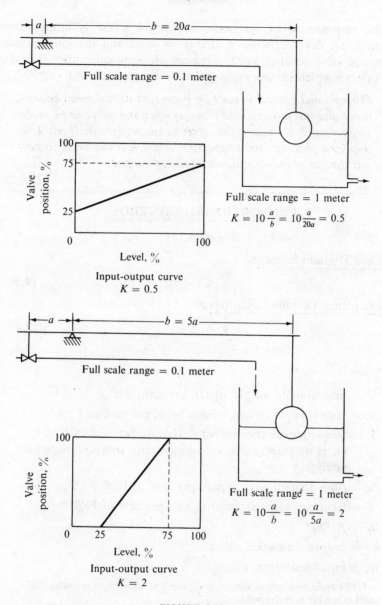

FIGURE 9.10

PROPORTIONAL CONTROLLERS WITH DIFFERENT GAIN VALUES

the size of the load changes and inversely proportional to the gain (K). Proportional control is used when the gain can be made large enough to counteract the effect of the largest load change on the proportional offset.

The response of the proportional control action is instantaneous. There is no delay between a change in level and the corresponding change in valve position. For this reason, proportional control is used on fast systems which require rapid action to maintain control.

Proportional control is used on processes with a small capacitance and fast moving load changes when the gain can be made large enough to reduce the offset to an acceptable level. This implies a process with a capacitance which is too small to permit the use of two–position or floating control.

PROPORTIONAL CONTROL

(Figure 9.11)

Time Domain Equation

$$v = K e + v_o \qquad (9.3)$$

Frequency Domain Equations

$$V = K E \qquad (9.4)$$

$$\text{Transfer Function} = V/E = K \qquad (9.5)$$

where:

$v =$ time domain output signal, per cent of F. S.*

$v_o =$ time domain output when $e = $ o, per cent of F. S.*

$V =$ frequency domain output deviation, per cent of F. S.*
(v_o is assumed to be zero to simplify frequency-domain analysis.)

$e =$ time-domain error signal, per cent of F. S.*

$E =$ frequency-domain error signal, per cent of F. S.*

$K = R_f/R_1$*

$R_f =$ output resistance, ohms

$R_1 =$ input resistance, ohms

  * The error and output signals in Figure 9.11 must have the same full-scale range for K as defined here.

An operational amplifier is frequently used to form an electronic proportional controller. Figure 9.11 is a schematic diagram of an electronic proportional controller. The time-domain and frequency-domain equations are included for convenience as a reference. Refer to Chapter Eight for the development of the time-domain equation and Chapter Four for the development of the frequency-domain equation.

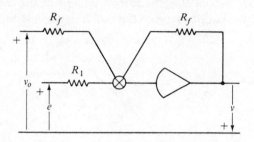

$$v = Ke + v_o; \quad K = \frac{R_f}{R_1}$$

FIGURE 9.11
AN ELECTRONIC PROPORTIONAL CONTROLLER

Example 9.2 An electronic proportional controller similar to Figure 9.11 uses a value of 10^4 ohms for output resistor R_f. Determine the value of R_1 for a gain of 0.2.

Solution:

$$K = R_f/R_1$$
$$0.2 = 10^4/R_1$$
$$R_1 = 10^4/0.2 = 5 \times 10^4 \text{ ohms}$$

Example 9.3 A proportional controller has a gain of 4. Determine the proportional offset required to maintain $V - V_o$ at 20 per cent.

Solution:

$$\text{Gain} = \frac{V - V_o}{E}$$

$$4 = \frac{20}{E}$$

$$E = 5 \text{ per cent}$$

An offset error of 5 per cent is required to maintain $V - V_o$ at 20 per cent.

9.5 Integral Control

The integral control mode changes the output of the controller by an amount proportional to the integral of the error signal. In other words, the change in controller output during the interval from 0 to t seconds is proportional to the net area under the error curve between 0 and t. Figure 9.12 illustrates the relationship between the error signal and the controller output. Notice that the rate of change of the controller output is proportional to the error signal (the rate of change is equal to the slope of the graph).

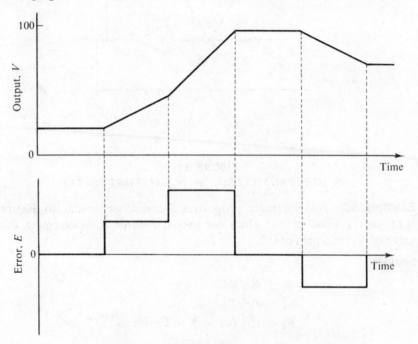

FIGURE 9.12

INTEGRAL-MODE RESPONSE TO AN ERROR SIGNAL

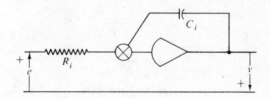

$$v = \frac{1}{T_i} \int_o^t e dt + v_o \; ; \; T_i = R_i C_i$$

FIGURE 9.13

AN ELECTRONIC INTEGRAL CONTROLLER

An integral-mode controller is quite simple to construct using the circuit diagram shown in Figure 9.13. See Chapter Eight for further details. Integral control is sometimes used alone, but most often it is combined with proportional control. This combination will be discussed next.

INTEGRAL CONTROL

(Figure 9.13)

Time Domain Equation

$$v = \frac{1}{T_i} \int_o^t e\, dt + v_o \qquad (9.6)$$

Frequency Domain Equations

$$V = \left[\frac{1}{T_i S}\right] E \qquad (9.7)$$

$$\text{Transfer Function} = V/E = \frac{1}{T_i S} \qquad (9.8)$$

where:

v = time-domain output signal, per cent of F. S.*

v_o = time-domain output at $t = 0$, per cent of F. S.*

V = frequency-domain output deviation, per cent of F. S.*
(v_o is assumed to be zero to simplify frequency-domain analysis.)

e = time-domain error signal, per cent of F. S.*

E = frequency-domain error signal, per cent of F. S.*

$T_i = R_i C_i$ = integral action time constant, seconds*

R_i = input resistance, ohms

C_i = output capacitance, farads

* The error and output signals in Figure 9.13 must have the same full-scale range for T_i as defined here.

9.6 Proportional-Plus-Integral Control

The integral mode is frequently combined with the proportional mode to provide an automatic reset action which eliminates the proportional offset. The combination is referred to as *proportional plus integral control* or *proportional plus reset control*. The integral mode provides the reset action by constantly changing the controller output until the error is reduced to zero. Figure 9.14 illustrates the step response of a proportional-plus-integral controller. The proportional mode provides a change in the controller output which is proportional to the error signal. The integral mode provides an additional change in the output which is proportional to the integral of the error signal. The integral action time constant (T_i) is the time required for the integral mode to match the change in output produced by the proportional mode.

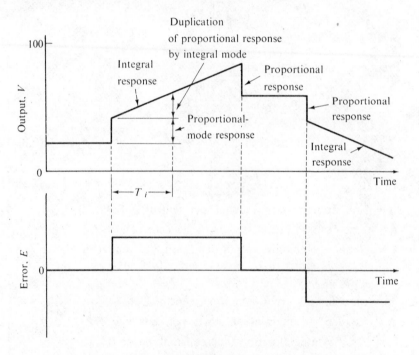

FIGURE 9.14

STEP RESPONSE OF A PROPORTIONAL-PLUS-INTEGRAL CONTROLLER

The disadvantage of the integral mode is that it increases the tendency for oscillation of the controlled variable. The gain of the proportional controller must be reduced when it is combined with the integral mode. This reduces the ability of the controller to respond to rapid load changes. If the process has a large dead-time lag, the error signal will not immediately reflect the actual error in the process. This delay often results in overcorrection by the integral mode—that is, the integral mode continues to change the controller output after the error is actually reduced to zero because it is acting on an "old" signal.

Proportional plus integral control is used on processes with large load changes when the proportional mode alone is not capable of reducing the offset to an acceptable level. The integral mode provides a reset action which eliminates the proportional offset.

PROPORTIONAL-PLUS-INTEGRAL CONTROL
(Figure 9.15)

Time-Domain Equation

$$v = Ke + \frac{K}{T_i} \int_o^t e\, dt + v_o \qquad (9.9)$$

Frequency-Domain Equations

$$V = KE + \left[\frac{K}{T_iS}\right] E \qquad (9.10)$$

$$\text{Transfer Function} = V/E = K\left[\frac{T_iS + 1}{T_iS}\right] \qquad (9.11)$$

where:

v = time-domain output signal, per cent of F. S.*

v_o = time-domain output at $t = 0$, per cent of F. S.*

V = frequency-domain output deviation, per cent of F. S.* (v_o is assumed to be zero to simplify frequency domain analysis.)

e = time-domain error signal, per cent of F. S.*

E = frequency-domain error signal, per cent of F. S.*

$T_i = R_iC_i$ = integral action time constant, seconds*

$K = R_i/R_1$ = gain*

R_1 = input resistance, ohms

R_i = output resistance, ohms

C_i = output capacitance, farads

  * The error and output signals in Figure 9.15 must have the same full-scale range for K and T_i as defined here.

An electronic proportional-plus-integral controller is illustrated in Figure 9.15. Equation (9.9) is the time-domain equation of the controller. The Ke-term in Equation (9.9) is the control action of the proportional mode. The $\dfrac{K}{T_i}\int edt$-term is the control action of the integral mode. Equation (9.10) is the frequency-domain equation, and Equation (9.11) gives the transfer function. The KE-term in Equation (9.10) is the proportional action, and the $\dfrac{K}{T_iS}$ term is the integral action.

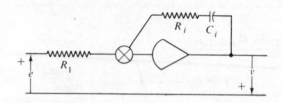

$$v = Ke + \frac{K}{T_i} \int_o^t edt + v_o$$

$$T_i = R_iC_i \; : \;\; K = R_i/R_1$$

FIGURE 9.15

AN ELECTRONIC PROPORTIONAL-PLUS-INTEGRAL CONTROLLER

Example 9.4 Determine the values of R_1 and R_i for an electronic proportional-plus-integral controller with a gain (K) of 2 and an integral action time constant (T_i) of 50 seconds. Use a 10^{-5} farad capacitor for C_i. Determine the time-domain equation and the transfer function.

Solution:

The proportional-plus-integral controller is defined by Equations (9.9), (9.10), and (9.11).

$$T_i = R_iC_i$$
$$50 = R_i \times 10^{-5}$$
$$R_i = 50 \times 10^5 = 5 \times 10^6 \text{ ohms}$$
$$K = R_i/R_1$$
$$2 = 5 \times 10^6/R_1$$
$$R_1 = 2.5 \times 10^6 \text{ ohms}$$

The resistance values are:

$$R_i = 5 \times 10^6 \text{ ohms}$$

$$R_1 = 2.5 \times 10^6 \text{ ohms}$$

The time-domain equation is given by Equation (9.9):

$$v = 2e + \frac{2}{50} \int_o^t e \, dt + v_o$$

The transfer function is given by Equation (9.11):

$$\frac{V}{E} = 2 \left[\frac{1 + 50S}{50S} \right]$$

Example 9.5 The controller in Example 9.4 has a value of v_o of 32 per cent at time $t = 0$. The graph of the error signal is given in **Figure 9.16–a.** Determine the value of the controller output (v) at the following times: $t = 0$ seconds, 10 seconds, 50 seconds, 75 seconds, and 100 seconds.

Solution:

The controller output is given by Equation (9.9).

$$v = Ke + \frac{K}{T_i} \int_o^t e \, dt + v_o$$

But $v_o = 32$ per cent, and from Example 9.4, $K = 2$ and $T_i = 50$ seconds.

$$v = 2\,e + \frac{2}{50} \int_o^t e \, dt + 32$$

The term $\displaystyle\int_o^t e \, dt$ is equal to the net area under the error curve between o and t seconds.

a.   At $t = 0$ seconds, $e = 0$ and $\int e \, dt = 0$.

$$v = 2(0) + \frac{1}{25}(0) + 32$$

$$v = 32 \text{ per cent}$$

PD ⟋‾‾ *PI* ⌐‾⌐‾⌐

b. At $t = 10$ seconds, $e = 0$ and $\int e\, dt = 0$.

$$v = 2\,(0) + \frac{1}{25}\,(0) + 32$$

$$v = 32 \text{ per cent}$$

c. At $t = 50$ seconds, $e = 10$ per cent and

$$\int e\, dt = (10 \text{ per cent}) (50 - 20 \text{ seconds}) = 300 \text{ per cent} - \text{seconds}$$

$$v = 2(10) + \frac{1}{25}\,(300) + 32$$

$$v = 20 + 12 + 32$$

$$v = 64 \text{ per cent}$$

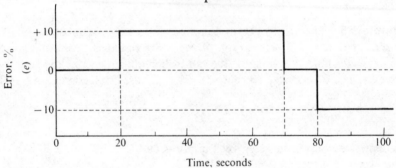

a) Graph of the error signal, e

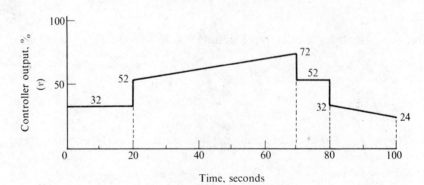

b) Graph of the controller output signal, v

FIGURE 9.16

ERROR AND CONTROLLER OUTPUT SIGNAL GRAPHS
FOR EXAMPLE 9.5

d. At $t = 75$ seconds, $e = 0$ and

$$\int e\, dt = (10)\,(70 - 20) = 500 \text{ per cent seconds}$$

$$v = 2\,(0) + \frac{1}{25}\,(500) + 32$$

$$v = 0 + 20 + 32$$

$$v = 52 \text{ per cent}$$

e. At $t = 100$ seconds, $e = -10$ per cent and

$$\int e\, dt = (10)\,(70 - 20) + (-10)\,(100 - 80)$$

$$\int e\, dt = 500 - 200$$

$$\int e\, dt = 300$$

$$v = 2(-10) + \frac{1}{25}\,(300) + 32$$

$$v = -20 + 12 + 32$$

$$v = 24 \text{ per cent}$$

A graph of the controller output is also given in Figure 9.16(b).

9.7 Derivative Control Mode

The derivative control mode changes the output of the controller pro-portional to the rate of change of the error signal. This change may be caused by a variation in the measured variable, the setpoint, or both. The derivative mode contributes to the output of the controller only while the error is changing. For this reason, the derivative control mode is always used in combination with the proportional, or proportional plus integral control modes.

The derivative control mode is never used alone. It is always used in combination with the proportional, or proportional plus integral modes.

The step and ramp responses of the ideal derivative control mode are given in Figure 9.17. At every instant, the output of the derivative con-trol mode is proportional to the slope or rate of change of the error signal. The step response indicates the reason that the ideal derivative control mode is never used. The error curve has an infinite slope when the step change occurs. The ideal derivative mode must respond with an infinite change in the controller output. In actual practice, the response of the derivative action to rapidly changing signals is limited. This greatly reduces the sensitivity of the controller to the unwanted noise spikes which frequently occur in actual practice.

The ideal and practical electronic differentiator circuits are shown in Figure 9.18. The time-domain equation and frequency-domain transfer

---

DERIVATIVE CONTROL MODE

(Figure 9.18)

Time-Domain Equation—(Ideal Differentiator)
(Figure 9.18–a)

$$v = T_d \frac{de}{dt} \qquad (9.12)$$

Frequency-Domain Equation—(Ideal Differentiator)
(Figure 9.18–a)

$$\text{Transfer Function} = \frac{V}{E} = T_d S \qquad (9.13)$$

Frequency-Domain Equation—(Practical Differentiator)
(Figure 9.18–b)

$$\text{Transfer Function} = \frac{V}{E} = \frac{T_d S}{(S R_1 C_d + 1)} \qquad (9.14)$$

where:

v = time-domain output signal, per cent of F. S.*

V = frequency-domain output signal, per cent of F. S.*

e = time-domain error signal, per cent of F. S.*

E = frequency-domain error signal, per cent of F. S.*

$T_d = R_d C_d$ = the derivative action time constant, seconds

R_1 = input resistance, ohms

R_d = output resistance, ohms

C_d = input capacitance, farads

* The error and output signals in Figure 9.18 must have the same full-scale range for T_d as defined here.

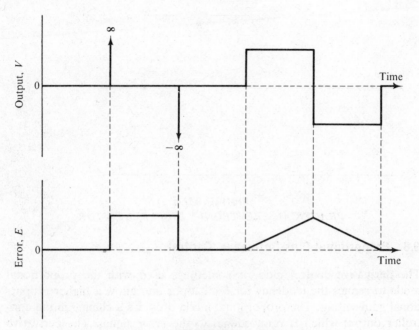

FIGURE 9.17
*THE STEP AND RAMP RESPONSE OF THE IDEAL
DERIVATIVE CONTROL MODE*

function are included here for future reference. In the transfer function of the practical differentiator [Equation (9.14)], the $(1 + R_1C_dS)$-term limits the response produced by rapidly changing signals.

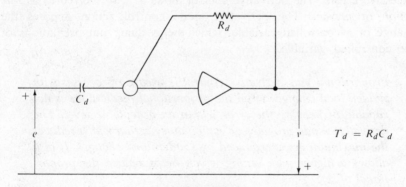

FIGURE 9.18-a
IDEAL ELECTRONIC DIFFERENTIATOR

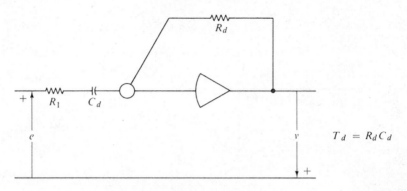

FIGURE 9.18-b
PRACTICAL ELECTRONIC DIFFERENTIATOR

9.8 Proportional-Plus-Derivative Control

The derivative control mode is sometimes used with the proportional mode to reduce the tendency for oscillations and allow a higher proportional gain setting. The proportional mode provides a change in the controller output which is proportional to the error signal. The derivative mode provides an additional change in the controller output which is proportional to the rate of change of the error signal. The derivative mode anticipates the future value of the error signal and changes the controller output accordingly. This anticipatory action makes the derivative mode useful in controlling processes with sudden load changes. For this reason, the derivative mode is usually used with proportional, or proportional plus integral control when the sudden load changes produce excessive errors. The derivative control mode is also referred to as *rate action* or *pre-act*. The derivative mode control action opposes the change of a controlled variable, which helps damp out oscillations of the controlled variable.

> *Proportional plus derivative control is used on processes with sudden load changes when the proportional mode alone is not capable of keeping the error within an acceptable level. The derivative mode provides an anticipatory action which reduces the maximum error produced by sudden load changes. It also allows a higher gain setting which helps reduce the proportional offset.*

An electronic proportional-plus-derivative controller is shown in Figure 9.19. Equation (9.15) is the defining time-domain equation. The *Ke-*

term in Equation (9.15) is the proportional mode action. The $KT_d\dfrac{de}{dt}$ term is the derivative mode action, and $\alpha T_d\dfrac{dv}{dt}$ is the term which limits the

---

PROPORTIONAL-PLUS-DERIVATIVE CONTROL

(Figure 9.19)

Time-Domain Equation

$$v + \alpha T_d\frac{dv}{dt} = Ke + KT_d\frac{de}{dt} + v_o \tag{9.15}$$

Frequency-Domain Equations

$$V = [KE + KT_dSE]\left[\frac{1}{\alpha T_dS + 1}\right] \tag{9.16}$$

$$\text{Transfer Function} = \frac{V}{E} = \frac{K(T_dS + 1)}{(\alpha T_dS + 1)} \tag{9.17}$$

where:

v = time-domain output signal, per cent of F. S.*

v_o = time-domain output signal with no control action, per cent of F. S.*

V = frequency-domain output deviation, per cent of F. S.* (v_o is assumed to be zero to simplify frequency-domain analysis.)

e = time-domain error signal, per cent of F. S.*

E = frequency-domain error signal, per cent of F. S.*

$T_d = R_dC_d$ = the derivative action time constant, seconds

R_1 = series input resistance, ohms

R_d = parallel input resistance, ohms

R_f = feedback resistance, ohms

C_d = input capacitance, farads

$K = \dfrac{R_f}{R_1 + R_d}$ = gain

$\alpha = \dfrac{R_1}{R_1 + R_d}$ = derivative limiter coefficient

    * The error and output signals in Figure 9.1 must have the same full-scale range for the equations as defined.

response produced by rapidly changing signals. Equation (9.16) is the time-domain equation, and Equation (9.17) gives the transfer function. In Equation (9.16), KE is the proportional action, KT_dSE is the derivative action, and $1/(\alpha T_dS + 1)$ is the limit on the derivative action.

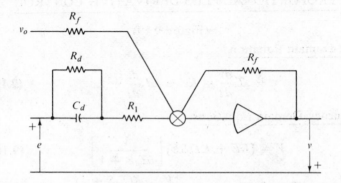

FIGURE 9.19

AN ELECTRONIC PROPORTIONAL-PLUS-DERIVATIVE CONTROLLER

Example 9.6 Determine the values of R_1, R_d, and R_f for an electronic proportional-plus-derivative controller with a gain (K) of 0.8, a derivative action time constant (T_d) of 1 second, and a value of α of 0.1. Use a 10^{-6} farad capacitor for C_d. Determine the time-domain equation and the transfer function.

Solution:

Equations (9.15), (9.16), and (9.17) apply to the proportional-plus-derivative controller.

1. $T_d = R_dC_d$

   $1 = R_d \times 10^{-6}$

   $\underline{R_d = 10^6 \text{ ohms}}$

2. $\alpha = \dfrac{R_1}{R_1 + R_d}$

   $0.1 = \dfrac{R_1}{R_1 + 10^6}$

   $0.1R_1 + 10^5 = R_1$

   $0.9R_1 = 10^5$

   $\underline{R_1 = 1.11 \times 10^5 \text{ ohms}}$

3. $K = \dfrac{R_f}{R_1 + R_d}$

$0.8 = \dfrac{R_f}{1.11 \times 10^5 + 10^6} = \dfrac{R_f}{1.11 \times 10^6}$

$R_f = 0.888 \times 10^6$ ohms

The time-domain equation is given by Equation (9.15):

$$\alpha\, T_d = (0.1)\,(1) = 0.1$$

$$K\, T_d = (0.8)\,(1) = 0.8$$

$$v + 0.1\,\frac{dv}{dt} = 0.8e + 0.8\,\frac{de}{dt} + v_o$$

The transfer function is given by Equation (9.17):

$$\frac{V}{E} = \frac{0.8\,(S + 1)}{(0.1\,S + 1)}$$

Example 9.7 The controller in Example 9.6 has a value of v_o of 40 per cent. The graph of the error signal is given in Figure 9.20. Determine the value of the controller output (v) at the following times: $t = 0$, $t = 20$, $t = 40$, $t = 60$ seconds. Assume that the $\alpha T_d \dfrac{dv}{dt}$-term is negligible.

Solution:

The controller output is given by Equation (9.15).

$$v + \alpha T_d \frac{dv}{dt} = Ke + K\, T_d \frac{de}{dt} + v_o$$

But $\dfrac{dv}{dt} = 0$ by assumption, $v_o = 40$ per cent, $K = 0.8$, and $T_d = 1$. With these values, Equation (9.15) is

$$v = 0.8\,e + 0.8\,\frac{de}{dt} + 40$$

The term $\dfrac{de}{dt}$ is equal to the slope of the error curve at any given instant of time.

a. At $t = 0$, $e = 0$, and $\dfrac{de}{dt} = 0$.

$$v = 0.8\,(0) + 0.8\,(0) + 40$$

$$\underline{v = 40 \text{ per cent}}$$

b. At $t = 20$ seconds, $e = 10$ per cent and

$$\frac{de}{dt} = \frac{20 \text{ per cent}}{20 \text{ seconds}} = 1 \text{ per cent/second}$$

$$v = 0.8 (10) + 0.8 (1) + 40$$

$$\underline{\underline{v = 48.8 \text{ per cent}}}$$

c. At $t = 40$ seconds, $e = 20$ per cent and $\dfrac{de}{dt} = 0$.

$$v = 0.8 (20) + 0.8 (0) + 40$$

$$\underline{\underline{v = 56 \text{ per cent}}}$$

d. At $t = 60$ seconds, $e = 10$ per cent, and

$$\frac{de}{dt} = \frac{-20}{20} = -1 \text{ per cent/second}$$

$$v = 0.8 (10) + 0.8 (-1) + 40$$

$$\underline{\underline{v = 47.2 \text{ per cent}}}$$

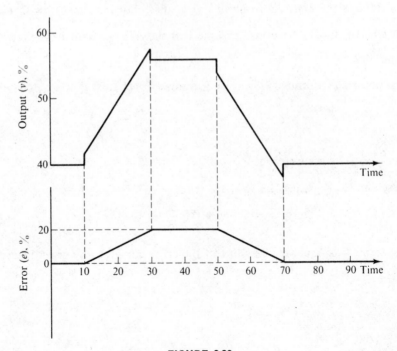

FIGURE 9.20

ERROR AND CONTROLLER OUTPUT GRAPHS FOR EXAMPLE 9.7

e.   At $t = 80$ seconds, $e = 0$, and $\dfrac{de}{dt} = 0$.

$$v = 0.8 \,(0) + 0.8 \,(0) + 40$$

$$\underline{\underline{v = 40 \text{ per cent}}}$$

A graph of the controller output is also given in Figure 9.20.

9.9 Proportional-Plus-Integral-Plus-Derivative Control

A proportional-plus-integral-plus-derivative controller is usually referred to as a *three-mode controller*. The integral mode is used to eliminate the proportional offset caused by large load changes. The derivative mode reduces the tendency toward oscillations and provides a control action which anticipates changes in the error signal. The derivative mode is especially useful when the process has sudden load changes.

> *Proportional-plus-integral-plus-derivative control is used on processes with sudden, large load changes when one or two mode control is not capable of keeping the error within acceptable limits. The derivative mode produces an anticipatory action which reduces the maximum error produced by sudden load changes. The integral mode provides a reset action which eliminates the proportional offset.*

Equation (9.18) is the defining equation for an ideal three-mode controller.

$$v = Ke + \frac{K}{T_i} \int_{o}^{t} e \, dt + KT_d \frac{de}{dt} + v_o \qquad \textbf{(9.18)}$$

Practical three-mode controllers use two modifications to the above equation. The first modification, mentioned in Sections 9.7 and 9.8, is the addition of the term $\alpha T_d \dfrac{dv}{dt}$ to the left-hand side of Equation (9.18). The purpose of this term is to reduce the derivative action produced by rapidly changing signals such as noise spikes.

The second modification is for the sake of economy. The implementation of the ideal three-mode controller is too expensive to be competitive because the derivative and reset terms must be formed in parallel. An additional summing amplifier is required to add the two signals in order to produce the response defined by Equation (9.18). The second modification consists of forming the integral and derivative terms in series. No matter which term is placed first, there is an interaction between the gain (K), the integral action time constant (T_i), and the derivative action time constant (T_d). Equation (9.19) is the defining equation of an interacting three-mode controller.

$$v = K \left(1 + \frac{T_d}{T_i}\right) \left(e + \frac{1}{T_d + T_i} \int_o^t e \, dt + \frac{T_d T_i}{T_d + T_i} \frac{de}{dt}\right) + v_o \quad \textbf{(9.19)}$$

Let k', T_i', and T_d' represent the equivalent ideal gain, integral time, and derivative time in Equation (9.19). Then

$$K' = 1 + T_d/T_i$$

$$T'_i = T_d + T_i$$

$$T_d' = \frac{T_d T_i}{T_d + T_i}$$

Substituting K', T_i', and T_d' into Equation (9.19), results in the ideal three-mode equation [Equation (9.18)].

$$v = K' \left(e + \frac{1}{T_i'} \int_o^t e \, dt + T_d' \frac{de}{dt}\right) + v_o$$

Figure 9.21 is a schematic diagram of an electronic three-mode controller in which the derivative action is formed at the input to the operational amplifier and the integral action is formed in the feedback circuit. Figure 9.22 is a schematic diagram of a three-mode controller in which the integral action is formed at the input and the derivative action is formed in the feedback circuit. The defining equations of both controllers are given by Equations (9.20), (9.21), and (9.22). The individual terms were explained in Sections 9.6 and 9.8. Refer to Appendix C for a development of the time-domain equations of the controller illustrated in Figure 9.22.

PROPORTIONAL-PLUS-INTEGRAL-PLUS DERIVATIVE CONTROL

(Figures 9.21 and 9.22)

Time-Domain Equation

$$v + \alpha T_d \frac{dv}{dt} = K\left(1 + \frac{T_d}{T_i}\right)\left(e + \frac{1}{T_d + T_i}\int_o^t e \, dt + \right.$$

$$\left. \frac{T_d T_i}{T_d + T_i} \frac{de}{dt}\right) + v_o \qquad \textbf{(9.20)}$$

Frequency-Domain Equations

$$V = K\left[1 + \frac{1}{T_iS}\right]\left[T_dS + 1\right]\left[\frac{1}{\alpha T_dS + 1}\right]E \qquad \textbf{(9.21)}$$

$$\text{Transfer Function} = \frac{V}{E} = K\left[\frac{T_iS + 1}{T_iS}\right]\left[\frac{T_dS + 1}{\alpha T_dS + 1}\right] \qquad \textbf{(9.22)}$$

Figure 9.21	Figure 9.22
$K = \dfrac{R_i}{R_1 + R_d}$	$K = C_i/C_1$
$\alpha = \dfrac{R_1}{R_1 + R_d}$	$\alpha = \dfrac{C_2}{C_1 + C_2}$
$T_i = R_iC_i$	$T_i = R_iC_i$
$T_d = R_dC_d$	$T_d = R_d(C_1 + C_2)$

where:

v = time-domain output signal, per cent of F. S.*

v_o = time-domain output signal at time $t = 0$, per cent of F. S.*

V = frequency-domain output deviation, per cent of F. S.* (v_o is assumed to be zero to simplify frequency-domain analysis)

e = time-domain error signal, per cent of F. S.*

E = frequency-domain error signal, per cent of F. S.*

Resistance values are in ohms.
Capacitance values are in farads.

* The error and output signals in Figures 9.21 and 9.22 must have the same full-scale range for the equations as defined.

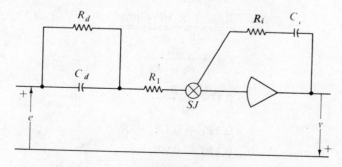

FIGURE 9.21
*AN ELECTRONIC THREE-MODE CONTROLLER WITH
DERIVATIVE INPUT AND INTEGRAL OUTPUT*

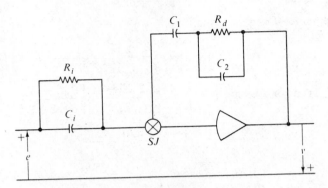

FIGURE 9.22

AN ELECTRONIC THREE-MODE CONTROLLER WITH
INTEGRAL INPUT AND DERIVATIVE OUTPUT

Example 9.8   Determine the values of R_1, R_i, R_d, and C_d for a 3-mode controller, as shown in Figure 9.21. The following controller parameters are desired.

$$K = 4$$
$$T_i = 7 \text{ seconds}$$
$$T_d = 0.5 \text{ second}$$
$$\alpha = 0.1$$

Use a 10^{-5}-farad capacitor for C_i. Determine the time-domain equation and the transfer function.

Solution:

Equations (9.20), (9.21), and (9.22) apply to the three-mode controller.

1.
$$T_i = R_i C_i$$
$$7 = R_i 10^{-5}$$
$$\underline{R_i = 7 \times 10^5 \text{ ohms}}$$

2.
$$\alpha = \frac{R_1}{R_1 + R_d}$$
$$0.1 = \frac{R_1}{R_1 + R_d}$$
$$0.1R_1 + 0.1R_d = R_1$$
$$0.1R_d = 0.9R_1$$
$$R_d = 9R_1$$

3.
$$K = \frac{R_i}{R_1 + R_d}$$

$$4 = \frac{7 \times 10^5}{R_1 + 9R_1}$$

$$10R_1 = 1.75 \times 10^5$$

$$\underline{\underline{R_1 = 1.75 \times 10^4 \text{ ohms}}}$$

4.
$$R_d = 9R_1 = 9(1.75) \times 10^4$$

$$\underline{\underline{R_d = 1.575 \times 10^5 \text{ ohms}}}$$

5.
$$T_d = R_d C_d$$

$$0.5 = 1.575 \times 10^5 C_d$$

$$C_d = 0.5/1.575 \times 10^5$$

$$\underline{\underline{C_d = 3.17 \times 10^{-6} \text{ farads}}}$$

6.
$$\alpha T_d = (0.1)(0.5) = 0.05$$

$$K\left(1 + \frac{T_d}{T_i}\right) = 4\left(1 + \frac{0.5}{7}\right) = 4(1.07) = 4.3$$

$$\frac{1}{T_d + T_i} = \frac{1}{0.5 + 7} = \frac{1}{7.5} = 0.133$$

$$\frac{T_d T_i}{T_d + T_i} = \frac{(0.5)(7)}{0.5 + 7} = \frac{3.5}{7.5} = 0.467$$

The time-domain equation is given by Equation (9.20).

$$v + 0.05\frac{dv}{dt} = 4.03\left(e + 0.133 \int e\, dt + 0.467\frac{de}{dt}\right) + v_o$$

7.   The transfer function is given by Equation (9.22).

$$\frac{V}{E} = 4\left[\frac{7S + 1}{7S}\right]\left[\frac{0.5S + 1}{0.05S + 1}\right]$$

Example 9.9   The controller in Example 9.8 has a value of $v_o = 10$ percent. The graph of the error signal is given in Figure 9.23. Determine the output of the controller at $t = 5, 10, 15$, and 25 seconds. Assume that the $_a T_d \frac{dv}{dt}$-term is negligible.

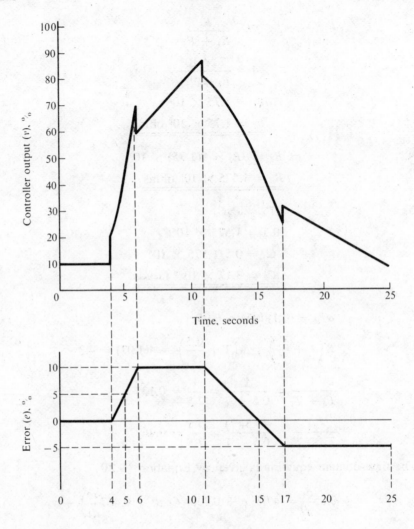

FIGURE 9.23

ERROR AND CONTROLLER OUTPUT GRAPHS FOR EXAMPLE 9.9

Solution:

The controller equation was determined in Example 9.8, Part 6. With $v_o = 10$ per cent and $\alpha T \dfrac{dv}{dt} = 0$, the equation is

$$v = 4.3\left(e + 0.133 \int_{o}^{t} e\, dt + 0.467\frac{de}{dt}\right) + v_o$$

1. At $t = 5$ seconds

$$e = 5 \text{ per cent}$$

$$\int e\, dt = \frac{1}{2}(1)(5) = 2.5 \text{ per cent—seconds}$$

$$\frac{de}{dt} = 10/(6 - 4) = 5 \text{ per cent/seconds}$$

$$v = 4.3[5 + 0.133(2.5) + 0.467(5)] + 10$$

$$v = 4.3[5 + 0.333 + 2.335] + 10$$

$$v = (4.3)(7.668) + 10 = 33 + 10$$

$$v = \underline{43 \text{ per cent}}$$

2. At $t = 10$ seconds

$$e = 10 \text{ per cent}$$

$$\int e\, dt = \frac{1}{2}(2)(10) + (4)(10) = 10 + 40 = 50$$

$$\frac{de}{dt} = 0$$

$$v = 4.3[10 + 0.133(50) + 0.467(0)] + 10$$

$$v = 4.3[10 + 6.65] + 10 = (4.3)(16.65) + 10$$

$$v = 71.7 + 10 = \underline{81.7 \text{ per cent}}$$

3. At $t = 15$ seconds

$$e = 0$$

$$\int e\, dt = \frac{1}{2}(2)(10) + (5)(10) + \frac{1}{2}(4)(10)$$

$$\int e\, dt = 10 + 50 + 20 = 80 \text{ per cent—seconds}$$

$$\frac{de}{dt} = -15/(17 - 11) = -2.5 \text{ per cent/seconds}$$

$$v = 4.3[0 + 0.133(80) + 0.467(-2.5)] + 10$$

$$v = 4.3(10.64 - 1.17) + 10 = (4.3)(9.47) + 10$$

$$v = 40.7 + 10 = \underline{50.7 \text{ per cent}}$$

4. At $t = 25$ seconds

$$e = -5 \text{ per cent}$$

$$\int e \, dt = 80 - \tfrac{1}{2}(2)(5) - (8)(5)$$
$$= 80 - 5 - 40 = 35 \text{ per cent—seconds}$$
$$\frac{de}{dt} = 0$$
$$v = 4.3[-5 + 0.133(35) + 0.467(0)] + 10$$
$$v = 4.3[-5 + 4.65] + 10 = -1.5 + 10 = \underline{\underline{8.5 \text{ per cent}}}$$

A graph of the controller output is included in Figure 9.23.

9.10 Digital Controllers

Digital controllers used for closed-loop control generally implement the proportional-plus-integral-plus-derivative control action. The controlled variable is measured at specific times separated by a time interval called the sample time, Δt. Each sample (or measurement) of the controlled variable is converted to a binary number for input to a digital computer or microprocessor. The computer subtracts each sample of the measured variable from the set point to determine a set of error samples.

$$E_1 = SP - C_{m1} = \text{the first error sample}$$
$$E_2 = SP - C_{m2} = \text{the second error sample}$$
$$E_3 = SP - C_{m3} = \text{the third error sample}$$
$$E_n = SP - C_{mn} = \text{the present error sample}$$

After computing each error sample, the computer follows a procedure called the PID algorithm to calculate a three-mode control action based on the error samples: $E_1, E_2, E_3, \ldots, E_n$. The PID algorithm has two versions, the positional version and the incremental version.

The positional algorithm determines the valve position, V_n, based on the error signals. Equation (9.23) is a simplified version of the positional algorithm.

$$V_n = KE_n + K\frac{\Delta t}{T_i} \sum_{j=1}^{j=n} E_j + K\frac{T_d \Delta E_n}{\Delta t} \qquad \textbf{(9.23)}$$

where:

$V_n =$ present valve position, per cent of F.S.
$K =$ controller gain
$E_n =$ present error sample, per cent of F.S.
$\Delta t =$ sample time, seconds
$T_i =$ integral action time constant, seconds

$$T_d = \text{derivative action time constant, seconds}$$
$$\Delta E_n = E_n - E_{n-1} = \text{the change in the error signal}$$

The incremental algorithm determines the change in the valve position, $\Delta V_n = V_n - V_{n-1}$, based on the error samples. The incremental algorithm can be determined by using Equation (9.23) to determine V_n and V_{n-1} and then subtracting to obtain Equation (9.24) as follows.

$$V_{n-1} = KE_{n-1} + K\frac{\Delta t}{T_i}\sum_{j=1}^{j=n-1}E_j + K\frac{T_d\Delta E_{n-1}}{\Delta t}$$

$$\Delta V_n = K\Delta E_n + K\frac{\Delta t}{T_i}E_n + K\frac{T_d}{\Delta t}(\Delta E_n - \Delta E_{n-1}) \qquad \textbf{(9.24)}$$

where:

$$\Delta V_n = V_n - V_{n-1}$$
$$\Delta E_n = E_n - E_{n-1}$$
$$\Delta E_n - \Delta E_{n-1} = E_n - 2E_{n-1} + E_{n-2}$$

The incremental algorithm is especially suited to incremental output devices such as stepper motors. The positional algorithm is more natural and has the advantage that the controller "remembers" the valve position. If the sample time, Δt, is much shorter than the integral time, T_i, then the positional algorithm will produce a behavior similar to an analog controller.

The integral and derivative modes in Equation (9.23) both present computational problems that can result in unsatisfactory results. The integral mode is given by the following term in Equation (9.23).

$$\text{Integral Term} = K\frac{\Delta t}{T_i}\sum_{j=1}^{j=n}E_j$$

For each sample, the integral mode must produce a change given by:

$$\text{Integral mode change} = \frac{K\Delta t}{T_i}E_j$$

If $K\Delta t$ is small compared with T_i, then the computer may ignore relatively large errors because of insufficient resolution. For example, consider a digital controller with a 12-bit word length. The largest value that can be represented by a 12-bit binary number is decimal 4,095, and the resolution is one part in 4,095. If $K = 0.5$, $t = 1$ second, and $T_i = 500$ seconds, then

$$K\Delta t/T_i = (0.5)(1)/(500) = 1/1000$$

Any value of E less than 1000 (i.e., 24 per cent of full scale) would result in an integral mode change less than 1, which would be ignored. This small change would then be lost unless a special provision is made to include the change in the calculations for the next sample. The end result is a permanent offset error that the integral mode is unable to eliminate.

One solution to this integral offset problem is to increase the precision by increasing the word length in the computer. A 16-bit word length has a precision of 1 part in 65,535, which would reduce the offset from 24 per cent to 1.5 per cent of full scale. Another solution is to add the unused portion of the sum of the error samples to the next error sample before computing the integral mode change. In the preceding example, an error of 900 in each of two successive samples would not produce an integral mode change because each sample is less than 1000. However, if the first sample is retained and added to the second sample, the sum of 1800 would produce an integral mode change of $1800/1000 = 1$ with a remainder of 800. The remainder would then be retained to be added to the next error sample.

The derivative mode is given by the following term in Equation (9.23).

$$\text{Derivative Term} = K \frac{T_d}{\Delta t} (E_n - E_{n-1})$$

If $K = 6$, $\Delta t = 1$ second, and $T_d = 100$ seconds, then

$$KT_d/\Delta t = (6)(100)/1 = 600$$

A slowly changing signal will result in a "jumpy" derivative mode action. For example, the following error samples would result from sampling a controlled variable that is increasing at a rate of 0.25 parts/second.

$$E_1 = 0, \quad E_2 = 0, \quad E_3 = 0, \quad E_4 = 1, \quad E_5 = 1,$$
$$E_6 = 1, \quad E_7 = 1, \quad E_8 = 2, \quad E_9 = 2, \quad E_{10} = 2,$$

etc. The resulting derivative mode control actions would be:

$$D_1 = 0, \quad D_2 = 0, \quad D_3 = 0, \quad D_4 = 600, \quad D_5 = 0,$$
$$D_6 = 0, \quad D_7 = 0, \quad D_8 = 600, \quad D_9 = 0, \quad D_{10} = 0$$

Thus every fourth sample would result in a derivative mode kick. This kick is considerably different than the steady output produced by an analog derivative mode with the same steady change in the controlled

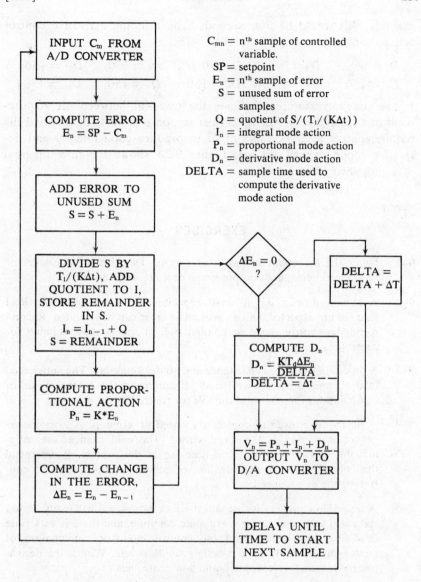

C_{mn} = nth sample of controlled variable.
SP = setpoint
E_n = nth sample of error
S = unused sum of error samples
Q = quotient of $S/(T_i/(K\Delta t))$
I_n = integral mode action
P_n = proportional mode action
D_n = derivative mode action
$DELTA$ = sample time used to compute the derivative mode action

FIGURE 9.24

FLOW DIAGRAM OF A SIMPLIFIED PID ALGORITHM

variable. The derivative mode kick can be reduced by increasing the sample time used to calculate the derivative mode action. The desired sample time is the interval between changes in the error. In the above example, this would be four seconds. The resulting derivative control actions would be:

$$D_1 = 0, \qquad D_2 = 0, \qquad D_3 = 0, \qquad D_4 = 150, \qquad D_5 = 150;$$
$$D_6 = 150, \qquad D_7 = 150, \qquad D_8 = 150, \qquad D_9 = 150, \qquad D_{10} = 150$$

The previous examples indicate the trade-off between the requirement of a small value of Δt for proper response to load changes and the requirement of a larger value of Δt to produce good integral and derivative mode control actions. Figure 9.24 shows the flow diagram of a simplified PID algorithm.

EXERCISES

9–1. Name the five common modes of control. Describe the operation of each mode by an input-output graph or a step-response curve.

9–2. A certain process has a small capacitance. Sudden moderate load changes are expected, and a small offset error can be tolerated. Recommend the control mode or combination of modes most suitable for controlling the process.

9–3. A process has a large capacitance and no dead-time lag. The anticipated load changes are relatively small. Recommend the control mode or combination of modes most suitable for controlling the process.

9–4. A dc motor is used to control the speed of a pump. A tachometer-generator is used as the speed sensor. The load changes are insignificant, and there is no dead-time lag in the process. Recommend the control mode or combination of modes most suitable for controlling the motor speed.

9–5. A liquid flow process is fast and the flow rate signal has many "noise spikes." Large load changes are quite common, and there is very little dead-time lag. Recommend the control mode or combination of modes most suitable for controlling the flow rate. Why is the derivative mode usually avoided in liquid flow controllers?

9–6. An electric heater is used to control the temperature of a plastic extruder. The process is a first-order lag with almost no dead-time lag. The time constant is very large. Under steady operation, the load changes are insignificant. Recommend the control mode or combination of modes most suitable for controlling the temperature of the extruder.

9–7. A dryer is a slow process with a very large dead-time lag. Sudden load changes are common, and proportional offset is undesirable. Recommend the control mode or combination of modes most suitable for controlling the process.

9–8.   The liquid level process in Example 9.1 is controlled by a two-position controller that opens and closes the inlet valve. Examine the effect on the oscillation in level (h) of each of the following changes.

   a.   The inlet flow rate with the valve open is 0.008 cubic meter/second instead of 0.004 cubic meter/second.
   b.   The dead-time lag is 20 seconds instead of 10 seconds.
   c.   The cross-sectional area of the tank is 1.0 square meters instead of 2.0 square meters.
   d.   The neutral zone is ±0.10 meter instead of ±0.05 meter.

9–9.   The solid flow rate control system shown in Figure 9.25 uses a gate to control the level of material on the belt. A single-speed, reversible motor is used to drive the cam that positions the gate. If the solid feed rate is below a predetermined value, the controller drives the gate up at a constant rate. If the feed rate is above a second predetermined value, the controller drives the gate down at a constant rate. Between the two predetermined values, the gate is motionless. Identify the mode of control used in this system.

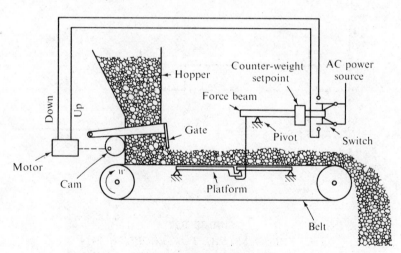

FIGURE 9.25
A SOLID FLOW RATE CONTROL SYSTEM

9–10.   Sketch the input-output curve for each of the following proportional controller conditions.

   a. $K = 10$, $v_o = 30$ per cent
   b. $K = 0.1$, $v_o = 60$ per cent
   c. $K = 5$, $v_o = 50$ per cent

9–11. An electronic proportional controller similar to Figure 9.11 used a value of 10^5 ohms for R_f. Determine the value of R_1 for each of the following controller gains.

    a. $K = 0.1$ b. $K = 0.5$ c. $K = 1.0$
    d. $K = 4.0$ e. $K = 8.0$ f. $K = 10.0$

9–12. Determine the proportional offset required to maintain $V - V_o$ at 10 per cent for proportional controllers with each of the following gain values.

    a. $K = 0.1$ b. $K = 0.5$ c. $K = 1.0$
    d. $K = 5.0$ e. $K = 10.0$

9–13. Determine the values of R_1 and R_i for an electronic proportional-plus-integral controller (Figure 9.15) with $K = 0.5$ and $T_i = 150$ seconds. Use a 10^{-5} farad capacitor for C_i. Determine the time-domain equation and the transfer function.

9–14. A proportional-plus-integral controller has a gain of 2, an integral time of 200 seconds, and a value of v_o of 40 per cent. The graph of the error signal is given in Figure 9.26. Determine the value of the controller output at the following times: $t = 100$ seconds, 200 seconds, 300 seconds, and 400 seconds.

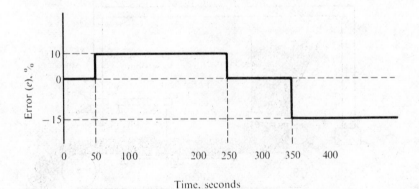

FIGURE 9.26
THE ERROR SIGNAL FOR PROBLEM 9–14

9–15. Determine the value of R_1, R_d, and R_f for an electronic proportional-plus-derivative controller with a gain (K) of 6, a derivative time (T_d) of 4 seconds, and a value of α of 0.1. Use a 10^{-6}-farad capacitor for C_d. Determine the time-domain equation and the transfer function.

9–16. A proportional-plus-derivative controller has the following parameter values: $K = 0.5$, $T_d = 4$ seconds, $\alpha = 0$, and $V_o = 65$ per cent. The graph of the error signal is given in Figure 9.27. Determine the value of the controller output at the following times: $t = 5$ seconds and $t = 15$ seconds.

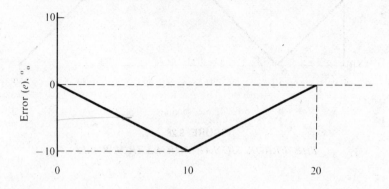

Time, seconds

FIGURE 9.27

THE ERROR SIGNAL FOR EXERCISE 9.16

9–17. Determine the values of R_1, R_i, R_d, and C_d for a three-mode controller, as shown in Figure 9.21. The following controller parameters are desired:

$$K = 0.75$$
$$T_i = 80 \text{ seconds}$$
$$T_d = 0.2 \text{ second}$$
$$\alpha = 0.1$$

Use a 10^{-5}-farad capacitor for C_i. Also determine the time-domain equation and the transfer function.

9–18. A three-mode controller has the following control parameters.

$$K = 1$$
$$T_i = 10 \text{ seconds}$$
$$T_d = 1 \text{ second}$$
$$v_o = 60 \text{ per cent}$$
$$\alpha = 0$$

Determine the controller output at $t = 5$, 20, and 40 seconds. The graph of the error signal is given in Figure 9.28.

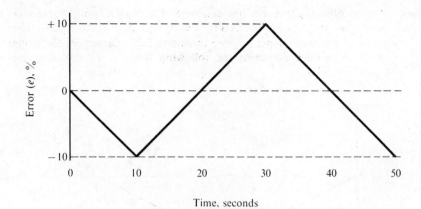

FIGURE 9.28

THE ERROR SIGNAL FOR EXERCISE 9.18

10

Final Control
Elements

10.1 Introduction

In every control system, the desired control action is accomplished by controlling the flow of energy to the process. Some type of final control element is used to regulate this energy input. Typical final control elements include: pneumatic control valves, electric motors, pumps, hydraulic motors, hydraulic cylinders, and electric relays or contactors. Various methods of power amplification are used to provide the power required to drive the final control element. Often, the final control element itself serves as a power amplifier. In other systems, a separate component called an *amplifier* provides a major portion of the required power amplification. Selected examples of final control elements and amplifiers are presented in this chapter.

10.2 Control Valves

General Considerations. The manipulated variable in a process control system is usually the flow rate of a fluid, and the pneumatic control valve is the most common final control element. A typical pneumatic control valve is illustrated in Figure 10.1. The input to the valve is a 3- 15-psi air-pressure signal which is applied to the top of the diaphragm. The diaphragm actuator converts the air pressure into a displacement of the valve stem. The valve body and trim varies the area through which the flowing fluid must pass.

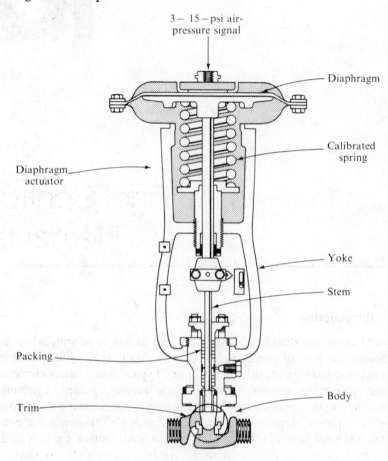

3 — 15 — psi air-pressure signal

Diaphragm

Diaphragm actuator

Calibrated spring

Yoke

Stem

Packing

Body

Trim

FIGURE 10.1

A PNEUMATIC CONTROL VALVE[1]

The input air-pressure signal may come directly from a pneumatic controller; it may come from a pneumatic valve positioner; or, it may come from an electropneumatic transducer. The valve positioner is a pneumatic amplifier that provides additional power to operate the valve. The electropneumatic transducer converts the milliamp signal from an

[1] Honeywell Specifications 5810–1401a, Figure 1, Model 1401 Diaphragm Operated. Honeywell Industrial Division, Honeywell, Inc.

electronic controller into the 3- 15-psi air-pressure signal required by the control valve.

The diaphragm actuator consists of a synthetic rubber diaphragm and a calibrated spring. The air pressure on the top of the diaphragm is opposed by the spring. Full travel of the valve stem is obtained with a change in the air-pressure signal from 3 to 15 psi. The graph of the valve-stem displacement versus the air-pressure signal is within a few per cent of a straight line. The actuator may be described as a *direct actuator* or a *reverse actuator,* depending on the location of the calibrated spring. The two types of actuator are illustrated in Figure 10.2.

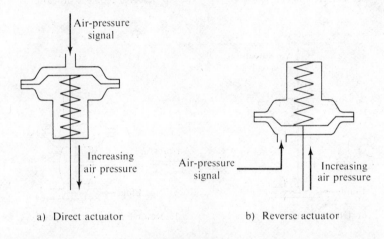

a) Direct actuator   b) Reverse actuator

FIGURE 10.2

DIRECT- AND REVERSE-VALVE ACTUATORS

A yoke attaches the diaphragm actuator to the valve body. The valve stem passes through a packing, which allows motion of the valve stem and prevents leakage of the fluid flowing around the stem. The trim consists of the valve plug and the valve seat or port. These two parts are often produced in matched pairs to provide tight shutoff.

The *valve position* (P) is the displacement of the plug from the fully seated position. The *valve action* is the direction in which valve plug moves as the air-pressure signal increases. If increasing air pressure closes the valve, the valve action is described as *air-to-close.* If increasing air pressure opens the valve, the action is described as *air-to-open.* The variations of valve action and actuator type are illustrated in Figure 10.3.

Valve Characteristics. The *characteristic* of a control valve is the relationship between the valve position and the flow rate through the valve.

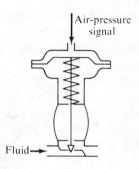

a) Direct acting air-to-close

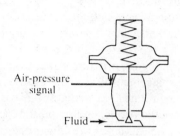

b) Reverse acting air-to-close

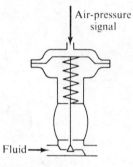

c) Direct acting air-to-open

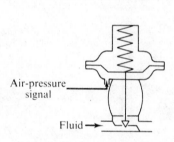

d) Reverse acting air-to-open

FIGURE 10.3

VALVE ACTION—ACTUATOR COMBINATIONS

The *inherent characteristic* of a valve is obtained when there is a constant pressure drop across the valve for all valve positions. The *installed characteristics* are obtained when the valve is installed in a system and the pressure drop across the valve depends on the pressure drop in the remainder of the system. Thus, the installed characteristics depend on both the inherent characteristics of the valve and the flow characteristics of the system in which the valve is installed.

The three most common valve characteristics—quick opening, linear, and equal percentage—are illustrated in Figure 10.4.

The *quick-opening* characteristic provides a large change in flow rate for a small change in value position. This characteristic is used for on-off or two-position control systems in which the valve must move quickly from open to closed or vice versa. A typical quick-opening characteristic is shown in Figure 10.4.

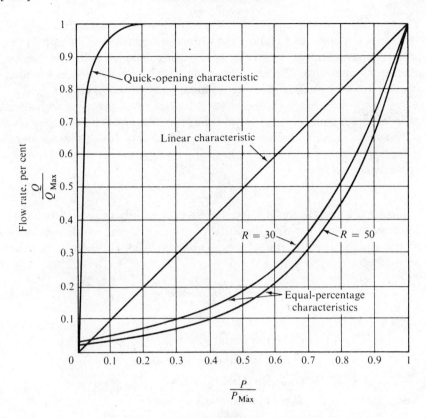

Valve position, per cent

FIGURE 10.4

CONTROL VALVE CHARACTERISTICS

A *linear characteristic* provides a linear relationship between the valve position (P) and the flow rate (Q). The linear characteristic is described by the following mathematical relationship.

$$\frac{Q}{Q_{MAX}} = \frac{P}{P_{MAX}} \tag{10.1}$$

where:

Q = valve flow rate

P = valve position (displacement of the plug from the closed position)

P_{MAX} = maximum valve position

Q_{MAX} = maximum flow rate

The flow rate and valve position are usually given in the English units of gallons per minute and inches. However, any convenient units may be used. The values of Q_{MAX} and P_{MAX} are usually included in the valve manufacturer's literature. The linear characteristic is usually used in installed systems in which most of the system pressure drop is across the valve.

An *equal-percentage characteristic* provides equal percentage changes in flow rate for equal changes in valve position. An equal-percentage valve is designed to operate between a minimum flow rate (Q_{MIN}) and a maximum flow rate (Q_{MAX}). The ratio Q_{MAX}/Q_{MIN} is called the *rangeability of the valve, R*.

$$R = Q_{MAX}/Q_{MIN} \qquad \textbf{(10.2)}$$

Most commercial control valves have a rangeability between 30 and 50. The simplest way to express the equal-percentage characteristic is in terms of the change in flow rate corresponding to a small change in valve position.

$$\frac{Q_2 - Q_1}{Q_1} = K\frac{P_2 - P_1}{P_{MAX}} \qquad \textbf{(10.3)}$$

where:

Q_1 = the initial flow rate

P_1 = the initial valve position

Q_2 = the final flow rate

P_2 = the final valve position

P_{MAX} = the maximum valve position

The value of the constant K depends on the rangeability and the relative size of the change in valve position, $(P_2 - P_1)/P_{MAX}$. For example, if $R = 50$ and $(P_2 - P_1)/P_{MAX} = 0.01 = 1$ per cent, then $K = 4$. In other words, a 1 per cent change in valve position will produce a 4 per cent change in flow rate. If the flow rate is 3 gallons/minute when the valve is 10 per cent open, then a change in valve position to 11 per cent will increase the flow rate by 4 per cent of 3—a total of 3.12 gallons/minute. When the valve is 50 per cent open, the flow rate will be 14.2 gallons/minute. A change in valve position to 51 per cent open will increase the flow rate by 4 per cent of 14.2—a total of 14.77 gallons/minute.

The exact mathematical relationship for the equal percentage valve is given by Equation (10.4).

$$\frac{Q}{Q_{MAX}} = \left(\frac{Q_{MAX}}{Q_{MIN}}\right)^{\left(\frac{P - P_{MAX}}{P_{MAX}}\right)} = R^{\left(\frac{P - P_{MAX}}{P_{MAX}}\right)} \qquad \textbf{(10.4)}$$

where:

$$Q = \text{valve flow rate}$$

$$Q_{MAX} = \text{maximum flow rate}$$

$$Q_{MIN} = \text{minimum flow rate}$$

$$R = \text{rangeability of the valve}$$

$$P = \text{valve position}$$

$$P_{MAX} = \text{maximum valve position}$$

Any convenient units of flow rate and position may be used in Equation (10.4).

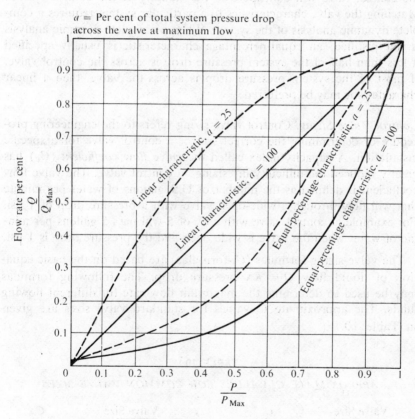

a = Per cent of total system pressure drop across the valve at maximum flow

Flow rate per cent $\dfrac{Q}{Q_{Max}}$

Valve position, per cent

FIGURE 10.5

INSTALLED CHARACTERISTICS OF LINEAR AND EQUAL-PERCENTAGE VALVES

The installed characteristics of a valve depend on how much of the total system pressure drop is across the valve. If 100 per cent of the system pressure drop is across the valve, then the installed characteristic is the same as the inherent characteristic. The installed characteristics of linear and equal-percentage valves are illustrated in Figure 12.5. As the percentage pressure drop across the valve decreases, the installed characteristic of the linear valve approaches the quick-opening characteristic, and the installed characteristic of the equal-percentage valve approaches the linear characteristic.

The proper selection of a control valve involves matching the valve characteristics to the characteristics of the system. When this is done, the control valve will contribute to the stability of the control system. Matching the valve characteristics to a particular system requires a complete dynamic analysis of the system. When a complete dynamic analysis is not justified, an equal-percentage characteristic is usually specified if less than half of the system pressure drop is across the control valve. If most of the system pressure drop is across the valve, then a linear characteristic may be preferred.

Control Valve Sizing. Control valve sizing refers to the engineering procedure of determining the correct size of a control valve for a specific installation. A capacity index called the *valve flow coefficient* (C_v) has greatly reduced the difficulty in "sizing" a control valve. The valve flow coefficient is defined as the number of U.S. gallons of water per minute that will flow through a wide-open valve with a pressure drop of 1 psi. For example, a control valve with a C_v of 5 will pass 5 gallons per minute of water when the valve is wide open and the pressure drop is 1 psi.

The valve-sizing formulas (C_v-formulas) are based on the basic equation of liquid flow $Q = K\sqrt{\text{pressure drop}}$. The following formulas may be used to determine the maximum flow rate for different flowing fluids. The approximate C_v-values for standard valve sizes are given in Table 10.1.

TABLE 10.1

APPROXIMATE C_v-VALUES FOR COMMON VALVE SIZES

Valve Size	C_v	Valve Size	C_v
¼	0.3	3	108
½	3	4	174
1	14	6	400
1½	35	8	725
2	55	10	1100

C_v FORMULAS

Liquids: $\quad Q_L = C_v \sqrt{(P_1 - P_2)/G_L}$ $\qquad$ **(10.5)**

Gases: $\quad Q_g = 960 C_v \sqrt{\dfrac{(P_1 - P_2)(P_1 + P_2)}{G_g(T + 460)}}$ $\qquad$ **(10.6)**

Steam: $\quad M = 90 C_v \sqrt{\dfrac{(P_1 - P_2)}{(V_1 + V_2)}}$ $\qquad$ **(10.7)**

where:

Q_L = the liquid flow rate, gallons per minute

Q_g = the gas flow rate, cubic feet per hour at 14.7 psia and 60°F

M = the steam flow, pounds per hour

C_v = the valve flow coefficient

P_1 = the valve inlet pressure, psia

P_2 = the valve outlet pressure, psia

G_L = the liquid specific gravity (density of the liquid divided by the density of water)

G_g = the gas specific gravity (density of the gas divided by the density of air with both gases at standard conditions)

T = gas temperature, °F

V_1 = Specific volume of the steam at the valve inlet, cubic feet per pound

V_2 = Specific volume of the steam at the valve outlet, cubic feet per pound

Control Valve Transfer Functions. A pneumatic control valve is basically an overdamped second-order system. The resonant frequency (ω_o) is usually between 1 and 10 radians/second. The transfer function is given by Equation (10.8), below.

CONTROL VALVE TRANSFER FUNCTION

$$\frac{P}{I} = \frac{K}{\dfrac{1}{\omega_o^2} S^2 + \dfrac{2\zeta}{\omega_o} S + 1} \qquad \textbf{(10.8)}$$

where:

P = the valve position, inches

I = the input air pressure signal, psi

K = the valve gain, inches/psi

S = the frequency parameter, 1/second

ω_o = the resonant frequency, radians/second
($1 \leq \omega_o \leq 10$)

ζ = the damping ratio, ($\zeta \geq 1$)

Example 10.1 Determine the size of the control valve required to control the flow rate of a liquid with a specific gravity of 0.92. The maximum flow rate is 320 gallons/minute. The available pressure drop across the valve at maximum flow is 60 psi. Allow a safety factor of 25 percent (i.e., $Q_L = 1.25\, Q_{MAX}$).

Solution:

Equation (10.5) may be used.

$$Q_L = C_v\sqrt{(P_1 - P_2)/G_L}$$
$$(1.25)(320) = C_v\sqrt{(60)/0.92}$$
$$400 = 8.08C_v$$
$$C_v = 49.5 \text{ minimum}$$

Therefore, a 2-inch valve is required (Table 10.1).

Example 10.2 Determine the size of the control valve required to control flow rate of a gas with a specific gravity of 0.85. The maximum flow rate is 20,000 standard cubic feet per hour (14.7 psia and 60°F). The gas temperature, inlet pressure, and outlet pressure are 500°F, 90 psia, and 70 psia, respectively. Allow a safety factor of 25 per cent.

Solution:

Equation (186) may be used.

$$Q_g = 960C_v\sqrt{\frac{(P_1 - P_2)(P_1 + P_2)}{G_g(T + 460)}}$$
$$(1.25)(20,000) = 960C_v\sqrt{\frac{(20)(160)}{(0.85)(960)}}$$
$$25,000 = 960\sqrt{3.92}\,C_v$$
$$C_v = 13.15$$

Therefore, a 1-inch valve is required (Table 10.1).

Example 10.3 Determine the size of the control valve required to control the flow rate of steam to a process. The maximum flow rate is 600 pounds/hour. The inlet and outlet steam conditions are given below.

$$P_1 = 40 \text{ psia}$$
$$V_1 = 10.5 \text{ cubic feet/pound}$$
$$P_2 = 30 \text{ psia}$$
$$V_2 = 14.0 \text{ cubic feet/pound}$$

Use a safety factor of 25 per cent.

Solution:

Equation (10.7) may be used.

$$M = 90C_v\sqrt{\frac{P_1 - P_2}{V_1 + V_2}}$$

$$(1.25)(600) = 90C_v\sqrt{\frac{10}{24.5}}$$

$$750 = \frac{90}{\sqrt{2.45}}C_v$$

$$C_v = 13.1$$

Therefore, a 1-inch valve is required (Table 10.1).

10.3 Armature-Controlled DC Motors

General Considerations. Armature-controlled dc motors are frequently used in velocity and position control systems to convert electric signals into a rotary or linear motion. They are extremely versatile drives, capable of reversible operation over a wide range of speeds, with accurate control of the speeds at all times. DC motors are available with horsepower ratings from 1/300 to over 700.

The operating characteristics of electric motors and generators depend on the following basic fact: A force is exerted on an electric charge whenever it moves through a transverse magnetic field. This phenomenon is responsible for the following two actions which occur in motors and generators.

1. A force is exerted on a conductor in a magnetic field whenever a current passes through the conductor.
2. A voltage is induced in a conductor whenever it is moved through a magnetic field.

The force exerted on a current-carrying conductor produces the torque that tends to rotate an electric motor. This force is also present in an electric generator whenever a current is passing through the armature winding. The direction of the force is perpendicular to both the direction of the current (i) and the direction of the magnetic field (B). The magnitude of the force is given by the following relationships.

$$f = i_c BL \sin \alpha \qquad (10.9)$$

where:

f = the force, newtons

i_c = the current, amperes

L = the length of the conductor, meters

B = the flux density of the magnetic field, webers/square meter

α = the angle between the direction of the current and the direction of the magnetic field, degrees

The voltage induced in a moving conductor produces the voltage output of an electric generator. This induced voltage is also present in an electric motor whenever the armature is rotating. If the conductor and the magnetic field are mutually perpendicular then the induced voltage is given by the following relationship.

$$e = LvB \sin \beta \qquad (10.10)$$

where:

e = the induced voltage, volts

L = the length of the conductor, meters

v = the velocity of the conductor, meters/second

B = the flux density, webers/square meter

β = the angle between the direction of motion of the conductor and the direction of the magnetic field, degrees.

DC Motor Construction. The schematic diagram in Figure 10.6 illustrates the fundamental parts of a dc motor. Four coils of wire are mounted in slots on a cylinder of magnetic material called the *armature*. The armature is mounted on bearings and is free to rotate in the magnetic field produced by the two *field poles*. The field poles may be permanent magnets or electromagnets, depending on the size of the motor (the smaller motors use permanent magnets). The ends of each coil are connected to adjacent segments of a segmented ring called the *commutator*. Electrical connection is made to the armature coils through carbon contacts called *brushes*.

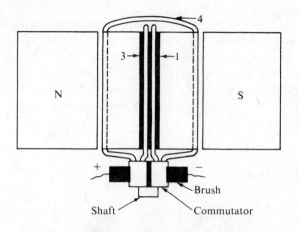

a) Top view

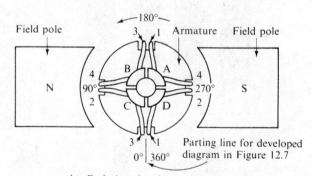

b) End view (brushes not shown)

FIGURE 10.6

*THE SCHEMATIC DIAGRAM OF A TWO-POLE,
FOUR-COIL DC MOTOR*

The construction illustrated in Figure 10.6 is typical of commercial dc motors. However, commercial motors have many more coils on the armature—usually 11 or more. Additional slots are provided in the armature, one for each additional coil. One segment is provided on the commutator for each coil. Thus an 11-coil dc motor has 11 slots in the armature and 11 segments on the commutator. Commercial motors are either two-pole, as shown in Figure 10.6, or four-pole. In a four-pole motor, the opposite field poles are placed 90° apart. The coil sides are placed one pole-span apart, or slightly more or less than one pole-span apart. The pole span is 180° in a two-pole motor and 90° in a four-pole motor.

Figure 10.7 is a developed diagram of the armature in Figure 10.6, and is made by unrolling the surface of the armature into a plane. The parting line for the development is along the centerline of the bottom slot in the armature (see Figure 10.6). The angles at the top of Figure 10.7 correspond to the angles around the armature in Figure 10.6. A graph of the flux density produced by the field poles is included at the bottom of Figure 10.7.

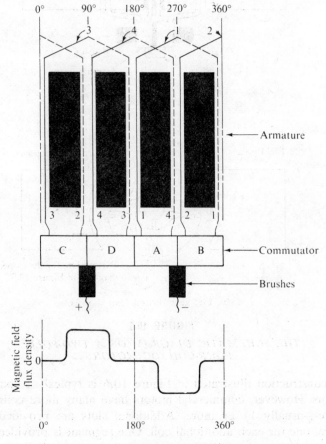

FIGURE 10.7

A DEVELOPED DIAGRAM OF THE ARMATURE WINDINGS OF THE TWO-POLE, FOUR-COIL DC MOTOR

Generator-Motor Action. The machine illustrated in Figures 10.6 and 10.7 is capable of operating as either a dc generator or a dc motor. Its operation is based on these two facts.

1. A voltage is induced in a conductor moving through a transverse magnetic field.
2. A force is exerted on a current-carrying conductor in a transverse magnetic field.

When the machine is used as a generator, the armature is rotated by an external prime mover. The motion of the coils through the magnetic field induces a voltage in the coil. The magnitude of the induced voltage is given by Equation (10.10). The polarity of the induced voltage reverses each time the coil passes from a north pole to a south pole, or vice versa. In a two-pole generator, the polarity reverses twice during each revolution, and in a four-pole generator it reverses four times during each revolution. The function of the commutator is to reverse the connection between the brushes and the coil at the same time that the polarity of the induced voltage reverses. In other words, the commutator is a rectifier that converts an ac voltage into a dc voltage.

The generator described above supplies current to a load; this current passes through the armature coils. As a result of this current, a force is exerted on the armature coils, according to Equation (10.9). This force produces a torque that opposes the motion of the armature. The prime mover must overcome this additional torque to maintain the armature speed.

When the machine is used as a motor, a current is produced in the armature coils by connecting an external voltage source to the brushes. The current produces a force, according to Equation (10.9) which tends to rotate the armature. The commutator reverses the current in each coil as it passes from a north pole to a south pole, or vice versa. This reversal in the direction of the current eliminates the reversal in the direction of the force which would otherwise occur. The rotation of the armature also results in an induced voltage, according to Equation (10.10). This voltage opposes the external voltage, and is called the *back emf* of the motor.

Commutation. A schematic diagram of the armature winding of the two-pole, four-coil motor is shown in Figure 10.8.

Notice in Figure 10.8-a that each coil carries one-half of the total armature current. As the armature rotates, both the polarity of the induced voltage and the direction of the current through the coil reverse

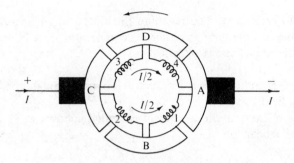

a) Before commutation

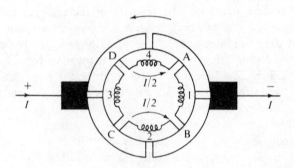

b) During commutation of coils 1 and 3

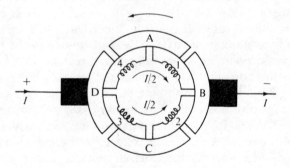

c) After commutation

FIGURE 10.8

*A SCHEMATIC DIAGRAM OF THE ARMATURE WINDING OF
A TWO-POLE, FOUR-COIL DC MOTOR*

each time a coil passes from one pole to the next. This reversal of current in the coil is called *commutation*. In Figure 10.8-b, coils 1 and 3 are undergoing commutation. Notice that both coils are shorted by the brushes in Figure 10.8-b. The reversal of the current in coils 1 and 3 can be observed by comparing the direction before commutation (Figure 10.8-a) with the direction after commutation (Figure 10.8-c).

Steady-State Operating Characteristics. The steady-state operating characteristic of a typical armature-controlled dc motor are illustrated in Figure 10.9. The first graph indicates a linear relationship between the armature current (i) and the armature torque (T). The slope of this line is called the *torque constant* (K_T). It indicates the change in torque ($\triangle T$) produced by a change in current ($\triangle i$).

$$\triangle T = K_T \triangle i \qquad (10.11)$$

The intercept on the i-axis is the value of the current required to overcome the friction torque in the motor.

The torque constant can be derived from Equation (10.9) $f = i_c BL \sin \alpha$), where f is the force on a conductor of length L. The torque on the armature is equal to the sum of the forces on each conductor times the mean radius of the armature, R. If N is the total number of conductors, and M is the fraction of conductors which are effective at any time, then

$$T = MN(f)R$$

From Equation (10.9),

$$T = MN(i_c BL \sin \alpha)R \qquad (10.12)$$

From Figure 10.8,

$$i_c = i/2$$

In dc motors, the angle α is always $90°$, so $\sin \alpha = \sin 90° = 1$ (α is the angle between the direction of the current and the direction of the magnet field). Equation (10.13) is the result of substitutions of the above values of i_c and α in Equation (10.12).

$$T = \left[\frac{NMBLR}{2} \right] i \qquad (10.13)$$

The term in brackets in Equation (10.13) is the torque constant K_T.

$$K_T = \left[\frac{NMBLR}{2} \right] \qquad (10.14)$$

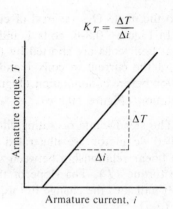

a) Graph of armature
current vs. torque

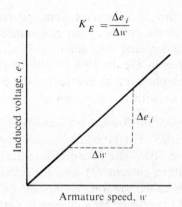

b) Graph of armature speed
vs. induced voltage

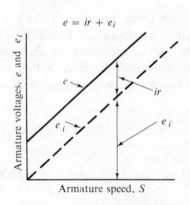

c) Graph of armature speed
vs. total and induced
voltages – no load

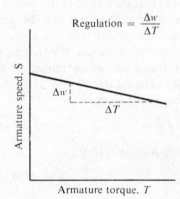

d) Graph of armature torque
vs. speed – constant
voltage (e)

FIGURE 10.9

*THE STEADY-STATE OPERATING CHARACTERISTICS OF
A TYPICAL DC MOTOR*

where:

T = the torque, newton meters

K_T = the torque constant, newton-meters/ampere

i = the armature current, amperes

N = the total number of conductors

M = the fraction of effective conductors

B = the magnetic flux density, webers/square meter

L = the length of each conductor, meters

R = the mean radius of the armature, meters

Figure 10.9-b shows a linear relationship between the armature speed (w) and the voltage induced in the armature coil (e_i). The slope of this line is called the *EMF constant* (K_E). It indicates the change in induced voltage ($\triangle e_i$) produced by a change in armature speed ($\triangle w$).

$$\triangle e_i = K_E \triangle w \qquad (10.15)$$

The EMF constant can be derived from Equation (10.10) ($e = LvB \sin \beta$). Notice that, in Figure 10.8, the coils are divided into two equal circuit paths. The total induced voltage is the sum of the voltages induced in each conductor in either path.

$$e_i = \frac{NM}{2}(e)$$

From Equation (10.10),

$$e_i = \frac{NM}{2}(LvB \sin \beta)$$

In dc motors, the angle β is 90°, so $\sin \beta = 1$.

$$e_i = NML(v)B/2$$

If w is the speed of the armature in radians per second, then each conductor travels Rw meters each second.

Thus,

$$v = Rw \text{ meters/second}$$

$$e_i = NML(Rw)B/2$$

$$e_i = \left[\frac{NMBLR}{2}\right]w \qquad (10.16)$$

The term in brackets in Equation (10.16) is the EMF constant, K_E.

$$K_E = \frac{NMBLR}{2} \qquad (10.17)$$

where:

> e_i = the induced voltage, volts
>
> w = the armature speed, radians per second
>
> K_E = the EMF constant, volts/radians per second
>
> $N, M, L, R,$ and B are defined for Equation (10.14).

It is interesting to note that $K_E = K_T$ [see Equations (10.14) and (10.17)].

Figure 10.9-c is a graph of the armature speed-vs.-armature voltage with no load on the motor. The total armature voltage (e) is made up of two components: the induced voltage (e_i), and the ir-drop across the armature resistance (r)—i.e., the voltage required to maintain the current i through resistance r. (Current i is required to produce a torque which will match the torque produced by the frictional forces in the motor.) The total armature voltage is the sum of the two voltage components.

$$e = ir + e_i \qquad \textbf{(10.18)}$$

Figure 10.9-d is a graph of the armature torque-vs.-armature speed with a constant armature voltage. This graph is a result of the combined effects of Equations (10.11), (10.15), and (10.18).

$$\triangle T = K_T \triangle i \qquad \textbf{(10.11)}$$

$$\triangle e_i = K_E \triangle w \qquad \textbf{(10.15)}$$

$$e = ir + e_i \qquad \textbf{(10.18)}$$

If the armature voltage (e) is constant, then according to Equation (10.18), any change in ir must be accompanied by an equal and opposite change in e_i.

$$\triangle e_i = -\triangle ir$$

From Equation (10.11),

$$\triangle i = \triangle T / K_T$$

From Equation (10.15),

$$\triangle e_i = K_E \triangle w$$

Thus,

$$\triangle e_i = K_E \triangle w = -\triangle ir = -\triangle Tr / K_T$$

or

$$K_E \triangle w = -\triangle Tr / K_T$$

$$\text{Regulation} = \frac{\triangle w}{\triangle T} = -\frac{r}{K_E K_T}$$

DC MOTOR EQUATIONS

$$e = ir + e_i \qquad (10.18)$$

$$e_i = K_E w \qquad (10.19)$$

$$T = K_T i - T_f \qquad (10.20)$$

$$REG = \triangle w / \triangle T = -\frac{r}{K_E K_T} \qquad (10.21)$$

$$P = wT \qquad (10.22)$$

where:

e = the armature voltage, volts

e_i = the induced voltage, volts

i = the armature current, amperes

r = the armature resistance, ohms

K_E = the EMF constant, volts/radian per second

w = the armature speed, radians per second

T = the output torque, newton meters

T_f = the friction torque, newton meters

K_T = the torque constant, newton meters/ampere

$\triangle w$ = a change in armature speed, radians per second

$\triangle T$ = a change in torque, newton meters

REG = the regulation, radians per second/newton meter

P = power, watts

DC Motor Transfer Functions. The schematic diagram and the block diagram of an armature-controlled dc motor are illustrated in Figure 10.10. A dc voltage is applied to the field winding (or the field is provided by permanent-magnet field poles). A variable voltage (e) is applied to the armature windings. The armature is represented by a resistor, an inductor, and an induced voltage source connected in series. The armature current (i) is defined by the following time domain equation.

$$e = L\frac{di}{dt} + ir + e_i$$

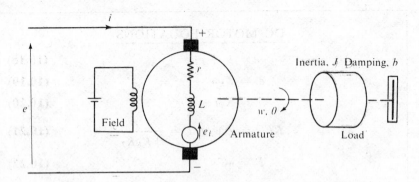

a) Schematic diagram

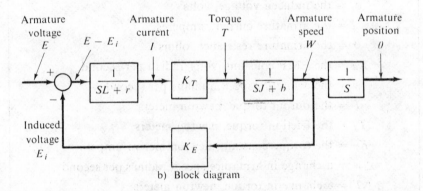

b) Block diagram

FIGURE 10.10

AN ARMATURE-CONTROLLED DC MOTOR

The corresponding frequency-domain equation is

$$E = LSI + Ir + E_i$$

or

$$I = \left[\frac{1}{SL + r}\right](E - E_i) \tag{10.23}$$

Equation (10.23) is represented on the block diagram by the summing junction, and the block between the summing junction and the current (I). The remaining system equations are each represented on the block diagram.

$$T = K_T I$$

$$W = \left[\frac{1}{JS + b}\right]T$$

$$E_i = K_E W$$

$$\theta = \left[\frac{1}{S}\right]W$$

The transfer functions are obtained by solving the system equations algebraically for the desired output-over-input ratio.

DC MOTOR TRANSFER FUNCTIONS— ARMATURE CONTROL

Velocity Transfer Function

$$\frac{W}{E} = \frac{\dfrac{K_T}{rb}}{\tau_m \tau_a S^2 + (\tau_m + \tau_a)S + \left(1 + \dfrac{K_E K_T}{rb}\right)} \qquad (10.24)$$

Position Transfer Function

$$\frac{\theta}{E} = \frac{1}{S}\left[\frac{W}{E}\right] \qquad (10.25)$$

where:

E = the armature voltage, volts

W = the armature speed, radians/second

θ = the armature position, radians

$\tau_m = J/b$, sec

$\tau_a = L/r$, sec

J = the moment of inertia of the load, kilograms-meter^2

b = the damping resistance, newton-meters/radian per second

L = the armature inductance, henrys

r = the armature resistance, ohms

K_T = the torque constant, newton meters/ampere

K_E = the EMF constant, volts/radian per second

S = the frequency parameter, 1/second

Example 10.4   An armature-controlled dc motor has the following ratings.

$$T_f = 1.2 \times 10^{-2} \text{ newton-meter}$$
$$K_T = 0.06 \text{ newton-meter/ampere}$$
$$I_{MAX} = 2 \text{ amperes}$$
$$K_E = 0.06 \text{ volt/radian per second}$$
$$w_{MAX} = 500 \text{ radians/second}$$
$$r = 1.2 \text{ ohms}$$

Determine the following.
1. The maximum output torque.
2. The maximum power output.
3. The regulation.
4. The maximum armature voltage.

Solution:

1.   The maximum output torque is obtained from Equation (10.20), when $i = I_{MAX}$.

$$T_{MAX} = K_T I_{MAX} - T_f = (0.06)(2) - 0.012$$
$$T_{MAX} = 0.108 \text{ newton-meter}$$

2.   The maximum power output is obtained from Equation (10.22) when w and T are both a maximum.

$$P_{MAX} = w_{MAX} T_{MAX} = (500)(0.108)$$
$$P_{MAX} = 54 \text{ watts}$$

3.   The regulation is obtained from Equation (10.21).

$$\text{Regulation} = -\frac{r}{K_E K_T} = -\frac{1.2}{(0.06)^2}$$
$$\text{Regulation} = -333 \text{ radians per second/newton-meter}$$

4.   The maximum armature voltage is obtained from Equations (10.18) and (10.19), when i and w are both a maximum.

$$e_i = K_E w$$
$$e = ir + e_i = ir + K_E w$$
$$E_{MAX} = I_{MAX} r + K_E w_{MAX}$$
$$E_{MAX} = (2)(1.2) + (0.06)(500)$$
$$E_{MAX} = 32.4 \text{ volts}$$

Example 10.5   The motor in Example 10.4 is operated at 300 radians per second with a load torque of 0.05 newton-meter. Determine the following.

1.   The armature voltage.
2.   The armature speed if the torque increases to 0.075 newton-meter and the armature voltage is not changed.

Solution:

1.   From Equation (10.20),

$$T = K_T i - T_f$$

$$i = \frac{T + T_f}{K_T} = \frac{0.05 + 0.012}{0.06}$$

$$i = 1.03 \text{ amperes}$$

From Equation (10.19),

$$e_i = K_E w = (0.06)(300) = 18 \text{ volts}$$

From Equation (10.18),

$$e = ir + e_i = (1.03)(1.2) + 18$$

$$e = 19.24 \text{ volts}$$

2.   The regulation was determined in Example 10.4.

$$\text{Regulation} = \frac{\triangle w}{\triangle T} = -333$$

$$\triangle T = 0.075 - 0.05 = 0.025 \text{ newton-meter}$$

$$\triangle w = (-333)(\triangle T) = (-333)(0.025)$$

$$\triangle w = -8.33 \text{ radians/second}$$

$$\text{Armature speed} = 300 - 8.33$$
$$= 291.67 \text{ radians/second}$$

Example 10.6   Determine the velocity and position transfer functions of the motor in Example 10.4. The following additional values are required.

$$J = 6.2 \times 10^{-4} \text{ kilogram-meter}^2$$

$$b = 1 \times 10^{-4} \text{ newton-meter/radian per second}$$

$$L = 0.020 \text{ henry}$$

Solution:

Equations (10.24) and (10.25) may be used.

$$\tau_m = J/b = 6.2 \times 10^{-4}/10^{-4} = 6.2 \text{ seconds}$$
$$\tau_a = L/r = 0.02/1.2 = 0.0167 \text{ seconds}$$
$$\tau_m\tau_a = 0.103$$
$$\tau_m + \tau_a = 6.22$$
$$1 + K_E K_T/(rb) = 1 + 30 = 31$$
$$K_T/(rb) = 0.06/(1.2 \times 10^{-4}) = 500$$

For velocity control,

$$\frac{W}{E} = \frac{500}{0.103S^2 + 6.22S + 31}$$

or, in the preferred form,

$$\frac{W}{E} = \frac{16.1}{3.32 \times 10^{-3}S^2 + 0.201S + 1}$$

For position control,

$$\frac{\theta}{E} = \frac{16.1}{S(3.32 \times 10^{-3}S^2 + 0.201S + 1)}$$

10.4 Two-Phase AC Motors

Two-phase ac motors are often used in control systems which require a low-power, variable-speed drive. The primary advantage of the ac motor over the dc motor is its ability to use the ac output of synchros, LVDTs, and other ac measuring means without demodulation of the error signal. An ac amplifier provides the gain for a proportional control mode. However, more elaborate control modes are difficult to implement with an ac signal. When additional control actions are required, the ac signal is usually demodulated, and the control action is inserted in the dc signal. The modified dc signal is then reconverted to an ac signal.

The schematic diagram of a two-phase ac motor is shown in Figure 10.11. The motor consists of an induction rotor, and two field coils located 90° apart. One field coil serves as a fixed reference field, the other as the control field. The amplified ac error signal is applied to the control field. This signal has a variable magnitude with a phase

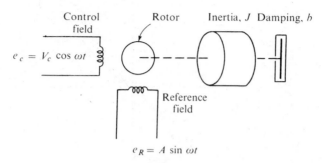

Load

FIGURE 10.11

A TWO-PHASE AC MOTOR

angle of either 0° or 180°. A constant ac voltage is applied to the reference field through a 90° phase-shift network. This signal has a constant magnitude and a phase angle of −90°. The two voltages are given below.

$$e_c = V_c \cos \omega t$$

$$e_R = A \cos (\omega t - 90°) = A \sin \omega t$$

where:

e_c = the control field voltage

e_R = the reference field voltage

ω = the operating frequency

V_c = the variable amplitude of the control voltage

A = the constant amplitude of the reference voltage

The sign and magnitude of V_c is determined by the sign and magnitude of the error signal. A negative error signal results in a negative value of V_c. This is usually interpreted as a 180° phase shift in e_c.

The linearized operating characteristics of a two-phase ac motor are shown in Figure 10.12. The actual operating line will depend on the speed-torque characteristics of the process. Two typical process load lines are indicated by the dotted lines. A change from one load line to another is an example of a process load change. The negative values of V_c in the third quadrant simply indicate that the motor reverses direction when V_c is negative.

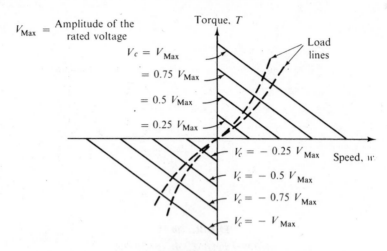

V_{Max} = Amplitude of the rated voltage

Torque, T

$V_c = V_{\text{Max}}$

$= 0.75\ V_{\text{Max}}$

$= 0.5\ V_{\text{Max}}$

$= 0.25\ V_{\text{Max}}$

Load lines

$V_c = -\ 0.25\ V_{\text{Max}}$ Speed, w

$V_c = -\ 0.5\ V_{\text{Max}}$

$V_c = -\ 0.75\ V_{\text{Max}}$

$V_c = -\ V_{\text{Max}}$

FIGURE 10.12

THE LINEARIZED OPERATING CHARACTERISTICS OF
A TWO-PHASE AC MOTOR

The torque equation of the two-phase motor may be obtained from Figure 10.12.

$$T = K_1 V_c - K_2 w \qquad (10.26)$$

In Equation (10.26), K_1 is a constant with units of newton-meters/volt, and K_2 is a constant with units of newton-meters/radians per second. The torque produced by the motor is applied to the inertia and friction of the load. The following frequency-domain equation defines the relationship between the motor torque and the motor speed.

$$T = (JS + b)w \qquad (10.27)$$

The velocity transfer function is obtained by equating Equations (10.26) and (10.27).

TWO-PHASE MOTOR TRANSFER FUNCTIONS

Velocity Transfer Function

$$\frac{W}{V_c} = \frac{K}{\tau S + 1} \qquad (10.28)$$

Position Transfer Function

$$\frac{\theta}{V_c} = \frac{K}{S(\tau S + 1)} \qquad (10.29)$$

where:

V_c = the control voltage amplitude, volts

W = the motor speed, radians/second

θ = the motor position, radians

$K = K_1/(b + K_2)$

$\tau = J/(b + K_2)$, sec.

K_1 = stall torque/rated voltage, newton-meters/volt

K_2 = stall torque/no load speed at the rated voltage, newton-meters/radian per second

J = the moment of inertia of the load, kilogram-meters^2

b = the damping resistance of the load, newton-meters/radian per second

Example 10.7   Determine the velocity and position transfer functions of a two-phase motor with the following data.

Rated voltage: 120 volts
Load inertia: 6×10^{-6} kilogram-meter^2
Load damping: 2×10^{-5} newton-meter/radian per second
Stall torque: 0.04 newton-meters at rated voltage (120 volts)
No load speed: 4000 rpm at rated voltage (120 volts)

Solution:

The transfer functions are given by Equations (10.28) and (10.29).

$$K_1 = 0.04/120 = 3.33 \times 10^{-4}$$

$$\text{no load speed} = 4000 \times 2\pi/60 = 419 \text{ radians/second}$$

$$K_2 = 0.04/419 = 9.56 \times 10^{-5}$$

$$(b + K_2) = 2 \times 10^{-5} + 9.56 \times 10^{-5} = 1.156 \times 10^{-4}$$

$$K_1/(b + K_2) = 3.33 \times 10^{-4}/1.156 \times 10^{-4} = 2.88$$

$$J/(b + K_2) = 6 \times 10^{-7}/1.156 \times 10^{-4} = 0.052$$

Velocity Transfer Function

$$\frac{W}{V_c} = \frac{2.88}{0.052S + 1}$$

Position Transfer Function

$$\frac{\theta}{V_c} = \frac{2.88}{S(0.052S + 1)}$$

10.5 Amplifiers

General Considerations. An amplifier is any device that receives an input signal and produces a proportional output signal at a higher power level. The term "amplifier" is usually associated with electronic components. However, pneumatic, hydraulic, and rotating amplifiers are also used extensively in control systems. Almost every position or velocity control system uses some type of amplifier. The power output of these amplifiers ranges from a few watts to megawatts. Selected examples of control system amplifiers are presented in this section. The emphasis is on the application of the amplifier, rather than on a detailed examination of the design of the amplifier.

AC Amplifiers. An application of an ac amplifier in a position control system is illustrated in Figure 10.13. A synchro transmitter and transformer are used as a combination measuring means and error detector. The synchro transmitter produces an ac output voltage with an amplitude proportional to the magnitude of the error between the two synchro rotors. The phase of the transmitter output is either 0° or 180°, depending on the sense (or sign) of the error.

   The ac amplifier magnifies the error signal, which is then applied to the control field of a two-phase ac motor. The magnitude and phase of this signal determine the speed and direction of the ac motor. The motor connections are such that the load is driven in a direction which tends to reduce the error signal. The gain of the amplifier provides a proportional control mode for the system.

DC Amplifiers. The application of a dc amplifier in a velocity control system is shown in Figure 10.14. A dc tachometer provides a negative voltage proportional to the speed of the load. A positive setpoint voltage is provided by the reference potentiometer. The two signals are applied to the inputs of an operational amplifier with a gain equal to R_2/R_1. The output of the operational amplifier (E_1) is given by the following equation.

$$E_1 = (E_{SP} - E_c)R_2/R_1$$

The power amplifier provides the additional voltage and power gain to drive the dc motor. The system as illustrated has a proportional control mode. Derivative and integral control modes can be added by the methods developed in Chapter Nine.

SCR Amplifiers. A silicone-controlled rectifier (SCR) amplifier is illustrated in Figure 10.15. The SCR is a rectifier which can be turned on by a control signal, but can be turned off only when the ac signal

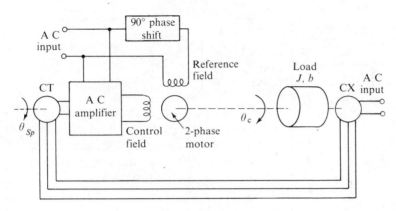

a) Schematic diagram

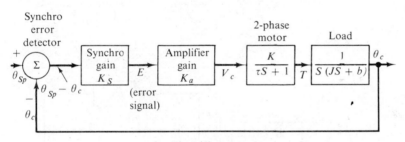

b) Block diagram

FIGURE 10.13

AN AC AMPLIFIER IN A POSITION CONTROL SYSTEM

reverses polarity. The two SCRs and two rectifiers are connected in a bridge circuit. When the SCR's are turned off, the bridge acts as an open switch. When the SCRs are turned on, the bridge acts as a full-wave rectifier. The output of the bridge is a pulsating voltage which is applied to the armature of a dc motor.

The SCR is turned on by a voltage pulse produced by the magnetic amplifier trigger circuit. The SCR is said to be 100 per cent on if the trigger occurs at the start of each ac half cycle. With the SCR 100 per cent on, the bridge acts as a full-wave rectifier (see Figure 10.15). The SCR is 75 per cent on if the trigger occurs 45° after the start of each ac half-cycle. With the SCR 75 per cent on, the first 25 per cent of each half-cycle is removed from the output waveform, as shown in Figure

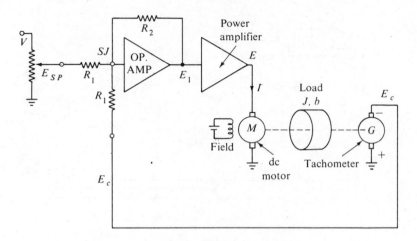

a) Schematic diagram

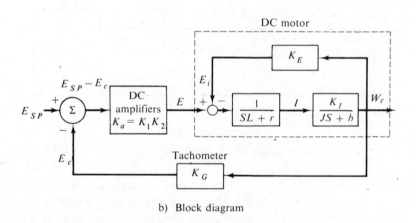

b) Block diagram

FIGURE 10.14

A DC AMPLIFIER IN A VELOCITY CONTROL SYSTEM

10.15. The SCR is 25 per cent on if the trigger occurs 135° after the start of each half-cycle. With the SCR 25 per cent on, the first 75 per cent of each half-cycle is removed from the output. The inertia of the motor and load effectively smooths out the pulsations in the armature voltage.

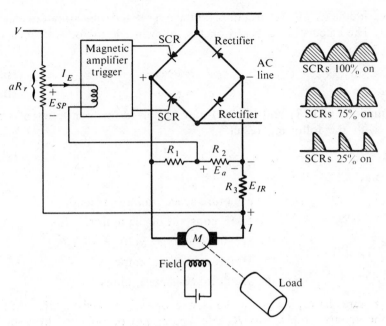

a) Schematic diagram

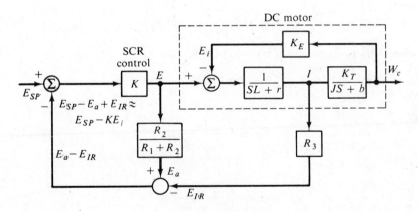

b) Block diagram

FIGURE 10.15

*AN SCR VELOCITY CONTROL SYSTEM WITH ARMATURE
VOLTAGE AND CURRENT FEEDBACK*

Another interesting feature of the system illustrated in Figure 10.15 is the nature of the measuring means. A tachometer is not used to measure the speed of the load. Instead, the speed is calculated from measurements of the armature voltage (e) and the armature current (i). The calculation is based on the following dc motor equations (see 10.3).

$$e = ir + e_i \qquad \textbf{(10.18)}$$

$$e_i = K_E w \qquad \textbf{(10.19)}$$

Equations (10.18) and (10.19) may be solved for w in terms of e and ir, with the following result.

$$e - ir = K_E w \qquad \textbf{(10.30)}$$

where:

w = the armature speed, radians/second

K_E = the EMF constant of the motor

e = the armature voltage, volts

i = the armature current, amperes

r = the armature resistance, ohms

In Figure 10.15, e_a is the voltage drop across resistor R_2 and e_{IR} is the voltage drop across R_3. Voltage e_a can be obtained by applying the voltage-divider rule to resistors R_1 and R_2.

$$e_a = \left[\frac{R_2}{R_1 + R_2} \right] e$$

Voltage e_{IR} is obtained by applying Ohm's law to R_3.

$$e_{IR} = iR_3$$

If

$$R_3 = \left(\frac{R_2}{R_1 + R_2} \right) r \qquad \textbf{(10.31)}$$

then

$$e_{IR} = \left(\frac{R_2}{R_1 + R_2} \right) ir$$

and

$$e_a - e_{IR} = \left(\frac{R_2}{R_1 + R_2} \right) (e - ir) \qquad \textbf{(10.32)}$$

A combination of Equations (10.30) and (10.32) produces the following result.

$$e_a - e_{IR} = K_m w \qquad \textbf{(10.33)}$$

where:

$$K_m = \frac{K_E R_2}{R_1 + R_2}, \text{ volts/radian per second}$$

w = the armature speed, rpm

e_a = the voltage across resistor R_2, volts

e_{IR} = the voltage across resistor R_3, volts

R_1, R_2, and R_3 are scaling resistors [R_3 is defined by Equation (10.31), R_1 and R_2 can be arbitrarily selected]

Equation (10.33) states that the voltage difference ($e_a - e_{IR}$) *is proportional to the armature speed w and can be used as a measurement of w.*

In Figure 10.15, Resistors R_2, R_3, and aR_r form an error-detector circuit in which the current i_E is proportional to the error signal.

$$i_E = \frac{e_{SP} - (e_a - e_{IR})}{R_2 + R_3 + aR_r} \tag{10.34}$$

Current i_E determines the timing of the magnetic amplifier trigger, which in turn determines the voltage applied to the dc motor.

Rotating Amplifiers. A dc generator is a rotating power amplifier which may be used to control the speed of a dc motor. An example of a single-stage rotating amplifier is shown in Figure 10.16. This method of control is also known as the *Ward Leonard method.*

The dc generator is driven by another prime mover, such as an induction motor. A low-power current in the field winding (I_f) controls the output voltage of the generator (E), according to the graph at the top of Figure 10.16. The remainder of the control system consists of a tachometer, a reference potentiometer, and a dc amplifier.

The transfer function of the amplifier and generator is a first-order lag, with a time constant determined by the field winding.

$$\frac{E}{I} = \frac{K_1}{\tau_1 S + 1} \tag{10.35}$$

where:

E = output voltage, volts

I = control field current, ampere

K_1 = the gain, volts/ampere

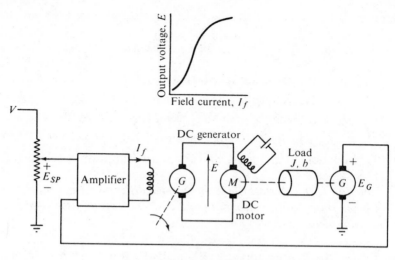

a) Schematic diagram

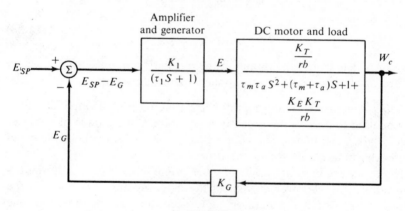

b) Block diagram

FIGURE 10.16

*A SINGLE-STAGE ROTATING AMPLIFIER IN A
VELOCITY CONTROL SYSTEM*

$\tau_1 = L_f/r_f$ = time constant, seconds

L_f = the inductance of the generator field, henrys

r_f = the resistance of the generator field, ohms

The single-stage rotating amplifier has several disadvantages. The time constant is too long for many applications. The power input is relatively large, and the power amplification is limited (a gain of 100 is typical). These disadvantages are greatly improved in several types of two-stage rotating amplifiers—such as Rototrol, Regulex, Metadyne, and Amplidyne—which have wide applications in control systems. These improved rotating amplifiers all produce a large amplification of a low-power signal with a short time constant. The Amplidyne will be used as an example of this type of amplifier.

The Amplidyne is a two-stage dc generator which has two sets of brushes and a center-tapped control field. The two brush sets are mutually perpendicular. One set is called the *direct brushes;* the other is called the *quadrature brushes.* The control field is located on the stator direct axis. The center tap is provided so that the field can be driven by a push-pull amplifier, as shown in Figure 10.17.

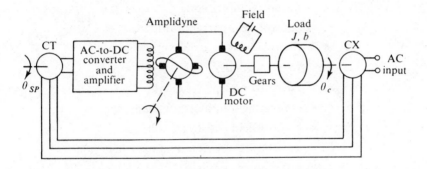

a) Schematic diagram

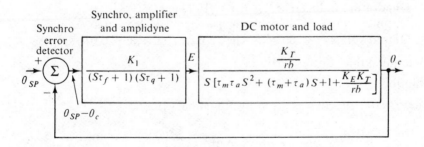

b) Block diagram

FIGURE 10.17

AN AMPLIDYNE IN A POSITION CONTROL SYSTEM

The quadrature brushes are short-circuited, and the control current in the field winding generates a very large quadrature current (i_q). The quadrature current, in turn, generates a large output voltage at the direct brushes. This output voltage is applied to the armature of a dc motor. The Amplidyne transfer function consists of a first-order lag for each stage. The first stage is the control field winding. The second stage is the quadrature winding. The transfer function is given by Equation (10.36), below.

where:

$$\frac{E}{I} = \frac{K}{(\tau_f S + 1)(\tau_q S + 1)} \qquad (10.36)$$

E = output voltage, volts

I = control field current, amperes

K = the steady-state gain, volts/ampere

$\tau_f = L_f/r_f$, seconds

$\tau_q = L_q/r_q$, seconds

L_f = control field inductance, henrys

r_f = control field resistance, ohms

L_q = quadrature circuit inductance, henrys

r_q = quadrature circuit resistance, ohms

Example 10.8 An Amplidyne has the following data.

Control field resistance, r_f = 1000 ohms
Control field inductance, L_f = 20 henrys
Quadrature resistance, r_q = 0.1 ohm
Quadrature inductance, L_q = 0.0015 henry
Output voltage = 145 volts when control current = 0.020 ampere
Output voltage = 60 volts when control current = 0.010 ampere

Determine the transfer function.

Solution:

The transfer function is given by Equation (10.36)

$$K = \frac{145 - 60}{0.020 - 0.010} = \frac{85}{0.010} = 8500 \text{ volts/ampere}$$

$$\tau_f = L_f/r_f = 20/1000 = 0.020 \text{ second}$$

$$\tau_q = L_q/r_q = 0.0005/0.1 = 0.005 \text{ second}$$

$$\text{Transfer Function} = \frac{8500}{(0.020S + 1)(0.005S + 1)}$$

EXERCISES

10–1. An equal-percentage valve has a rangeability of 20. Use Equation (10.4) to calculate the value of Q/Q_{MAX} for values of P/P_{MAX} from 0 to 1 in increments of 0.1.

10–2. Determine the size of the control valve required to control the flow rate of water. The maximum flow rate is 850 gallons/minute, and the available pressure drop across the valve is 50 psi. Allow a safety factor of at least 30 per cent.

10–3. Determine the size of the control valve required to control the flow rate of a gas with a specific gravity of 0.85. A maximum flow rate of 2500 standard cubic feet per hour is required. The gas temperature is 420° F, the inlet pressure is 40 psia, and the outlet pressure is 20 psia. Allow a minimum safety factor of 25 per cent.

10–4. Determine the size of a control valve required to control the flow rate of steam to a process. The maximum flow rate is 320 pounds/hour. The inlet and outlet steam conditions are given below. Use a safety factor of 30 per cent.

$$P_1 = 60 \text{ psia}$$
$$V_1 = 7.175 \text{ cubic feet/pound}$$
$$P_2 = 40 \text{ psia}$$
$$V_2 = 10.72 \text{ cubic feet/pound}$$

10–5. An armature-controlled dc motor has the following ratings.

$$T_f = 3.0 \times 10^{-2} \text{ newton-meter}$$
$$K_T = 0.15 \text{ newton-meter/ampere}$$
$$I_{MAX} = 2.4 \text{ amperes}$$
$$K_E = 0.15 \text{ volt/radian per second}$$
$$w_{MAX} = 5000 \text{ revolutions per minute}$$
$$r = 0.8 \text{ ohm}$$

Determine the following.
1. The maximum output torque.
2. The maximum power output.
3. The regulation.
4. The maximum armature voltage.

10–6. The motor in Exercise 10–5 is operated at 3000 revolutions per minute with a load torque of 0.15 newton-meter. Determine the following.
1. The armature voltage.
2. The armature speed if the torque increases to 0.30 newton-meter and the armature voltage is not changed.

10–7. Determine the velocity and position transfer functions of the motor in Example 10–5. The following additional values are required.

$$J = 10.2 \times 10^{-4} \text{ kilogram-meter}^2$$
$$b = 1.2 \times 10^{-4} \text{ newton-meter/radian per second}$$
$$L = 0.045 \text{ henry}$$

10–8. Determine the velocity and position transfer functions of a two-phase motor with the following data.

Rated voltage: 120 volts
Load inertia: 15×10^{-6} kilogram-meter^2
Load damping: 6×10^{-6} newton-meter/radian per second
Stall torque: 0.12 newton-meter
No load speed: 5000 rpm

10–9. The SCR control in Figure 10.15 is to be used to control a dc motor with an armature resistance (r) of 0.4 ohm. The feedback voltage e_a must be 20 per cent of the armature voltage e for proper operation of the control circuit. A total value of 10K ohms for R_1 and R_2 is required. Determine the resistance values required for R_1, R_2, and R_3.

10–10. An Amplidyne has the following data.

control field resistance, $r_f = 10,000$ ohms
control field inductance, $L_f = 25$ henrys
quadrature resistance, $r_q = 0.08$ ohm
quadrature inductance, $L_q = 0.0008$ henry
output voltage = 110 volts when control current = 0.020
output voltage = 55 volts when control current = 0.010

Determine the transfer function.

Part Three

ANALYSIS AND DESIGN OF CONTROL SYSTEMS

11

Basic Elements
of a System

11.1 Introduction

The analysis and design of closed-loop control systems involve the determination of how each component in the system responds to sinusoidal input signals of different frequencies (i.e., its frequency response). A linear component responds to a sinusoidal input signal by changing only the amplitude and the phase of the signal, but not the frequency. The transfer function of a component gives us the amount of these changes in terms of the ratio of the output amplitude over the input amplitude and the phase difference between the input and the output signals. Figure 11.1 illustrates these two changes.

Four basic elements of a component determine its transfer function and, hence, its frequency response. The four elements are:

1. Resistance
2. Capacitance
3. Inertia or inductance
4. Dead time

The four basic elements are common to many different types of systems, such as electrical, mechanical, thermal, liquid flow, and gas flow.

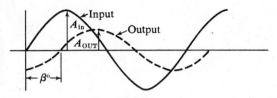

β = the phase difference
A_{OUT}/A_{in} = the amplitude ratio

FIGURE 11.1

THE TWO WAYS A SIGNAL IS CHANGED BY A COMPONENT

For each type of system, the basic elements are defined in terms of a quantity of material, energy, or distance and a driving force or potential. For example, in a liquid system, the liquid pressure is the potential parameter and the volume of liquid is the quantity parameter. The potential and quantity parameters are summarized in Table 11.1 for the five most common types of systems.

The determination of the basic elements of the components in a control system is one of the first steps in the design of a closed-loop control system. The objective of this chapter is to provide the information necessary to calculate the basic elements of the five most common types of systems using the text as a guide. The equations for each element are included in a box to make them easy to locate. All terms are defined within the box, and the units are included for each term. The method of explanation by example is used to describe how to complete the calculations. An example of each calculation is provided for *use as a guide* in calculating the elements. Section 11.2 includes a general discussion of the four basic elements. The remaining five sections cover electrical systems (11.3), liquid flow systems (11.4), gas flow systems (11.5), thermal systems (11.6), and mechanical systems (11.7).

TABLE 11.1

THE THREE PARAMETERS OF VARIOUS SYSTEM TYPES

Type of system	Parameter		
	Quantity	Potential	Time
Electrical	Charge	EMF or Voltage	Second
Liquid	Volume	Pressure	Second
Gas	Mass	Pressure	Second
Thermal	Heat	Temperature	Second
Mechanical	Distance	Force	Second

11.2 The Four Basic Elements

Resistance. Resistance is an opposition to motion or the flow of material or energy. It occurs in many forms: for instance, electrical resistance, thermal resistance, flow resistance, and mechanical resistance. Resistance is measured in terms of the change in potential required to produce a unit change in the quantity transferred each second. The exact nature of the change in the quantity transferred each second is given in Table 11.2.

TABLE 11.2

*CHANGE IN QUANTITY PARAMETER USED
TO DEFINE RESISTANCE*

Type of system	Change in quantity parameter	
	Description	Name (SI units)
Electrical	Quantity of charge transferred each second	Current (amperes)
Liquid	Quantity of liquid transferred each second	Flow rate (cubic meters/second)
Gas	Quantity of gas transferred each second	Flow rate (kilograms/second)
Thermal	Quantity of heat transferred each second	Flow rate (joules/second)
Mechanical	Distance traveled each second	Velocity (meters/second)

Electrical resistance is the change in voltage required to produce a change in electric current of 1 ampere. Liquid flow resistance is the change in pressure required to produce a change in liquid flow rate of 1 cubic meter per second. Gas flow resistance is the change in pressure required to produce a change in gas flow rate of 1 kilogram per second. Thermal resistance is the change in temperature required to produce a change in heat flow rate of 1 joule per second. Mechanical resistance is the change in force required to produce a change in the velocity of 1 meter per second.

> *Resistance is measured in terms of the change in potential required to make a unit change in electric current, liquid flow rate, gas flow rate, heat flow rate, or velocity.*

Capacitance. Capacitance is the quantity of material, energy, or distance required to make a unit change in potential. It is a dynamic quantity that expresses the relationship between a change in quantity and the corresponding change in potential. Capacitance should not be confused with capacity, which is the total material- or energy-holding ability of a device. If we say a jug holds 1 liter, we are stating its capacity. If we say that 100 cubic meters of liquid must be added to a tank to increase the pressure at the bottom by 1 newton per square meter, we are stating the tank's capacitance.

> *Capacitance is measured in terms of the change in material, energy, or distance required to make a unit change in potential.*

Inertia, Inertance, or Inductance. This element is measured in terms of the potential required to produce a unit change in the flow rate each second. It is usually an important element in electrical and mechanical systems, sometimes important in liquid and gas systems, and usually ignored in thermal systems. The term *inductance* is used with electrical systems, the term *inertance* is used with fluid systems, and the term *inertia* is used with mechanical systems.

> *Inertia, inertance, or inductance is measured in terms of the change in potential required to produce a unit rate of change of electric current liquid flow rate, gas flow rate or velocity.*

Dead Time. Dead time is the time interval between the time a signal appears at the input of a component and the time a corresponding response appears at the output. A pure dead-time element does not change the magnitude of the signal; only the timing of the signal is changed. Dead time occurs whenever mass or energy is transported from one point to another. It is the time required for the mass or energy to travel from the input location to the output location. Dead time is also called *transport time, pure delay* or *distance-velocity lag*. Since control becomes increasingly more difficult as dead time increases, dead time is considered the most difficult characteristic of a process.

If V is the velocity of the mass or energy and D is the distance traveled, then the dead-time lag is equal to the distance divided by the velocity.

$$t_d = D/V$$

The effect of dead time is to delay the input signal by the dead-time lag (t_d). As an example, consider the system with a dead-time lag of 5 seconds. Let $f_i(t)$ represent the input signal and $f_o(t)$ represent the output signal. The (t) is a symbol that indicates that f_i and f_o have different values at different times. If $f_i(t)$ is 5 at time $t = 0$ seconds, then $f_o(t)$ will be 5 at $t = 5$ seconds. If $f_i(t)$ is 7 at $t = 5$ seconds, then $f_o(t)$ will be 7 at time $t = 10$ seconds. Several more time intervals are indicated below.

t	$f_i(t)$	$f_o(t)$	relationship between f_i and f_o
0	5		
5	7	5	$f_o(5) = f_i(0)$
10	8	7	$f_o(10) = f_i(5)$
15	6	8	$f_o(15) = f_i(10)$
20		6	$f_o(20) = f_i(15)$

It is interesting to note that in each case, $f_o(t) = f_i(t - 5)$ or, more generally,

$$f_o(t) = f_i(t - t_d)$$

This equation is often used to represent a dead-time process. In simple terms, it expresses the concept that the output at any time t is the same as the input was t_d seconds before time t; that is at time $t - t_d$.

11.3 Electrical Systems

Electrical Resistance. Electrical resistance is that property of a material which impedes the flow of electrons. Good conductors have a low resistance; insulators have a very high resistance. Electrical resistance is described by Ohm's law:

ELECTRICAL RESISTANCE EQUATION

$$R = \frac{E}{I} = \frac{\Delta E}{\Delta I} \tag{11.1}$$

where:

R = electrical resistance, ohms

E = applied voltage, volts

I = resulting current, amperes

ΔE = a change in applied voltage, volts

ΔI = a corresponding change in current, amperes

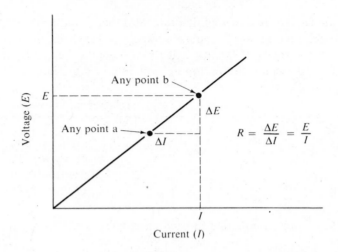

FIGURE 11.2
ELECTRICAL RESISTANCE FROM A VOLT-AMPERE GRAPH

Expressed as $E = IR + O$, Ohm's law is the equation of a straight line through the origin with a slope of R (see Figure 11.2). No matter how small the change in voltage (ΔE), there is a corresponding change in current (ΔI) such that the change in voltage divided by the change in current is equal to the resistance, as stated in Equation (11.1).

Example 11.1 The volt-ampere graph of Figure 11.2 applies to an electrical component. An applied voltage of 24 volts results in a current of 12 milliamperes. Determine the resistance of the component.

Solution:

The volt-ampere graph is a straight line, so Ohm's law applies.

$$R = E/I = 24/0.012 = 2000 \text{ ohms}$$

Example 11.2 A light bulb is an example of an electric component with a curved volt-ampere graph. The electrical resistance of these nonlinear components must be determined using the increment of current ($\triangle I$) produced by a small increment of voltage ($\triangle E$) at the operating point. Determine the resistance of a light bulb at 6 volts from the following information.

5.95 volts results in 0.500 amperes
6.05 volts results in 0.504 amperes

Solution:

$$\triangle E = 6.05 - 5.95 = 0.10 \text{ volts}$$
$$\triangle I = 0.504 - 0.500 = 0.004 \text{ amperes}$$
$$R = \triangle E / \triangle I = 0.10/0.004 = 25 \text{ ohms}$$

Electrical Capacitance. Electrical capacitance is the quantity of electric charge (in coulombs) required to make a unit change in the electric potential (in volts). The unit of electrical capacitance is the *farad*.

ELECTRICAL CAPACITANCE

$$\text{Capacitance} = C = \frac{\triangle q}{\triangle e}, \text{ farads} \qquad \textbf{(11.2)}$$

where:

$\triangle q$ = a change in charge, coulombs

$\triangle e$ = a change in potential, volts

A simple manipulation of Equation (11.2) results in the following relationship between the voltage and current in a capacitance.

$$\triangle q = C(\triangle e)$$
$$\frac{\triangle q}{\triangle t} = I = C\frac{\triangle e}{\triangle t} \qquad \textbf{(11.3)}$$

where:

$\triangle t$ = the time required to make the $\triangle q$-change in charge, seconds

I = the average current during the time interval $\triangle t$, amperes

If $\triangle t$ is reduced until it approaches zero, then $\triangle q / \triangle t$ becomes dq/dt, the instantaneous rate of change of charge. Also $\triangle e / \triangle t$ becomes de/dt, the instantaneous rate of change of potential.

$$\frac{dq}{dt} = i = C\frac{de}{dt} \qquad \textbf{(11.4)}$$

where:

i = the instantaneous current, amperes

Example 11.3 A current pulse with an amplitude of 0.1 milliampere and a duration of 0.1 second is applied to an electrical capacitor. The voltage across the capacitor is increased from 0 to +25 volts by the current pulse. Determine the capacitance (C) of the capacitor.

Solution:

$$I = C\frac{\Delta e}{\Delta t}$$

$$0.1 \times 10^{-3} = C\frac{25 - 0}{0.1}$$

$$C = (0.1 \times 10^{-3})(0.1)/25$$

$$C = 100 \ 10^{-7}/25$$

$$C = 0.4 \text{ microfarad}$$

Electrical Inductance. Electrical inductance is measured in terms of the voltage required to produce a unit change in electric current each second. It is usually defined by the following equation.

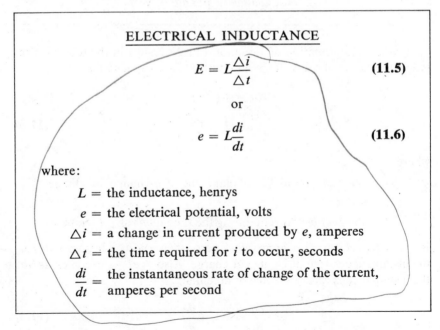

ELECTRICAL INDUCTANCE

$$E = L\frac{\Delta i}{\Delta t} \qquad \textbf{(11.5)}$$

or

$$e = L\frac{di}{dt} \qquad \textbf{(11.6)}$$

where:

L = the inductance, henrys

e = the electrical potential, volts

Δi = a change in current produced by e, amperes

Δt = the time required for i to occur, seconds

$\dfrac{di}{dt}$ = the instantaneous rate of change of the current, amperes per second

Example 11.4 A voltage pulse with an amplitude of 5 volts and a duration of 0.02 second is applied to an inductor. The current through the inductor is increased from 1 to 2.1 amperes by the voltage pulse. Assume that the resistance of the inductor is negligible, and determine the inductance (L).

Solution:

Equation (11.5) may be used to determine L.

$$e = L\frac{\Delta i}{\Delta t}$$

$$5 = \frac{L(2.1 - 1)}{0.02} = L(1.1/0.02)$$

$$L = (5)(0.02/1.1) = 0.0908$$

$$L = 0.0908 \text{ henrys}$$

Electrical Dead Time. Although electrical signals travel at tremendous speeds (2×10^8 to 3×10^8 meters/second), the transport time of an electrical signal constitutes a dead-time lag which has important consequences in some systems. In digital computer circuits, the delay on a transmission line is sometimes used to deliberately delay a signal in order to accomplish the desired logic function. In most control systems, however, the effect of electrical dead time is negligible because the delay is so small compared to the delay in other parts of the system.

The velocity of a signal on a transmission line is called the *velocity of propagation* (V_p). As mentioned before, V_p varies between 2×10^8 and 3×10^8 meters/second. The dead-time lag of the line is equal to the distance the signal travels (D) divided by the velocity of propagation (V_p)

DEAD-TIME LAG

$$t_d = D/V_p \hspace{3cm} \text{(11.7)}$$

where:

t_d = the dead-time lag, seconds

D = the distance between the input and the output, meters

V_p = the velocity of travel of the signal, meters/second

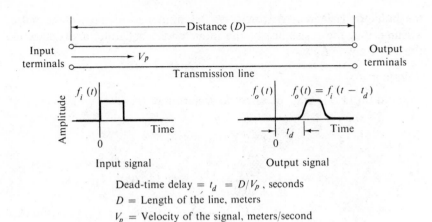

Dead-time delay $= t_d = D/V_p$, seconds
D = Length of the line, meters
V_p = Velocity of the signal, meters/second

a) An electrical dead-time element – a transmission line

FIGURE 11.3
AN ELECTRICAL DEAD-TIME ELEMENT—A TRANSMISSION LINE

Example 11.5   a.   Determine the dead-time lag of a 600-meter long transmission line if the velocity of propagation is 2.3×10^8 meters/second.

            b.   Determine the dead-time lag of a signal from a space vehicle which is located 2000 kilometers from the earth station receiving the signal. The signal travels at 3×10^8 meters/second.

Solution:

a.   Equation (11.7) applies.

$$t_d = 600/2.3 \times 10^8 = 2.61 \times 10^{-6} \text{ second}$$

b.   Equation (11.7) applies.

$$t_d = 2 \times 10^6/3 \times 10^8 = 0.67 \times 10^{-2} \text{ second}$$
$$t_d = 6.7 \text{ milliseconds}$$

11.4 Liquid Flow Systems

Liquid Flow Resistance. Liquid resistance is that property of pipes, valves, or restrictions which impedes the flow of a liquid. It is measured in terms of the change in pressure required to make a unit change in flow rate. Liquid resistance is determined by the relationship between the pressure drop and flow rate as expressed by a flow equation. There are two different types of flow: laminar and turbulent. Each has a different flow equation and, hence, a different liquid resistance. Laminar flow occurs when the fluid velocity is relatively low and the liquid flows in layers. A colored dye injected in the center of a laminar flow will move with the liquid and remain concentrated in the center. Turbulent flow occurs when the fluid velocity is relatively high and the liquid does not flow in layers. A colored dye injected in the center of turbulent flow is soon mixed throughout the flowing fluid.

The type of flow that occurs depends on four parameters: the density of the fluid (ρ), the inside diameter of the pipe (D), the dynamic viscosity of the fluid (μ), and the average velocity of the flowing fluid (V). These four parameters are arranged in a *dimensionless grouping* called the Reynolds number.

$$\text{Reynolds number} = \rho V D / \mu \tag{11.8}$$

Laminar flow occurs when the Reynolds number has a value less than 2000. Turbulent flow occurs when the Reynolds number has a value greater than 4000. A transition between laminar and turbulent flow occurs when the Reynolds number has a value between 2000 and 4000. Graphs of pressure vs flow rate for laminar and turbulent flow are shown in Figures 11.4 and 11.5.

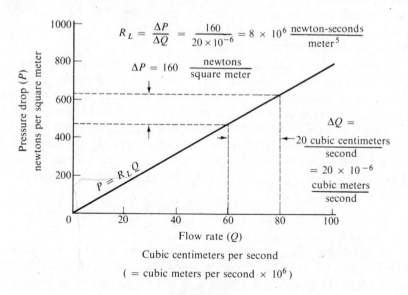

$$R_L = \frac{\Delta P}{\Delta Q} = \frac{160}{20 \times 10^{-6}} = 8 \times 10^6 \frac{\text{newton-seconds}}{\text{meter}^5}$$

$$\Delta P = 160 \frac{\text{newtons}}{\text{square meter}}$$

$$\Delta Q = 20 \frac{\text{cubic centimeters}}{\text{second}}$$

$$= 20 \times 10^{-6} \frac{\text{cubic meters}}{\text{second}}$$

$$P = R_L Q$$

Flow rate (Q)

Cubic centimeters per second

(= cubic meters per second × 10^6)

FIGURE 11.4

A PRESSURE VERSUS FLOW RATE GRAPH FOR LAMINAR FLOW

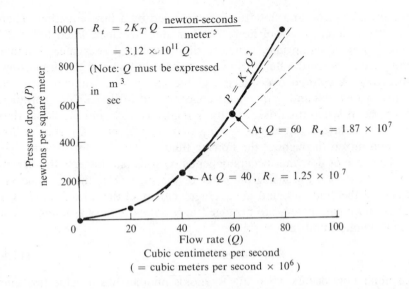

FIGURE 11.5

A PRESSURE VERSUS FLOW RATE GRAPH FOR TURBULENT FLOW

In both Figures 11.4 and 11.5, the slope of the graph is the liquid flow resistance. The laminar flow graph is a straight line and the resistance has a constant value for all flow rates. The turbulent flow graph is curved; thus the turbulent flow resistance is not constant and increases as the flow rate is increased.

The slope of the turbulent flow graph (Figure 4.4) is the turbulent flow resistance (R_t). However, since the graph is a curved line, the slope is not constant; thus the turbulent flow resistance is not a constant. The slope at any point on the curve is equal to the slope of a straight line that is tangent to the curve at that point. An approximate value for the slope can be obtained by determining the change in pressure ($\triangle P$) that corresponds to a small change in flow rate ($\triangle Q$). The approximate slope is equal to $\triangle P / \triangle Q$, and this is also an approximate value of R_t.

$$R_t \approx \frac{\triangle P}{\triangle Q}$$

As $\triangle Q$ becomes smaller, the approximate value of R_t becomes closer to the true value. The limit as $\triangle Q$ approaches zero is equal to the true value of R_t. In calculus, this limit is called *the derivative of P with respect to Q*. The derivative symbol is obtained by replacing the $\triangle$ by d, as indicated below.

$$\lim_{\triangle Q \to 0} \frac{\triangle P}{\triangle Q} = \frac{dP}{dQ} = R_t$$

The flow equation for laminar flow in a round pipe is given by Equation (11.9) (called the Hagen–Poiseuille law). The laminar flow resistance, R_L, is a constant that depends on three parameters, namely μ, l, and D. The SI units for the flow parameters and some English conversions are given in Table 11.4.

LIQUID LAMINAR FLOW EQUATIONS

$(\rho V D / \mu < 2000)$ [see Equation (B)]

$$P = R_L Q \qquad\qquad (11.9)$$

$$R_L = \left[\frac{128\mu l}{\pi D^4} \right] \qquad\qquad (11.10)$$

$$Q = \pi D^2 V / 4 \qquad\qquad (11.11)$$

where:

R_L = the laminar flow resistance, newton-seconds/meter^5

P = the difference between the inlet pressure and the outlet pressure, newtons/square meter

Q = the fluid flow rate, cubic meters/second

μ = the dynamic viscosity of the fluid, newton-seconds/square meter

l = the length of the pipe, meters

V = the average fluid velocity, meters/second

D = the inside diameter of the pipe, meters

The flow equation for turbulent flow in a round pipe is given by Equation (11.12) (called the Fanning equation). The turbulent flow resistance, R_t, can be determined by differentiating the turbulent flow equation as follows.

Turbulent Flow Equation:

$$P = K_t Q^2 \qquad\qquad (11.12)$$

Turbulent Flow Resistance:

$$R_t = \frac{dP}{dQ} = 2K_t Q$$

LIQUID TURBULENT FLOW EQUATIONS

$(\rho VD/\mu > 4000)$ [see Table 11.3 or Equation (11.8)]

$$P = K_t Q^2 \qquad \textbf{(11.12)}$$

$$R_t = 2K_t Q \qquad \textbf{(11.13)}$$

$$K_t = \left[\frac{8\rho fl}{\pi^2 D^5}\right] \qquad \textbf{(11.14)}$$

$$Q = \pi D^2 V/4 \qquad \textbf{(11.15)}$$

where:

R_t = the turbulent flow resistance, newton-seconds/meter^5

P = the difference between the inlet pressure and the outlet pressure, newtons/square meter

K_t = the turbulent flow constant

Q = the fluid flow rate, cubic meters/second

ρ = the fluid density, kilograms/cubic meter

l = the length of the pipe, meters

D = the inside diameter of the pipe, meters

f = the friction factor (a dimensionless number that depends on the Reynolds number and the internal surface of the pipe—see Table 11.3)

V = the average velocity, meters/second

TABLE 11.3

VALUES OF THE FRICTION FACTOR (f)

Type	Diameter centimeters	Reynolds number*					
		4×10^3	10^4	10^5	10^6	10^7	10^8
Smooth tubing	1–2	0.039	0.030	0.018	0.014	0.012	0.012
	2–4	0.039	0.030	0.018	0.013	0.011	0.010
	4–8	0.039	0.030	0.018	0.012	0.010	0.009
	8–16	0.039	0.030	0.018	0.012	0.009	0.008
Commercial steel pipe	1–2	0.041	0.035	0.028	0.026	0.026	0.026
	2–4	0.040	0.033	0.024	0.023	0.023	0.023
	4–8	0.039	0.030	0.022	0.020	0.019	0.019
	8–16	0.039	0.030	0.020	0.018	0.017	0.017

* Reynolds number $= \rho VD/\mu$

Reference: Moody, L. F. "Friction Factors for Pipe Flow," *A.S.M.E. Transactions,* 66, no. 8 (November 1944): 671.

TABLE 11.4

PARAMETERS INVOLVED IN LIQUID FLOW EQUATIONS

Symbol	Name	SI units	Conversion
g $\left[9.81 \dfrac{\text{meters}}{\text{second}}\right]$	gravitational acceleration	meter per square second	1 m/sec² = 3.28 ft/sec²
ρ	density or mass per unit volume	kilogram per cubic meter	1 kg/m³ = 0.0624 lb/ft³
w $[w = \rho g]$	specific weight (weight per unit volume)	newton per cubic meter	1 n/m³ = 0.00637 lb/ft³
h	pressure head	meter	1 meter = 3.28 feet
P $[P = \rho g h]$	pressure	newton per square meter	1 n/m² = 1.45 × 10⁻⁴ lb/in²
μ	dynamic viscosity	newton-second per square meter	
D	inside diameter	meter	1 meter = 39.37 inches
V	average velocity	meter per second	1 m/sec = 3.28 ft/sec
l	length	meter	1 meter = 3.28 feet
Q	flow rate	cubic meter per second	1 m³/sec = 35.34 ft³/sec
R		newton-second per meter⁵	

* $g = 9.81$ meters/second squared
  $= 32.2$ feet/second squared

Example 11.6 Oil at a temperature of 15° Celsius flows through a horizontal tube which is 1 centimeter in diameter with a velocity of 2 meters per second. The line is 10 meters long. Determine if the flow is laminar or turbulent by calculating the Reynolds number. Select the correct equations and calculate the resistance and pressure drop of the tube.

Solution:

From Table 2, Appendix A,

$\rho = 880$ kilograms per cubic meter
$\mu = 0.160$ newton-seconds per square meter

From Equation (11.8),

$$\text{Reynolds number} = \rho V D / \mu = \frac{(880)(2)(0.01)}{0.160} = 110$$

The Reynolds number is less than 2000, so the flow is laminar and Equations (11.9), (11.10), and (11.11) apply.

$$R_L = \left[\frac{128 \mu l}{\pi D^4} \right] = \frac{(128)(0.16)(10)}{\pi (0.01)^4} = 6.51 \times 10^9 \; n - s/m^5$$

$$Q = \pi D^2 V / 4 = \pi (0.01)^2 (2) / 4 = 1.57 \times 10^{-4} \; m^3/\text{sec}$$

$$P = R_L Q = (6.51 \times 10^9)(1.57 \times 10^{-4}) = 1.04 \times 10^6 \; n/m^2$$

Example 11.7 Water at $15°$ Celsius flows through a commercial steel pipe which is 0.1525 meters in diameter with a velocity of 3.5 meters/second. The line is 40 meters long. Determine if the flow is laminar or turbulent by calculating the Reynolds number. Select the correct equations and calculate the resistance and pressure drop of the pipe.

Solution:

From Table 2, Appendix A,

$$\rho = 1000 \text{ kilograms per cubic meter}$$
$$\mu = 0.001 \text{ newton-second per square meter}$$

$$\text{Reynolds number} = \rho V D / \mu = (1000)(3.5)(0.1525) / 0.001$$
$$= 5.35 \times 10^5$$

The Reynolds number is greater than 4000, so the flow is turbulent and Equations (11.12), (11.13), (11.14), and (11.15) apply. From Table 11.3, $f = 0.019$ (by interpolation),

$$K_t = \left[\frac{8 \rho f l}{\pi^2 D^5} \right] = \frac{(8)(10^3)(0.019)(40)}{\pi^2 (1.525 \times 10^{-1})^5} = 7.5 \times 10^6$$

$$Q = \pi D^2 V / 4 = \pi (1.525 \times 10^{-1})^2 (3.5) / 4$$
$$= 6.4 \times 10^{-2} \text{ cubic meter/second}$$

$$R_t = 2 K_t Q = (2)(7.5 \times 10^6)(6.4 \times 10^{-2})$$
$$= 9.6 \times 10^5 \text{ newton-seconds/meter}^5$$

$$P = K_t Q^2 = (7.5 \times 10^6)(6.4 \times 10^{-2})^2$$
$$= 3.1 \times 10^4 \text{ newtons/square meter}$$

The flow through a control valve is turbulent and depends on the orifice size and the position of the valve. The equation for water flow through a valve is shown in the following adaptation of the turbulent flow equation.

LIQUID CONTROL VALVE EQUATIONS
(ENGLISH UNITS)

$$Q = 0.088C_v\sqrt{h} \qquad \text{(11.16)}$$

where:

Q = the flow rate in cubic feet per minute

C_v = the valve flow coefficient

h = the difference in pressure across the valve in feet of head (see Table 4.3)

The resistance of the control valve may be found by squaring both sides of the above equation and proceeding as before.

$$Q^2 = 0.0077C_v{}^2h$$

$$2QdQ = 0.0077C_v{}^2dh$$

$$R_v = \frac{dh}{dQ} = \frac{2Q}{0.0077C_v{}^2} \text{ min/ft}^2 \qquad \text{(11.17)}$$

The SI version of the control valve equation is given in Equations (11.18) and (11.19) below.

LIQUID CONTROL VALVE EQUATIONS (SI UNITS)

$$Q = 7.53 \times 10^{-5}C_v\sqrt{H} \qquad \text{(11.18)}$$

$$R_t = \frac{dH}{dQ} = 3.53 \times 10^8 Q/C_v{}^2 \qquad \text{(11.19)}$$

where:

Q = flow rate in cubic meters per second

C_v = the valve flow coefficient

H = the pressure across the valve in meters of head

R_t = turbulent flow resistance in seconds per meter^2

Example 11.8 Find the valve resistance when $Q = 10$ ft^3/min and $C_v = 20$.

Solution:

$$R = \frac{dh}{dQ} = \frac{(2)(10)}{0.0077(20)^2} = 6.5 \text{ min/ft}^2 \text{ [from Equation (11.17)]}$$

Liquid Flow Capacitance. Liquid flow capacitance is defined in terms of the change in volume of liquid in a tank required to make a unit change in pressure at the outlet of the tank.

$$C = \frac{\triangle v}{\triangle p}, \frac{\text{cubic meters}}{\text{newton per square meter}} \tag{11.20}$$

where:

$\triangle v$ = a change in volume, cubic meters

$\triangle p$ = the corresponding change in pressure, newtons per square meter

The change in pressure in the tank depends on two things: the change in level of the liquid ($\triangle h$), and the density of the liquid. The relationship is expressed by the following equation.

$$\triangle p = \rho g \triangle h, \text{newtons/square meter} \tag{11.21}$$

where:

ρ = the density of the liquid, kilograms/cubic meter

g = the acceleration due to gravity = 9.8 meters/second^2

$\triangle h$ = the change in liquid level, meters

The change in level in the tank is equal to the change in volume divided by the average area (A).

$$\triangle h = \triangle v/A, \text{meters} \tag{11.22}$$

By substitution of Equations (11.21) and (11.22) into Equation (11.20),

$$\triangle p = \frac{\rho g \triangle v}{A}$$

$$C = \frac{\triangle v}{\triangle p} = \frac{\triangle v A}{g \rho \triangle v} = \frac{A}{\rho g}$$

LIQUID FLOW CAPACITANCE

$$C = \frac{A}{\rho g}$$

where:

C = liquid flow capacitance, $\dfrac{\text{cubic meters}}{\text{newton per square meter}}$

A = the cross-sectional area of the tank at the liquid interface, square meters

ρ = the density of the liquid (see Table 3, Appendix A), kilograms/cubic meter

g = 9.81 meters/second²

Example 11.9 A liquid tank has a diameter of 1.83 meters and a height of 10 feet. Determine the capacitance of the tank for each of the following fluids.

a. Water
b. Oil
c. Kerosene
d. Gasoline

Solution:

Equation (11.23) may be used to determine C.

$$C = \frac{A}{\rho g}$$

$$A = \pi D^2/4 = \pi (1.83)^2/4 = 2.63 \text{ square meters}$$

$$g = 9.81 \text{ meters/second}^2$$

$$C = 2.63/(9.81\rho) = 0.268/\rho$$

The density (ρ) may be obtained from Table 2, Appendix A.

a. Water: $\rho = 1000$

$$C = 0.268/1000 = 2.68 \times 10^{-4} \frac{\text{cubic meter}}{\text{newton per square meter}}$$

b. Oil: $\rho = 880$

$$C = 0.268/880 = 3.04 \times 10^{-4} \frac{\text{cubic meter}}{\text{newton per square meter}}$$

c. Kerosene: $\rho = 800$

$$C = 0.268/800 = 3.35 \times 10^{-4} \ \frac{\text{cubic meter}}{\text{newton per square meter}}$$

d. Gasoline: $\rho = 740$

$$C = 0.268/740 = 3.62 \times 10^{-4} \ \frac{\text{cubic meter}}{\text{newton per square meter}}$$

Liquid or Gas Flow Inertance. Liquid or gas flow inertance is measured in terms of the pressure difference between two points in the pipe required to produce a unit change in flow rate each second. The pressure

LIQUID OR GAS FLOW INERTANCE

$$P = I\frac{\Delta Q}{\Delta t}$$

or

$$P = I\frac{dQ}{dt}$$

$$I = \rho L/A \qquad\qquad (11.24)$$

where:

I = the inertance, $\dfrac{\text{newtons per square meter}}{\text{cubic meters per second}^2}$

P = the pressure difference between two points in the pipe, newtons/square meter

ρ = the fluid density, kilograms/cubic meter

L = the distance between the two pressure points, meters

A = the cross-sectional area of the pipe, square meters

Q = the fluid flow rate, cubic meters/second

ΔQ = a change in flow rate produced by P, cubic meters/second

Δt = the time required for Q to occur, seconds

$\dfrac{dQ}{dt}$ = the rate of change of liquid flow rate, cubic meters per second

difference acting on the cross-sectional area of the pipe produces a force (F) equal to the pressure drop (P) times the area (A).

$$F = P \times A$$

This force will accelerate the fluid between the two points, according to Newton's law of motion.

$$F = PA = M\frac{\triangle v}{\triangle t} \qquad (11.25)$$

The mass of fluid in the pipe (M) is equal to the density of the fluid (ρ) times the volume of fluid between the two points. The volume is equal to the area of the pipe (A) times the distance between the two points (L).

$$M = \rho AL \qquad (11.26)$$

Combining Equations (11.25) and (11.26), we have

$$PA = \rho AL\frac{\triangle v}{\triangle t} = \rho L\left[\frac{(A)(\triangle v)}{\triangle t}\right]$$

The area of the pipe (A) times the change in fluid velocity $(\triangle v)$ is equal to the change in fluid flow rate $(\triangle Q)$; i.e., $(A)(\triangle v) = \triangle Q$. Substituting $\triangle Q$ for $(A)(\triangle v)$ and dividing both sides by A produces the final form.

$$P = \left[\frac{\rho L}{A}\right]\frac{\triangle Q}{\triangle t} = I\frac{\triangle Q}{\triangle t}$$

Example 11.10 Determine the liquid flow inertance of water in a pipe which has a diameter of 2.1 centimeters and a length of 65 meters.

Solution:

Equation (11.24) applies: $I = \rho L/A$

$\rho = 1000$ kilograms/cubic meter (Table 2, Appendix A)

$A = \pi D^2/4 = \pi(0.021)^2/4 = 3.46 \times 10^{-4}$ square meter

$I = (1000)(65)/3.46 \times 10^{-4} = 1.88 \times 10^8 \dfrac{\text{newtons per meter}^2}{\text{cubic meters per second}^2}$

Liquid Flow Rate Dead Time. Dead time in a liquid system occurs when the liquid is transported from one point to another in a pipe line. An example of liquid flow dead time is shown in Figure 11.6. Hot and cold

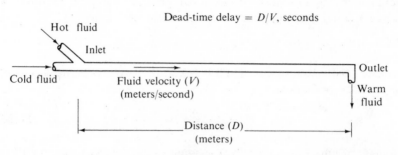

Dead-time delay $= D/V$, seconds

Hot fluid

Inlet

Cold fluid

Fluid velocity (V)
(meters/second)

Outlet

Warm fluid

Distance (D)
(meters)

FIGURE 11.6

A LIQUID-FLOW, DEAD-TIME ELEMENT—A LONG PIPE

fluids are combined in a Y connection to produce a warm fluid which flows a distance D to the outlet of the pipe. The input to the system is the ratio of hot fluid to cold fluid. The output is the temperature of the fluid at the outlet end. The dead-time lag is the distance traveled (D) divided by the average velocity (V) of the fluid. Equation (11.7) may be used to compute t_d. The fluid flow rate (Q) is given by Equation (11.27).

$$t_d = D/V \qquad\qquad\qquad \textbf{(11.7)}$$

$$V = Q/A = 4Q/(\pi d^2) \qquad\qquad \textbf{(11.27)}$$

where:

$D =$ the distance traveled, meters

$V =$ the average velocity of the fluid, meters/second

$Q =$ the fluid flow rate, cubic meters/second

$A =$ the area of the pipe, square meters

$d =$ pipe inside diameter, meters

Example 11.11 Liquid flows in a pipe which is 200 meters long and has a diameter of 6 centimeters. The flow rate is 0.0113 cubic meters per second. Determine the dead-time lag.

Solution:

Equations (11.7) and (11.27) apply.

$$V = 4Q/(\pi d^2) = 4(0.0113)/(\pi 0.06^2) = 4 \text{ meters/second}$$

$$t_d = D/V = 200/4 = 50 \text{ seconds}$$

11.5 Gas Flow Systems

Gas Flow Resistance. Gas flow resistance is that property of pipes, valves, or restrictions which impedes the flow of a gas. It is measured in terms of the change in pressure required to produce a change in gas flow rate of 1 kilogram per second. Gas flow in a pipe may be laminar or turbulent. In laminar flow, the pressure drop varies directly with the viscosity of the fluid and the fluid velocity. In turbulent flow, the pressure drop varies directly with the density of the fluid and the square of the fluid velocity. In practice, gas flow is almost always turbulent, and the commonly used equations apply to turbulent flow.

If the pressure drop is less than 5–10 per cent of the initial gas pressure, the Fanning equation for incompressible flow may be used for gas flow. This is the equation which is used for turbulent liquid flow with the volume flow rate (Q) replaced by the mass flow rate (w).

LOW PRESSURE DROP GAS FLOW EQUATIONS

$$\text{Pressure drop} = P = \left[\frac{8fl}{\pi^2 D^5 \rho}\right] w^2 \qquad \textbf{(11.28))}$$

$$R_g = \left[\frac{8fl}{\pi^2 D^5 \rho}\right]$$

where:

R_g = gas flow resistance, newton-seconds/kilogram-square meter

P = the difference between the inlet and outlet pressure, newtons/square meter

f = the friction factor (a dimensionless number that depends on the Reynolds number and the internal surface of the pipe—see Table 11.3)

l = the length of the pipe, meters

D = the inside diameter of the pipe, meters

ρ = the fluid density, kilograms/cubic meter

w = the gas mass flow rate, kilograms/second

The Fanning equation does not account for the expansion of a gas as the pressure decreases. For this reason it does not apply when the pressure drop is greater than 5–10 per cent of the initial gas pressure.

The amount of heat added to the gas also affects the expansion. In most gas flow systems, the amount of heat added is just enough to maintain the temperature of the gas constant. This is called an *isothermal* (constant temperature) expansion. The relationship between the pressure and the unit volume of the gas is given by the perfect gas law.

$$Pv = 10^3 RT/M$$

where:

P = the absolute pressure of the gas, newtons/square meter

v = the unit volume of the gas, cubic meters/kilogram

R = the universal gas constant = 8.314 joules/K-mole

M = the molecular weight of the gas (see Table 4, Appendix A)

T = the absolute temperature, Kelvin

In the isothermal expansion, T is a constant. Since R and M are also constant, this means that Pv is a constant and the unit, volume (v) increases by the same proportion that the pressure (P) decreases. The following flow equation applies to the constant temperature flow of a perfect gas with a large pressure drop.

LARGE PRESSURE DROP GAS FLOW EQUATION

$$w = 0.358 \sqrt{\frac{(p_1^2 - p_2^2)d^5 M}{fTL}} \qquad (11.29)$$

where:

w = the mass flow rate, kilograms per second

p_1 = the inlet absolute pressure, newtons per square meter

p_2 = the outlet absolute pressure, newtons per square meter

d = the inside diameter of the pipe, meters

M = the molecular weight of the gas (Table 4, Appendix A)

f = the friction factor (see Table 4-4)

T = the absolute temperature, Kelvin

L = the length of the pipe, meters

The gas resistance may be expressed in terms of a change in the inlet pressure (p_1) or a change in the outlet pressure (p_2). An exact derivation of each resistance expression is beyond the scope of this text. However, a graphical method may be used to determine the resistance as illustrated in Example 11.12. The gas resistance is defined as the change in pressure required to produce a unit change in mass flow rate. On a graph of pressure versus mass flow rate, the resistance is equal to the slope of the curve.

Example 11.12 Carbon dioxide gas at 15°Celsius flows through a commercial steel pipe 2.1 centimeters in diameter and 50 meters long. The outlet pressure (p_2) is 1×10^5 newtons/square meter. Use the large pressure drop gas flow equation (11.29) to determine the inlet pressure (p_1) for mass flow rates (w) of 0.2, 0.4, 0.6, 0.8, and 1 kilogram per second. Plot a graph of p_1 versus w and find the resistance at $w = 0.6$ kilograms per second from the slope of the curve.

Solution:

$$w = 0.358 \sqrt{\frac{(p_1{}^2 - p_2{}^2)d^5 M}{fTL}}$$

$$p_1{}^2 - p_2{}^2 = \left[\frac{w^2 fTL}{(0.358)^2 d^5 M}\right]$$

$$p_1 = \sqrt{\frac{w^2 ftL}{(0.358)^2 d^5 M} + p_2{}^2}$$

$$T = 15°C = 288°K$$

$$d^5 = (2.1 \times 10^{-2})^5 = 4.09 \times 10^{-9}$$

$$L = 50$$

$$M = 44 \quad \text{(from Table 4, Appendix A)}$$

$$p_2{}^2 = (1 \times 10^5)^2 = 10^{10}$$

$$\frac{TL}{(0.358)^2 d^5 M} = \frac{(288)(50)}{(0.128)(4.09 \times 10^{-9})(44)} = 6.25 \times 10^{11}$$

$$p_1 = \sqrt{6.25 \times 10^{11} w^2 f + 10^{10}}$$

From Table 4, Appendix A, $\mu = 1.46 \times 10^{-5} \dfrac{\text{newton-second}}{\text{square meter}}$.

The following alternate form of the Reynolds number equation may be used for gas flow.

$$\text{Reynolds number} = 4w/\pi\mu D$$

Reynolds number $= \dfrac{4w}{\pi\,(1.46 \times 10^{-5})\,(2.1 \times 10^{-2})} = 4.15 \times 10^{6}w$

w	0.2	0.4	0.6	0.8	1.0
Reynolds number	8.3×10^{5}	1.66×10^{6}	2.49×10^{6}	3.32×10^{6}	4.15×10^{6}
f (Table 4.4)	0.023	0.023	0.023	0.023	0.023
$w^2 f$	9.2×10^{-4}	3.68×10^{-3}	8.3×10^{-3}	1.47×10^{-2}	2.3×10^{-2}
$6.25 \times 10^{11} w^2 f$	5.75×10^{8}	2.3×10^{9}	5.2×10^{9}	9.2×10^{9}	1.44×10^{10}
$6.25 \times 10^{11} w^2 f + 10^{10}$	1.06×10^{10}	1.23×10^{10}	1.52×10^{10}	1.92×10^{10}	2.44×10^{10}
p_1	1.03×10^{5}	1.11×10^{5}	1.23×10^{5}	1.39×10^{5}	1.56×10^{5}

The inlet pressure (p_1) versus mass flow rate (w) curve is shown in Figure 11.7. The gas flow resistance is equal to the slope of the p_1 versus w curve. At $w = 0.6$ kilogram/second, the resistance ($R_{0.6}$) is equal to the slope of the dotted line in Figure 11.7.

$$R_{0.6} = \frac{\triangle p_1}{\triangle w} = \frac{1.52 \times 10^5}{0.72} = 2.11 \times 10^5 = \frac{\text{newton-seconds}}{\text{kilogram-square meter}}$$

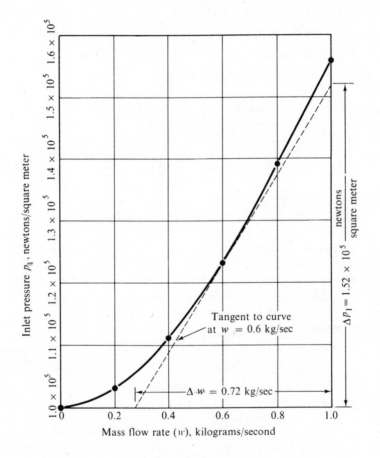

FIGURE 11.7

GRAPHICAL DETERMINATION OF GAS FLOW RESISTANCE

Gas Flow Capacitance. Gas flow capacitance is defined in terms of the change in the mass of gas in a pressure vessel required to produce a unit change in pressure while the temperature remains constant. The perfect gas law may be used to determine the capacitance equation.

$$PV = \left(\frac{10^3 m}{M}\right) RT \tag{11.30}$$

where:

   P = the absolute pressure of the gas, newtons/square meter

   V = the volume of the gas, cubic meters

   m = the mass of the gas, kilograms

   M = the molecular weight of the gas (see Table 4, Appendix A)

   R = the universal gas constant = 8.314 joule/K-mole

   T = the absolute temperature, Kelvin

Solving Equation (11.30) for m, the mass of the gas,

$$(10^3 RT)m = (VM)P$$

$$m = \left[1.2 \times 10^{-4} M\left(\frac{V}{T}\right)\right] P \tag{11.31}$$

Equation (11.31) applies to a pressure vessel where V is the volume of the vessel, T is the absolute temperature of the gas, and M is the molecular weight of the gas. For a given pressure vessel and gas, V and M are constants. The temperature (T) must also be held constant to determine the gas capacitance. When M, V, and T are all constant, the term in brackets is constant. Under these conditions, Equation (11.31) is linear, and the relationship between a change in mass (Δm) and the corresponding change in pressure (ΔP) is given by

$$\Delta m = [1.2 \times 10^{-4} MV/T]\Delta P$$

Gas flow capacitance = $\Delta m/\Delta P = [1.2 \times 10^{-3} MV/T]$

GAS FLOW CAPACITANCE

$$C = 1.2 \times 10^{-4} MV/T \tag{11.32}$$

where:

   C = the gas flow capacitance, kilograms/newton per meter^2

   M = the molecular weight of the gas (see Table 4, Appendix A)

   V = the pressure vessel volume, cubic meters

   T = the gas temperature, Kelvin

Example 11.13 A pressure tank has a volume of 0.75 cubic meters. Determine the capacitance of the tank if the gas is nitrogen at 20°C.

Solution:

Equation (11.32) may be used to determine C.

$$C = 1.2 \times 10^{-4} \, MV/T$$

$M = 28.016$ for nitrogen (Table 4, Appendix A)

$T = 20\ °C = 293$ Kelvin

$V = 0.75$ cubic meter

$C = (1.2 \times 10^{-4})\ (28.016)\ (0.75)/(293)$

$C = 8.6 \times 10^{-6}$ kilogram/newton per meter^2

Gas Flow Inertance. Gas flow inertance is covered with liquid flow inertance in Section 11.4.

11.6   Thermal Systems

Thermal Resistance. Thermal resistance is that property of a substance which impedes the flow of heat. It is measured in terms of the change in temperature required to produce a change in heat flow rate of 1 joule per second. Figure 11.8 shows a composite wall separating two fluids which are at different temperatures. The fluids may be either liquids or gases. Heat flows through the wall from the hotter fluid to the cooler fluid. The resistance of the wall (R) and the difference between the two fluid temperatures ($T_o - T_i$) determines the flow rate of heat (Q) from the outside fluid to the inside fluid.

$$Q = \frac{1}{R}(T_o - T_i) \qquad \text{watts} \tag{11.33}$$

If the heat flow rate (Q) is negative because T_i is greater than T_o, it simply means that the heat is going from the inside fluid to the outside fluid.

    The thermal resistance of the composite wall in Figure 11.8 is caused by two factors—the thermal conductivity (K) of the two wall sections, and the film coefficient (h) of the two fluid films. A film is a stagnant layer of fluid between the surface of the solid and the main body of the fluid. It acts as an insulating blanket over the solid surface. In many heat transfer systems, the film coefficients are a major portion of the thermal resistance. The total resistance of the composite wall is the sum of the resistances of each of the four components.

Thermal resistance = outside film resistance + outside wall resistance + inside wall resistance + inside film resistance.

THERMAL RESISTANCE EQUATIONS

$$R = \left[\begin{array}{l} \text{the sum of all film resistances } (R_h) \text{ and} \\ \text{wall section resistances } (R_k) \text{ between the} \\ \text{inside fluid and the outside fluid} \end{array} \right] \quad \textbf{(11.34)}$$

$$R_k = \frac{x}{Ak} \qquad \textbf{(11.35)}$$

$$R_h = \frac{1}{Ah} \qquad \textbf{(11.36)}$$

where:

R = the total thermal resistance, Kelvin/watt

R_k = the resistance of a wall section, Kelvin/watt

R_h = the resistance of a film, Kelvin/watt

x = the thickness of a wall section, meters

k = the thermal conductivity of the wall material, watts/ meter-Kelvin (see Tables 1, 2, and 4 in Appendix A)

A = the heat transfer surface area, square meters

h = the film coefficient, watts/meter^2—Kelvin
[see Equations (11.35)−(11.43)]

The resistance of each wall section is given by Equation (11.35). The thermal conductivity (k) is a property of the material in the wall section. The thermal conductivity of selected solids, liquids, and gases are included in Tables 1, 2, and 4 in Appendix A. The resistance of the fluid films is given by Equation (11.36). Equations (11.37) through (11.45) may be used to determine the value of the film coefficient (h).

The film coefficient (h) depends on several properties and conditions of the film, such as:

1. The thermal conductivity.
2. The specific heat of the fluid.
3. The viscosity of the fluid.
4. The velocity of the main body of the fluid relative to the surface of the solid.

Experiments were used to determine general empirical equations for computing the film coefficient of any fluid with either forced or natural convection. These general equations are too complex for consideration here. The equations may be simplified considerably by restricting the

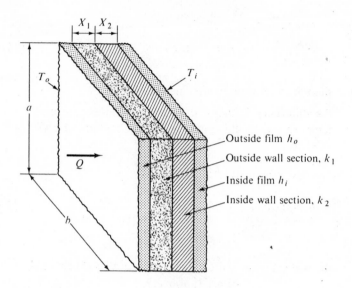

A = heated surface area = $a \times b$, square meters

R = thermal resistance = $\dfrac{1}{Ah_o} + \dfrac{X_1}{AK_1} + \dfrac{X_2}{AK_2} + \dfrac{1}{Ah_i} \quad \dfrac{\text{kelvin}}{\text{watt}}$

T_o = outside surface temperature, Kelvin or ° Celsius

T_i = inside surface temperature, Kelvin or ° Celsius

Q = heat flow rate = $\dfrac{1}{R}(T_o - T_i)$, watts

FIGURE 11.8

*THERMAL RESISTANCE AND HEAT FLOW RATE
FOR A COMPOSITE WALL*

application to a particular fluid over a limited range. Here, the equations have been converted to SI units, and in some cases, are simplified by linearization over a specified range.[1]

1. Natural convection in still air.
   a. For horizontal surfaces facing down.

$$h = 1.32\, T_d^{0.25} \tag{11.37}$$

   b. For horizontal surfaces facing up.

$$h = 2.5\, T_d^{0.25} \tag{11.38}$$

---

[1] Equations (11.37)–(11.45) are based on empirical equations in J. Kenneth Salisbury, *Kent's Mechanical Engineers' Handbook: Power,* 12th edition (New York: John Wiley & Sons, 1950), pp. 3–17—3–20.

c.   For vertical surfaces.

$$h = 1.78 \, T_d^{0.25} \qquad\qquad\qquad \textbf{(11.39)}$$

where

$h =$ the film coefficient, watts/meter^2, Kelvin

$T_d =$ the difference between the surface temperature and the temperature of the main body of the fluid, Kelvin or °Celsius

2.   Natural convection in still water.

$$h = 2.26 \, (T_w + 34.4) \, \sqrt{T_d}, \ \frac{\text{watts}}{\text{meter}^2\text{---Kelvin}} \qquad \textbf{(11.40)}$$

where:

$t_w =$ the water temperature, °Celsius (h and T_d are defined in Part 1, above.)

Note: Equation (11.40) does not apply when the surface temperature is above the boiling point of the water.

3.   Natural convection in oils.

$$h = 7.0 \, T_d^{\ 0.25}/\mu^{0.4}, \ \frac{\text{watts}}{\text{meter}^2\text{---Kelvin}} \qquad \textbf{(11.41)}$$

where:

$\mu =$ the dynamic viscosity, newton-second/square meter (h and T_d are defined in Part 1, above.)

4.   Forced convection for air inside pipes and against smooth surfaces.
a.   Air velocities below 4.6 meters/second.

$$h = 4.55 + 4.1v, \ \text{watts/meter}^2\text{---Kelvin} \qquad \textbf{(11.42)}$$

where:

$v =$ the air velocity, meters/second

b.   Air velocities above 4.6 meters/second.

$$h = 7.75v^{0.75}, \ \frac{\text{watts}}{\text{meter}^2\text{---Kelvin}} \qquad \textbf{(11.43)}$$

where:

$v =$ the air velocity, meters/second

5. Forced convection, turbulent water flow in straight pipes.
   a. Water temperature of 15° Celsius.

$$h = 2800v + 700, \frac{\text{watts}}{\text{meter}^2\text{—Kelvin}} \qquad \textbf{(11.44)}$$

where:

v = average water velocity in pipe, meters/second

   b. Water temperatures between 0° C and 100° C.

$$h = 850(1.35 + 0.02T_w)(2v + 0.5),$$
$$\text{watts/meter}^2\text{—Kelvin} \qquad \textbf{(11.45)}$$

where:

T_w = the water temperature, ° Celsius

v = the average water velocity in the pipe, meters/second

Note: Equation (11.45) is restricted to pipe inside diameters from ½ inches to 2 inches.

Example 11.14   In Figure 11.8, the inside wall section is a 1-centimeter-thick steel plate. The outside wall section is a 2-centimeter-thick layer of insulation. The inside air velocity is 0.5 meters/second and the outside air velocity is 4.0 meters/second. The air velocity determines the film coefficient, according to Equation (11.42). The wall dimensions are: $a = 2$ meters and $b = 3$ meters. Determine the thermal resistance of the wall section.

Solution:

From Table 1, Appendix A,

$$k_1 = 45 \text{ watts/meter-Kelvin} \quad \text{for steel}$$
$$k_2 = 0.036 \text{ watts/meter-Kelvin} \quad \text{for insulation}$$

From Figure 11.8,

$$A = a \times b = 2 \times 3 = 6 \text{ square meters}$$

From Equation (11.42),

$$h_i = 4.55 + 4.1v = 4.55 + 4.1 (0.5)$$
$$4.55 + 2.05 = 6.6 \text{ watts/square meter-Kelvin}$$
$$h_o = 4.55 + 4.1v = 4.55 + 4.1 (4)$$
$$4.55 + 16.4 = 20.95 \text{ watts/square meter-Kelvin}$$

$$R_t = \frac{1}{Ah_o} + \frac{x_1}{Ak_1} + \frac{x_2}{Ak_2} + \frac{1}{Ah_i}$$

$$R_t = \frac{1}{(6)(20.95)} + \frac{0.01}{(6)(45)} + \frac{0.02}{(6)(0.036)} + \frac{1}{(6)(6.6)}$$

$$R_t = 0.008 + 0.000037 + 0.0925 + 0.0252$$

$$R_t = 0.126 \text{ Kelvin/watt}$$

Example 11.15 Determine the thermal resistance of each of the following film conditions for an area of 1 square meter.

a. Natural convection in still air on a vertical surface where $T_d = 20$ K.
b. Natural convection in still water where $T_d = 30$ K and $T_w = 20°$C.
c. Natural convection in oil (Table 2, Appendix A) where $T_d = 16$ K.
d. Forced convection in air with a velocity of 6 meters/second.
e. Forced convection in water with a temperature (T_w) of 40° C and a velocity of 4 meters/second.

Solutions:

a. Equation (11.39) applies, $h = 1.78\ T_d^{0.25}$

$$h = 1.78\ (20)^{0.25} = 1.78\ (2.11) = 3.76\ \frac{\text{watts}}{\text{meter}^2\text{—K}}$$

$$R = \frac{1}{Ah} = \frac{1}{(1)(3.76)} = 0.266 \text{ Kelvin/watt}$$

b. Equation (11.40) applies, $h = 2.26\ (T_w + 34.4)\ \sqrt{T_d}$

$$h = 2.26\ (20 + 34.4)\ \sqrt{30}$$

$$2.26\ (54.4)\ (5.5) = 6.75 \times 10^2\ \frac{\text{watts}}{\text{meter}^2\text{—Kelvin}}$$

$$R = \frac{1}{Ah} = \frac{1}{(1)(6.75 \times 10^2)} = 1.48 \times 10^{-3}\ \text{K/watt}$$

c. Equation (11.41) applies, $h = 7.0\ T_d^{0.25}/\mu^{0.4}$

$$\mu = 0.160 \text{ newton-second/square meter}$$

$$h = 7.0\ (16)^{0.25}/(0.16)^{0.4} = (7.0)\ (2)/(0.48)$$

$$= 29.2 \text{ watts/meter}^2\text{—Kelvin}$$

$$R = \frac{1}{Ah} = \frac{1}{(1)(29.2)} = 0.0343 \text{ Kelvin/watt}$$

d. Equation (11.43) applies, $h = 7.75v^{0.75}$

$$h = 7.75 \ (6)^{0.75} = 7.75 \ (3.83) = 29.6 \ \frac{\text{watts}}{\text{meter}^2\text{—K}}$$

$$R = \frac{1}{Ah} = \frac{1}{29.6} = 0.0338 \ \text{Kelvin/watt}$$

e. Equation (11.45) applies, $h = 850 \ (1.35 + 0.02T_w) \ (2v + 0.5)$

$$h = 850 \ [1.35 + 0.02 \ (40) \ ] \ [2(4) + 0.5]$$
$$= 850 \ (1.35 + 0.8) \ (8 + 0.5)$$
$$= 850 \ (2.15) \ (8.5) = 15550 \ \frac{\text{watts}}{\text{meter}^2\text{—Kelvin}}$$

$$R = \frac{1}{Ah} = \frac{1}{(1) \ (1.555 \times 10^4)}$$
$$= (6.44) \ 10^{-6} \ \text{Kelvin/watt}$$

Thermal Capacitance. Thermal capacitance is defined in terms of the change in heat required to make a unit change in temperature. The specific heat of a material is defined as the amount of heat required to raise the temperature of one kilogram of the material by one degree Kelvin. The thermal capacitance is simply the product of the specific heat times the mass of the material.

THERMAL CAPACITANCE

$$C = mS_h \qquad\qquad \textbf{(11.46)}$$

where:

C = the thermal capacitance, joule/Kelvin

m = the mass of material, kilograms

S_h = the specific heat of the material, joule/kilogram-Kelvin

Example 11.16 Determine the thermal capacitance of 8.31 cubic meters of water.

Solution:

Equation (11.46) may be used to determine C.

$$C = mS_h$$

The mass of water (m) is equal to the density (ρ) times the volume of water $(8.31\ m^3)$. From Table 2, Appendix A, the density of water is 1000 kilograms/cubic meter, and the specific heat is 4190 joules/kilogram-Kelvin.

$$C = (1000)\ (8.31)\ (4190)$$
$$C = 3.48 \times 10^7\ \text{joules/Kelvin}$$

11.7 Mechanical Systems

Mechanical Resistance. Mechanical resistance (or friction) is that property of a mechanical system which impedes motion. It is measured in terms of the change in force required to produce a unit change in the velocity of 1 meter per second.

An automobile shock absorber and a dashpot are examples of mechanical resistance devices, whose operation is illustrated in Figure 11.9-a. The cylinder is stationary and the piston rod is attached to the moving part (M). When part M moves, the fluid in the cylinder must move through the orifice around the piston from one side to the other. The flow rate of the fluid through the orifice is proportional to the velocity of part M. A difference of pressure $(p_2 - p_1)$ is required to produce the fluid flow through the orifice. This difference in pressure produces a restraining force that opposes the motion. The restraining force is equal to the difference in pressure times the area of the piston $F = (p_2 - p_1)A$.

If the fluid flow rate through the orifice is small, the flow is laminar and the force is proportional to the velocity (Figure 11.9-b). If the flow rate is large, the flow is turbulent and the force is proportional to the square of the velocity (Figure 11.9-c). The friction effects of the seal are neglected in the preceding statements. The seal produces a constant friction force that is independent of velocity. Mechanical resistance that produces a force proportional to the velocity is called *viscous friction,* and the type that produces a force independent of velocity is called *coulomb friction.* The combination of coulomb friction and viscous friction is illustrated in Figure 11.9–d.

Pure viscous friction is the simplest to treat mathematically, and Figure 11.9–b is usually used as a first approximation to mechanical resistance. The resistance is equal to the force (F) divided by the velocity (V) at any point [see Equation (11.47)].

In Figure 11.9–c, the resistance varies with the velocity. At any point on the curve, the resistance is equal to the tangent to the curve at that point [see Equation (11.48)].

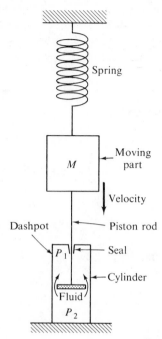

a) A mechanical part with resistance
provided by a dashpot

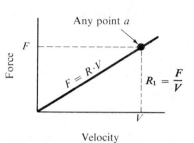

b) Resistance due to laminar
flow in a dashpot – viscous friction.

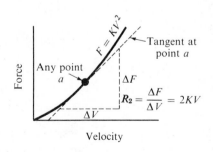

c) Resistance due to turbulent
flow in a dashpot.

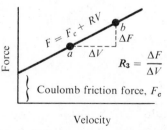

d) Resistance due to both
viscous friction and coulomb friction.

FIGURE 11.9

EXAMPLES OF MECHANICAL RESISTANCE

In Figure 11.9–d, the resistance is a constant, but it cannot be determined by dividing the force (F) by the velocity (V) at any point. It can be determined by dividing the change in force $(\triangle F)$ by the change in velocity (ΔV) between two points on the curve [see Equation (11.49)].

MECHANICAL RESISTANCE

$$R_1 = F/V \tag{11.47}$$

$$R_2 = \frac{dF}{dV} \tag{11.48}$$

$$R_3 = \frac{\triangle F}{\triangle V} \tag{11.49}$$

where:

R_1 = resistance of dashpots with laminar flow (Figure 11.9-b), newton-seconds/meter

R_2 = resistance of dashpots with turbulent flow (Figure 11.9-c), newton-seconds/meter

R_3 = resistance of dashpots with coulomb and viscous friction (Figure 11.9-d), newton-seconds/meter

F = the force, newtons

V = the velocity, meters/second

$\triangle F$ = the change in force between two points on the F-vs.-V curve

$\triangle V$ = the change in velocity between the same two points

$\dfrac{dF}{dV}$ = the slope of the F-vs.-V curve

Example 11.17 A dashpot is used to provide mechanical resistance in a packaging machine. The flow is laminar, so the viscous friction equation shown in Figure 11.9–a applies. A test was conducted in which a force of 98 newtons produced a velocity of 24 meters/second. Determine the mechanical resistance (R_m).

Solution:

$$R_m = F/V = 98/24 = 4.08 \text{ newton-seconds/meter}$$

Example 11.18 A mechanical system consists of a sliding load (coulomb friction) and a shock absorber (viscous friction). The force-versus-velocity curve is shown in Figure 11.9–d. The following data were obtained from the system.

Run	Force (F)	Velocity (V)
a	7.1 newtons	10.5 meters/second
b	9.6 newtons	15.75 meters/second

Determine the resistance (R) and the coulomb friction force (F_c). Write an equation for the applied force (F) in terms of the velocity (V).

Solution:

$$R = \frac{\Delta F}{\Delta V} = \frac{F_b - F_a}{V_b - V_a} = \frac{9.6 - 7.1}{15.75 - 10.5} = 2.5/5.25 = 0.476$$

$R = 0.476$ newton-second/meter

$F = F_c + 0.476V$　　(From Figure 11.9-d)

$7.1 = F_c + (0.476)(10.5)$　　(From Run a)

$F_c = 7.1 - 5.0 = 2.1$ newtons

The equation for the applied force is

$$F = 2.1 + 0.476\,V$$

where:

$$F = \text{the applied force, newtons}$$
$$V = \text{the velocity, meters/second}$$

Mechanical Capacitance. Mechanical capacitance is defined as the change in the displacement of a spring required to make a unit change in force. The reciprocal of the capacitance is called the spring constant, K.

MECHANICAL CAPACITANCE

$$C = \Delta X / \Delta F = 1/K \qquad \textbf{(11.50)}$$

where:

$C = $ the mechanical capacitance, meters/newton

$K = $ the spring constant, newtons/meter

$\Delta X = $ the change in deflection, meters

$\Delta F = $ the corresponding change in force, newtons

Some useful conversion factors:

1 pound per inch = 175 newtons per meter
1 pound per foot = 14.6 newtons per meter
1 inch per pound = 0.0057 meters per newton
1 foot per pound = 0.0685 meters per newton

Example 11.19 A spring is to be used to provide mechanical capacitance in a system. A force of 100 newtons compresses the spring by 30 centimeters. Determine the mechanical capacitance.

Solution:

Equation (47) may be used to determine C.

$$C = \triangle X / \triangle F = 0.30 / 100 = 0.003 \text{ meters per newton}$$

Mechanical Inertia (Mass.) Mechanical inertia is measured in terms of the force required to produce a unit change in acceleration. It is defined by Newton's law of motion, and the term *mass* is used for the inertia element.

MECHANICAL INERTIA (MASS)

$$F = M \frac{\triangle v}{\triangle t} \qquad\qquad (11.51)$$

or

$$F = M \frac{dv}{dt} \qquad\qquad (11.52)$$

where:

M = the mass, kilograms
F = the force, newtons
$\triangle v$ = a change in velocity produced by F, meters/second
$\triangle t$ = the time required for $\triangle v$ to occur, seconds
$\dfrac{\triangle v}{\triangle t}$ = the average acceleration during interval $\triangle t$, meters/second squared
$\dfrac{dv}{dt}$ = the instantaneous acceleration, meters/second squared

---

Example 11.20 Automobile A has a mass of 1500 kilograms. Determine the force required to accelerate A from 0 to 27.5 meters/second in 6 seconds. Automobile B requires a force of 8000 newtons to accelerate from 0 to 27.5 meters/second in 6 seconds. Determine the mass of B.

Solution:

Equation (11.51) may be used for both problems.

$$F = M\frac{\Delta v}{\Delta t}$$

Automobile A:

$$F = (1500)\,(27.5)/6 = 6875 \text{ newtons}$$

Automobile B:

$$8000 = M(27.5 - 0)/6$$
$$M = (8000)\,(6)/(27.5)$$
$$M = 1745 \text{ kilograms}$$

Solid Flow Rate Dead Time. A belt conveyor is frequently used to transport solid material in a process (see Figure 11.10). The dead-time lag is the time it takes for the belt to travel from the inlet end to the outlet end. Equation (11.7) may be used to compute the dead-time lag The effect of the dead time is to delay the input flow rate by the dead-time lag (t_d).

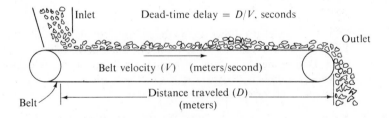

FIGURE 11.10
A SOLID FLOW, DEAD-TIME ELEMENT—A BELT CONVEYOR

Example 11.21 a. A 30-meter long belt conveyor has a belt speed of 3 meters/second. Determine the dead-time lag between the input and output ends of the belt.

b. Write the equation for the output mass flow rate $f_o(t)$ in terms of the input mass flow rate $f_i(t)$.

Solution:

a. Equation (11.7) may be used.

$$t_d = D/V = 30/3 = 10 \text{ seconds}$$

b. The following equation from Section 11.2 may be used:

$$f_o(t) = f_i(t - t_d)$$

EXERCISES

11-1. An electrical component has a linear volt-ampere graph. An applied voltage of 40 volts produces a current of 0.6 amperes. Determine the resistance of the component.

11-2. The following data were obtained in a test of a non-linear electrical resistor.

Volts	0	5	10	15	20	25
Amperes	0	0.363	0.63	0.87	1.10	1.31

Plot a volt-ampere graph. Determine the increment of current ($\triangle I$) produced by each 5-volt increment of voltage. Use these increments to determine the resistance when $E = 2.5, 7.5, 12.5, 17.5,$ and 22.5 volts (i.e., use $R = \triangle E/\triangle I$). Use the calculated values of R to plot an R-vs.-E curve.

11-3. Find the resistance of a control valve when $Q = 1.1$ cubic meters/second and $C_v = 16$.

11-4. Oil at 15° Celsius flows through smooth tubing 0.0127 meters in diameter with a velocity of 2.7 meters/second. The line is 8 meters long. Determine if the flow is laminar or turbulent by calculating the Reynolds number. Select the correct equations and calculate the resistance and pressure drop of the tube.

11-5. Kerosene at 15° Celsius flows through smooth tubing 0.0254 meters in diameter with a velocity of 11.3 meters/second. The line is 17 meters long. Determine if the flow is laminar or turbulent and calculate the resistance and pressure drop of the tube.

11-6. Nitrogen gas at 15° Celsius flows through a commercial steel pipe 2.54 centimeters in diameter and 32 meters long. The outlet pressure (p_2) is 1.2×10^5 newtons/square meter. Assume that the large pressure drop gas flow equation applies. Determine the inlet pressure p_1 for mass flow rates of 0.2, 0.4, 0.6, 0.8 and 1 kilogram per second. Plot a graph of p_1-vs.-w and find the resistance at $w = 0.8$ kilogram per second from the slope of the curve.

11–7. In Figure 11.8, the inside wall section is an aluminum plate 0.8 centimeter thick. The outside wall section is a 3-centimeter-thick layer of insulation. The inside air velocity is 0.35 meter/second. The outside air velocity is 5.2 meters/second. The wall dimensions are: $a = 4$ meters and $b = 2$ meters. Determine the thermal resistance of the wall section.

11–8. Determine the thermal resistance of each of the following film conditions for an area of 2.2 square meters.

   a. Natural convection in still air on a vertical surface where $T_d = 32$ K.
   b. Natural convection in still water where $T_d = 24$ K and $T_w = 36°$ C.
   c. Natural convection in oil where $T_d = 64$ K.
   d. Forced convection in air with a velocity of 9.4 meters/second.
   e. Forced convection in water with a water temperature of 28° C and a velocity of 3.4 meters/second.

11–9. A dashpot operates in the laminar flow region. A test is conducted in which a force of 210 newtons produces a velocity of 2 meters/second. Determine the mechanical resistance (R_m).

11–10. A mechanical system contains a sliding load (coulomb friction) and a shock absorber (viscous friction). A force of 22 newtons produces a velocity of 3 meters/second, and 68 newtons produces a velocity of 16 meters/second. Determine the resistance (R) and the coulomb friction force (F_c). Write an equation for the applied force (F) in terms of the velocity (V).

11–11. An electric current of 0.3 milliampere is applied to a capacitor for a duration of 1.0 seconds. The voltage across the capacitor is increased from 0 to 18 volts by the current pulse. Determine the capacitance (C) of the capacitor.

11–12. A liquid tank has a diameter of 2.3 meters and a height of 4.1 meters. Determine the capitance of the tank for each of the following fluids.
   a. Water
   b. Oil
   c. Kerosene
   d. Gasoline
   e. Turpentine

11–13. A pressure tank has a volume of 2.1 cubic meters. Determine the capacitance of the tank for each of the following gases.
   a. Nitrogen at 40°C
   b. Carbon dioxide at 30°C
   c. Oxygen at 20°C
   d. Carbon monoxide at 50°C

11–14. Determine the thermal capacitance of 3.75 cubic meters of ethyl alcohol.

11–15. A spring is used to provide mechanical capacitance. A force of 2000 newtons compresses the spring by 0.12 meter. Determine the mechanical capacitance.

11–16. A voltage pulse with an amplitude of 11.3 volts and a duration of 0.12 second is applied to an inductor. The current through the inductor is increased from 0 to 0.85 amperes by the voltage pulse. Assume that the resistance of the inductor is negligible and determine the inductance (L).

11–17. A mechanism consists of a linear acceleration cam. The cam has two sections: a and b. Section a accelerates the load from 0 to 0.65 meter/second in 0.24 second. Section b deaccelerates the load from 0.65 to 0 meter/second in 0.18 second. Determine the inertial force in a and b if the load has a mass of 3.8 kilograms.

11–18. Determine the liquid flow inertance of oil in a pipe which has a diameter of 0.0762 meter and a length of 1000 meters.

11–19. Determine the dead-time lag of a 20,000-meter-long transmission line if the velocity of propagation is 2.7×10^8 meters/second.

11–20. Determine the dead-time lag of a belt conveyor 18 meters long. The belt speed is 2.2 meters/second. Write an equation for the output mass flow rate $f_o(t)$ in terms of the input mass flow rate $f_i(t)$.

11–21. Determine the dead-line lag in a pipe which is 61 meters long and has a diameter of 2.1 centimeters. The flow rate is 0.00183 cubic meter per second.

12

Process
Characteristics

12.1 Introduction

A process or component is characterized by the relationship between the input signal and the output signal. It is this input/output (I/O) relationship that determines the design requirements of the controller. If the I/O relationship of the process is completely defined, the designer can specify the optimum controller parameters. If the I/O relationship is poorly defined, then the designer must provide a large adjustment of the controller parameters so the optimum settings can be determined during start-up of the system. The objective of this chapter is to provide the information necessary to determine the I/O relationship of the following types of processes:

a. The integral or ramp process (Section 12.2).
b. The first-order lag process (Section 12.3).
c. The dead-time process (Section 12.4).
d. The second-order lag process (Section 12.5).
e. The first-order lag plus dead-time lag process (Section 12.6).

The I/O relationship of a process may be defined by any or all of the following.

a. The step-response graph.
b. The time-domain equation.
c. The analog-computer diagram.
d. The transfer function.
e. The frequency-response graph.

The *step-response graph* is the time graph of the output signal following a step change in the input signal from one value to another. Figures 12.1 and 12.5 are examples of step-response graphs. The *time-domain equation* expresses the size-vs.-time relationship between the input signal and the output signal. They are expressed in terms of the basic elements defined in Chapter Eleven, and frequently contain integral or derivative terms. Some examples of time-domain equations are included in Figures 12.3, 12.7, 12.13, and 12.16. The *analog-computer diagram* may be obtained from the time-domain equation as described in Chapter Eight. Figures 12.2 and 12.6 illustrate analog-computer diagrams that describe two different processes. The *transfer function* of a process is obtained by transforming the time-domain equation into the frequency-domain algebraic equation and then solving for the output H over the input M. The transfer function defines the changes in the size and timing between the input and output signals at each frequency. The *frequency-response graphs* are another way of describing the size- and timing-vs.-frequency relationship. There are two graphs: one gives the size relationship between the input signal and the output signal at each frequency. The other gives the timing relationship between the input signal and the output signal at each frequency.

The characteristics of a process depend on its basic elements (resistance, capacitance, inertance, or dead time), not on the type of process (thermal, electrical, mechanical, etc.). Two completely different types of processes may have the same characteristics. That is, they may have the same step response, defining equations, analog-computer diagrams, transfer functions, and frequency response. This is a definite advantage, because an understanding of each characteristic may be extended to all types of systems with the same characteristic. The format used to present the basic elements is continued in this chapter. Each section begins with a general description of the step-response graph, time-domain equation, analog-computer diagram, and transfer function of the process. The frequency response is covered separately in Chapter Thirteen. The equations for each type of process are included in a box with all terms and their units. An example calculation for each type of system is provided for use as a guide in calculating the parameters that characterize the process.

12.2 The Integral or Ramp Process

General Description. The integral or ramp process is a single-capacity process in which the outflow of material or energy is independent of the amount of material or energy stored in the capacitive element. The quantity of stored material or energy remains constant only if the inflow rate is exactly equal to the outflow rate. If the inflow rate is greater than the outflow rate, the quantity stored will increase at a rate proportional to the difference. If the inflow rate is less than the outflow rate, the quantity stored will decrease at a rate proportional to the difference.

The step response of an integral process is illustrated in Figure 12.1. Before the step change, the input flow rate is equal to the output flow rate and the level is maintained constant. After the step change, the input flow rate is greater than the output flow rate, and the level increases at a constant rate. The term *ramp process* is derived from the ramp-like shape of the output response graph. The step response of an integral process is measured by the *integral time,* which is the number of seconds (or minutes) required for the output to reach the same percentage change as the input (see Figure 12.1).

A liquid tank will be used as an example to develop the time-domain equation of the integral process. The change in level is equal to the change in the volume of liquid in the tank divided by the cross-sectional area of the tank at the liquid surface, A. The change in volume of liquid in the tank during a time interval from t_1 to t_2 is equal to the average

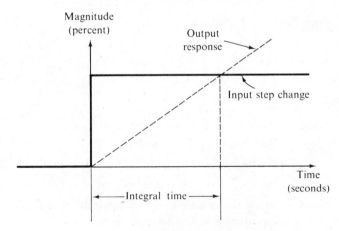

FIGURE 12.1

THE STEP RESPONSE OF AN INTEGRAL PROCESS

difference between the input flow rate (m) and the output flow rate (q) times the time interval $(t_2 - t_1)$.

$$\text{change in level} = \frac{1}{A}(m - q)_{avg} (t_2 - t_1)$$

In mathematics, the quantity $(m - q)_{avg} (t_2 - t_1)$ is equal to the integral of the difference $(m - q)$ over the interval from t_1 to t_2. That is, the integral of $(m - q)$ over the interval from t_1 to t_2 is equal to the change in volume in the tank during the interval from t_1 to t_2. The integral is designated by the following symbol.

$$\text{integral of } (m - q) \text{ from } t_1 \text{ to } t_2 = \int_{t_1}^{t_2} (m - q)dt$$

The level at any time t_2 is equal to the level at any previous time t_1 plus the change in level during the time interval from t_1 to t_2. The symbol $h(t_2)$ is used to represent the level at time t_2, and $h(t_1)$ represents the level at time t_1. Equation (12.1) uses the symbols $h(t_2)$, $h(t_1)$, and $\int(m - q)dt$ to define the relationship between the level h, the input flow rate m, and the output flow rate q.

THE INTEGRAL PROCESS

The Time-Domain Equation

$$h(t_2) = \frac{1}{A} \int_{t_1}^{t_2} (m - q)dt + h(t_1) \qquad \textbf{(12.1)}$$

where:

$h(t_2)$ = the level at time t_2, meters

$h(t_1)$ = the level at time t_1, meters

t_1 = the beginning of the time interval, seconds

t_2 = the end of the time interval, seconds

A = the cross-sectional area of the tank at the liquid interface, square meters

m = the input flow rate, cubic meters/second

q = the output flow rate, cubic meters/second

The Transfer Function

$$\frac{H(S)}{M(S)} = \frac{1}{AS} \qquad\qquad (12.2)$$

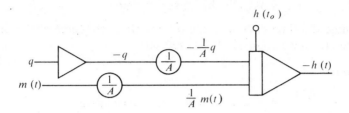

$$h(t) = \frac{1}{A}\int_{t_o}^{t}[m(t) - q]\,dt + h(t_o)$$

FIGURE 12.2

THE ANALOG-COMPUTER DIAGRAM OF AN INTEGRAL PROCESS

Example 12.1   A liquid-level integral process   A liquid-level integral process is shown in Figure 12.3. A positive displacement pump provides a constant output flow rate, q. Equation (12.1) defines the relationship between the tank level, $h(t)$, and the input flow rate, $m(t)$. The system has the following parameters.

$$h(t) = \frac{1}{A}\int_{t_o}^{t}[m(t) - q]\,dt + h(t_o)$$

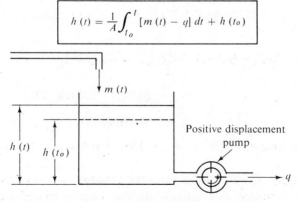

FIGURE 12.3

A LIQUID-LEVEL INTEGRAL PROCESS

tank height $= 4$ meters
tank diameter $= 1.5$ meters
level at time $t_o = h(t_o) = 0.9$ meter
output flow rate $= q = 0.006$ cubic meter/second
input flow range; 0 to 0.01 cubic meters/second
average input flow rate between t_o and $t_o + 100$ seconds $=$
    $m_{avg} = 0.008$ cubic meter/second

Determine the time domain equation, the transfer function, the integral time, and the level at time $t_o + 100$ seconds.

Solution:

$$A = \pi(1.5)^2/4 = 1.76 \text{ square meters}$$

$$h(t) = \frac{1}{1.76} \int_{t_o}^{t} [m(t) - 0.006]dt + h(t_o) \qquad \text{time-domain equation}$$

$$\frac{H(S)}{M(S)} = \frac{1}{1.76S} \qquad \text{transfer function}$$

The integral time is the number of seconds required for the level to reach the same percentage change as the input. Any convenient percentage change may be used for the calculation.

A 10 per cent change in input $= 0.001$ cubic meter/second
A 10 per cent change in level $= 0.4$ meter

Let $T_i =$ the integral time

$$(0.001)T_i = 0.4(1.76)$$

$$T_i = (0.4)(1.76)/0.001 = 704 \text{ seconds} \qquad \text{the integral time}$$

$$\text{Change in level} = \frac{1}{A}(m_{avg} - q)(t_o + 100 - t_o)$$

$$= \frac{1}{1.76}(0.008 - 0.006)(100)$$

$$= 0.114 \text{ meters}$$

The level at time $t_o + 100 = h(t_o + 100) = 0.9 + 0.114 = 1.014$ meter

Example 12.2 A sheet-loop integral process A *sheet-loop* integral process is illustrated in Figure 12.4. The loop is placed between the output of one processing stage and the input of the next stage. It provides a tension-free means of matching the sheet velocity in the two processing units. The supply-belt velocity, $v_i(t)$, is the input flow rate; the take-away belt velocity, $v_o(t)$ is the output flow rate. Since the change in the loop height, $d(t)$, is one-half of the change in length of sheet in the loop, $A = 2$. The time-domain equation is given in Figure 12.4.

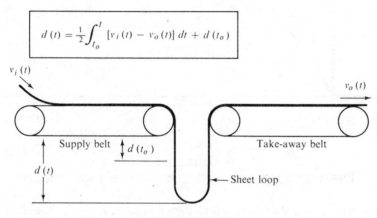

$$d(t) = \frac{1}{2}\int_{t_o}^{t} [v_i(t) - v_o(t)]\, dt + d(t_o)$$

FIGURE 12.4
A SHEET-LOOP INTEGRAL PROCESS

12.3   The First-Order Lag Process

General Description. The first-order lag process is a single-capacity process in which the output flow rate is proportional to the level. For each input flow rate, there is a corresponding level that will produce an output flow rate equal to the input. The first order lag process is a self-regulating process because it automatically produces an output flow rate to match each input flow rate. In contrast, the integral process is a non-self-regulating process.

The step response of the first-order lag process is shown in Figure 12.5. Before the step change, the input flow rate is equal to the output flow rate and the level is maintained constant. The step change consists of increasing the input flow rate. Let M represent the percentage increase in the input flow rate. The output flow rate is proportional to the level which does not change immediately. The input is greater than the output, so the level will increase at a rate proportional to the difference. As the level increases, the difference between the input and the output decreases. This in turn reduces the rate at which the level increases. The result is the output response curve of Figure 12.5.

The step response of a first-order lag is measured by the *time constant,* which is the number of seconds (or minutes) required for the output to

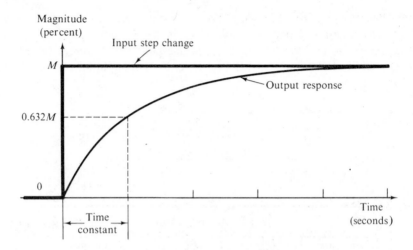

FIGURE 12.5

THE STEP RESPONSE OF A FIRST-ORDER LAG PROCESS

reach 63.2 per cent of the total change. During each additional interval equal to the time constant, the output will reach 63.2 per cent of the remaining change. For example, during the second time interval, the output will increase by $0.632(0.368M) = 0.232M$. Thus, after two time constants, the output will reach $0.632M + 0.232M = 0.864M$. After five time constants, the output will reach $0.993M$.

The time-domain equation of a first-order lag liquid system was developed in Section 4.2 and is given below.

THE FIRST-ORDER LAG PROCESS

The Time-Domain Equation

$$\tau\frac{dh}{dt} + h = Gm \qquad (12.3)$$

where:

τ = the time constant, seconds

h = the output signal

$\dfrac{dh}{dt}$ = the rate of change of the output signal

m = the input signal

G = a gain constant that depends on the type of system

The Transfer Function

$$\frac{H(S)}{M(S)} = \frac{G}{\tau S + 1} \qquad\qquad (12.4)$$

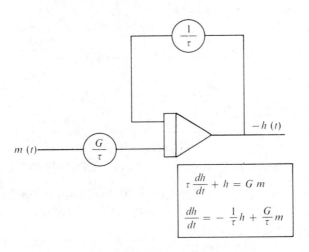

FIGURE 12.6

*THE ANALOG-COMPUTER DIAGRAM OF A
FIRST-ORDER LAG PROCESS*

Example 12.3 A liquid-level, first-order, lag process A liquid-level, first-order, lag process is illustrated in Figure 12.7. The time-domain equation is given by Equation (12.5).

$$RC\frac{dh}{dt} + h = \left(\frac{R}{\rho g}\right)m \qquad\qquad (12.5)$$

where;

    R = the liquid resistance, newton-seconds/meter[5]

    C = the liquid capacitance, $\dfrac{\text{cubic meters}}{\text{newtons per square meter}}$

    h = the tank level, meters

    m = the input flow rate, cubic meters/second

    ρ = the liquid density, kilograms/cubic meter

    g = 9.81 meters/second squared

An oil tank similar to Figure 12.7 has a diameter of 1.83 meters and a height of 10 meters. The outlet at the bottom has a length of

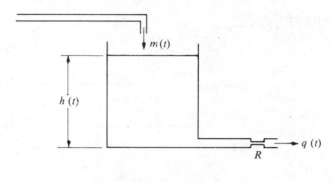

$$RC \frac{dh}{dt} + h = \left(\frac{R}{\rho g}\right) m$$

FIGURE 12.7

A LIQUID-LEVEL, FIRST-ORDER, LAG PROCESS

1 meter and a diameter of 0.025 meter. The oil temperature is 15° C and the outlet is laminar. Determine each of the following.

a. The capacitance of the tank.
b. The resistance of the outlet.
c. The RC time constant.
d. The time-domain equation.
e. The transfer function.

Solution:

a. Equation (11.23) in Chapter Eleven may be used to compute the tank capacitance.

$$C = \frac{A}{\rho g}$$

$A = \pi D^2/4 = \pi(1.83)^2/4 = 2.62$ square meters

$\rho = 880$ kilograms/cubic meter (Appendix A, Table 2)

$g = 9.81$ meters/second^2

$C = 2.62/(880 \times 9.81) = 3.04 \times 10^{-4} \dfrac{\text{meter}^3}{\text{newton per meter}^2}$

b. Equation (11.10) in Chapter Eleven may be used to compute the outlet resistance.

$$R_L = \left[\frac{128 \mu l}{\pi D^4}\right]$$

$\mu = 0.160$ newton second/square meter

$l = 1$ meter

$D^4 = (0.025)^4 = (2.5 \times 10^{-2})^4 = 39 \times 10^{-8}$

$R_L = \dfrac{128 \times 0.160 \times 1}{\pi 39 \times 10^{-8}} = 1.67 \times 10^7$ newton-seconds/meter^5

c. The time constant, $\tau = RC$

$\tau = (1.67 \times 10^7)(3.04 \times 10^{-4}) = 5.07 \times 10^3$ seconds

$\tau = 5,070$ seconds (or 84.7 minutes)

d. The time-domain equation is obtained from Equation (12.5).

$$RC\frac{dh}{dt} + h = \left(\frac{R}{\rho g}\right) m$$

$$\frac{R}{\rho g} = \frac{1.67 \times 10^7}{880 \times 9.81} = 1.94 \times 10^3$$

The time-domain equation is

$$5.07 \times 10^3 \frac{dh}{dt} + h = 1.94 \times 10^3\, m$$

e. The transfer function is obtained from Equation (12.4).

$$\frac{H(S)}{M(S)} = \frac{1.94 \times 10^3}{5.07 \times 10^3 S + 1}$$

Example 12.4 An electrical, first-order, lag process Three *electrical,* first-order, lag circuits are shown in Figure 12.8. In circuit (a), the output is the voltage (e) across the parallel RC circuit, and the input is the current (i) provided by the current source. The current through the resistor is e/R and the current through the capacitor is $C\frac{de}{dt}$. The time-domain equation is obtained by applying Kirchhoff's current law ($i = i_R + i_C$), and then substituting the above expressions for i_R and i_C.

$$RC\frac{de}{dt} + e = Ri \qquad\qquad (12.6)$$

In circuit (b), the output is the voltage (e_o) across the resistor in a series RL circuit. The input is the voltage (e_i) provided by the voltage source. The time-domain equation is obtained by applying Kirchhoff's voltage

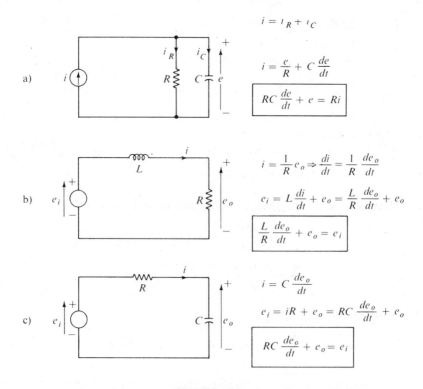

FIGURE 12.8

ELECTRICAL, FIRST-ORDER, LAG CIRCUITS

law ($e_i = e_L + e_o$), and then substituting the equivalent expression for e_L in terms of the output voltage $\left(e_L = \dfrac{L}{R}\dfrac{de_o}{dt}\right)$.

$$\frac{L}{R}\frac{de_o}{dt} + e_o = e_i \qquad\qquad (12.7)$$

In circuit (c), the output is the voltage (e_o) across the capacitor in a series RC circuit. The input is the voltage (e) provided by the voltage source. The time-domain equation is

$$RC\frac{de_o}{dt} + e_o = e_i \qquad\qquad (12.8)$$

An electrical circuit similar to Figure 12.8–c has an 8.2×10^3 ohm resistance value and a 60×10^{-6} farad capacitance value. Determine the RC time constant and the transfer function.

Solution:

a. Time constant, $\tau = RC$

$$\tau = (8.2 \times 10^3)(60 \times 10^{-6}) = 4.92 \times 10^{-1} \text{ second}$$
$$\tau = 0.492 \text{ second}$$

b. The transfer function is given by Equation (12.4).

$$\frac{E_o(S)}{E_i(S)} = \frac{1}{0.492S + 1}$$

Example 12.5 A thermal, first-order, lag process A thermal, first-order, lag process is shown in Figure 12.9. A jacketed kettle is used to heat a liquid. The mixer maintains a uniform temperature throughout the liquid. The input is the jacket temperature (θ_j). The output is the liquid temperature (θ_L). The amount of heat (Δq) transferred to the liquid depends on the resistance (R) of the wall between the steam and the liquid, the temperature difference ($\theta_j - \theta_L$) between the jacket and the liquid, and the time interval ($\triangle t$).

$$\Delta\theta_L = \frac{\Delta q}{C} = \frac{1}{RC}(\theta_j - \theta_L)\Delta t$$

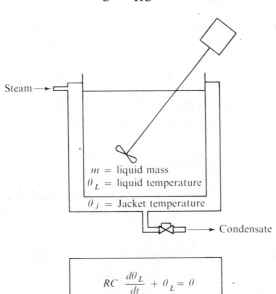

m = liquid mass
θ_L = liquid temperature
θ_j = Jacket temperature

$$RC\frac{d\theta_L}{dt} + \theta_L = \theta$$

FIGURE 12.9
A THERMAL FIRST-ORDER LAG PROCESS

$$RC\frac{\Delta\theta_L}{\Delta t} + \theta_L = \theta_j$$

or as $\Delta t \rightarrow 0$

$$RC\frac{d\theta_L}{dt} + \theta_L = \theta_j \qquad \textbf{(12.9)}$$

where:

R = the thermal resistance, Kelvin/watt

C = the thermal capacitance, joule/Kelvin

θ_L = the liquid temperature, Kelvin or ° Celsius

θ_j = the jacket temperature, Kelvin or ° Celsius

An oil-bath thermal system similar to Figure 12.9 has an inside diameter of 1 meter and a height of 1.2 meters. The inside film coefficient is 62 watts/meter²-Kelvin. The wall resistance and outside film coefficient are negligible. Determine each of the following: the thermal resistance, the thermal capacitance, the time constant, the time-domain equation, and the transfer function.

Solution:

a.   Equations (11.34) and (11.36) in Chapter Eleven may be used to compute the thermal resistance. Let h_f = film coefficient and h = height.

$$R = \frac{1}{Ah_f}$$

$$A = \pi D^2/4 + \pi Dh = \pi D(D/4 + h)$$

$$A = \pi(1)(1/4 + 1.2) = \pi(1.45) = 4.55 \text{ meters}^2$$

$$R = \frac{1}{Ah_f} = \frac{1}{(4.55)(62)} = 3.55 \times 10^{-3} \text{ Kelvin/watt}$$

b.   Equation (11.46) in Chapter Eleven may be used to compute the thermal capacitance.

$$C = mS_h$$

$$S_h = 2180 \text{ joule/kilogram-Kelvin (Appendix A)}$$

$$\rho = 880 \text{ kilograms/cubic meter (Appendix A)}$$

$$m = \rho h\pi D^2/4 = (880)(1.2)(\pi)(1)^2/4$$

$$m = 830 \text{ kilograms}$$

$$C = (830)(2180) = 1.81 \times 10^6 \text{ joule/Kelvin}$$

c.   Time constant $= \tau = RC$

$$\tau = (3.55 \times 10^{-3})(1.81 \times 10^6)$$

$$\tau = 6.43 \times 10^3 \text{ seconds (or 107 minutes)}$$

d.   Equation (12.9) may be used to determine the time-domain equation.

$$6.43 \times 10^3 \frac{d\theta_1}{dt} + \theta_1 = \theta_j$$

e.   The transfer function is

$$\frac{\theta_1(S)}{\theta_j(S)} = \frac{1}{6.43 \times 10^3 S + 1}$$

Example 12.6   A gas-pressure, first-order, lag process   The *gas-pressure*, first-order, lag process is shown in Figure 12.10. This process is almost identical to the liquid flow process. The input to the process is the inlet gas flow rate (m). The output is the gas pressure in the vessel (P). The rate of change of the pressure in the vessel is equal to the difference between the input and output flow rates $(m - q)$ divided by the gas capacitance (C).

$$\frac{dP}{dt} = \frac{m - q}{C}$$

The output flow rate q is equal to the pressure (P) divided by the outlet resistance (R).

$$q = P/R$$

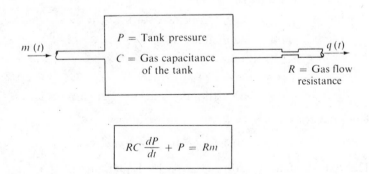

FIGURE 12.10
A GAS-PRESSURE, FIRST-ORDER, LAG PROCESS

A substitution of P/R for q in the differential equation results in the time-domain equation for the gas flow process.

$$RC\frac{dP}{dt} + P = Rm \qquad\qquad (12.10)$$

where:

$\quad R =$ the gas flow resistance, newton-seconds/kilogram-meter^2

$\quad C =$ the gas capacitance, kilograms/newton per square meter

$\quad P =$ the pressure in the vessel, newtons/square meter

$\quad m =$ the input flow rate, kilograms/second

A carbon dioxide (CO_2) gas-pressure process similar to Figure 12.10 has the following parameters.

$\qquad$ Pressure vessel volume: 1.4 cubic meters
$\qquad$ Temperature: 530 Kelvin
$\qquad$ Gas flow resistance: $2 \times 10^5 \dfrac{\text{newton-second}}{\text{kilogram-meter}^2}$

Determine each of the following:

a. $\quad$ The capacitance of the pressure vessel.
b. $\quad$ The time constant.
c. $\quad$ The time-domain equation.
d. $\quad$ The transfer function.

Solution:

a. $\quad$ Equation (11.32) in Chapter Eleven may be used to determine the capacitance of the pressure vessel.

$$\begin{aligned}
C &= 1.2 \times 10^{-4} MV/T \\
M &= 44 \text{ for } CO_2 \text{ (Appendix A)} \\
V &= 1.4 \text{ cubic meters} \\
T &= 530 \text{ Kelvin} \\
C &= (1.2 \times 10^{-4})(44)(1.4)/530 \\
&= 1.4 \times 10^{-5} \text{ kilogram/newton per meter}^2
\end{aligned}$$

b. $\quad$ The time constant, $\tau = RC$

$$\tau = (2 \times 10^5)(1.4 \times 10^{-5}) = 2.8 \text{ second}$$

c. $\quad$ The time-domain equation is given by Equation (12.10).

$$RC\frac{dP}{dt} + P = Rm$$

$$2.8\frac{dP}{dt} + P = 2 \times 10^5 m$$

d.   The transfer function is

$$\frac{P(S)}{M(S)} = \frac{2 \times 10^5}{2.8S + 1}$$

Example 12.7   A blending, first-order, lag process   A blending, first-order, lag process is illustrated in Figure 12.11. A constant flow rate of Q cubic meters per second passes through the tank. The input signal is the concentration (c_i) of component A in the incoming fluid. The output signal is the concentration (c_o) of component A in the fluid in the tank (and the outgoing fluid). A material balance is used to determine the time-domain equation. In a time interval Δt, an amount of liquid equal to $Q(\Delta t)$ is added at the inlet, and an equal amount is removed at the outlet. The amount of component A added to the tank in the inlet fluid is equal to $Q(\Delta t)c_i$. The amount of component A removed in the outgoing fluid is $Q(\Delta t)c_o$. The difference is the increase in component A in the tank which is equal to $V(\Delta c_o)$.

$$\text{amount of buildup} = \text{amount inputed} - \text{amount removed}$$

$$V(\Delta c_o) = Q(\Delta t)c_i - Q(\Delta t)c_o$$

$$\frac{V}{Q}\frac{\Delta c_o}{\Delta t} + c_o = c_i$$

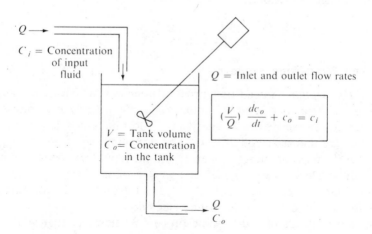

$Q \longrightarrow$

C_i = Concentration of input fluid

Q = Inlet and outlet flow rates

$$\left(\frac{V}{Q}\right)\frac{dc_o}{dt} + c_o = c_i$$

V = Tank volume
C_o = Concentration in the tank

Q
C_o

FIGURE 12.11

A BLENDING FIRST-ORDER LAG PROCESS

or as $\triangle t \rightarrow 0$

$$\left(\frac{V}{Q}\right)\frac{dc_o}{dt} + c_o = c_i \qquad\qquad \textbf{(12.11)}$$

where:

V = the volume of the tank, cubic meters

Q = the liquid flow rate, cubic meters/second

c_o = the concentration in the tank

c_i = the concentration in the incoming fluid

A blending tank similar to Figure 12.11 has a tank volume of 3.1 cubic meters and a flow rate Q of 0.0031 cubic meter per second. Determine the time constant, time-domain equation, and transfer function.

Solution:

a.   The time constant, $\tau = V/Q$

$$\tau = 3.1/0.0031 = 1000 \text{ seconds}$$

b.   The time-domain equation is given by Equation (12.11).

$$10^3\frac{dC_o}{dt} + C_o = C_i$$

c.   The transfer function is

$$\frac{C_o(S)}{C_i(S)} = \frac{1}{10^3 S + 1}$$

12.4   The Dead-Time Process

General Description. A dead-time process is one in which mass or energy is transported from one point to another. The output signal is identical to the input signal except for a time delay. The time delay is called the *dead-time lag* and is denoted by t_d. The dead-time lag is the time required for the mass or energy to travel from the input location to the output location. Dead time was included as one of the basic elements in Chapter Eleven (see Sections 11.2, 11.3, 11.4 and 11.7) for further discussion and examples of dead-time elements.

The step response of a dead-time process is shown in Figure 12.12. Before the step change, the output signal and the input signal are equal. The step change increases the input signal to a new value at time $t = 0$ seconds. The output signal remains at the original value until time $t = t_d$

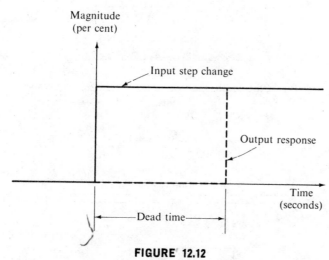

FIGURE 12.12

THE STEP RESPONSE OF A DEAD-TIME PROCESS

when it also increases to the new value. The graph of the output signal is a duplication of the input graph moved to the right by t_d seconds.

The step response of a dead-time process is measured by the dead time lag—i.e., the number of seconds (or minutes) that elapse between an input change and the corresponding output change. The time-domain equation of the dead-time process was developed in Section 11.2. Figure 12.13 illustrates two examples of dead-time processes.

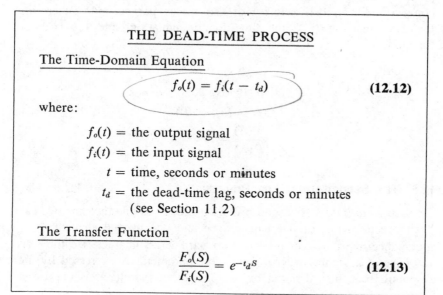

THE DEAD-TIME PROCESS

The Time-Domain Equation

$$f_o(t) = f_i(t - t_d)$$ **(12.12)**

where:

$f_o(t)$ = the output signal

$f_i(t)$ = the input signal

t = time, seconds or minutes

t_d = the dead-time lag, seconds or minutes (see Section 11.2)

The Transfer Function

$$\frac{F_o(S)}{F_i(S)} = e^{-t_d S}$$ **(12.13)**

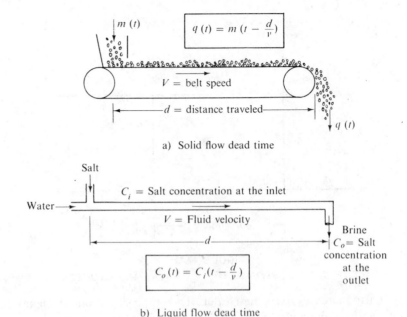

a) Solid flow dead time

b) Liquid flow dead time

FIGURE 12.13
SOLID AND LIQUID FLOW DEAD-TIME PROCESSES

Example 12.8 A solid-flow, dead-time, lag process A dead-time process similar to Figure 12.13–a consists of a 12-meter long belt conveyor with a belt velocity of 0.6 meter per second. Determine the dead-time lag (t_d), the time-domain equation, and the transfer function.

Solution:

$$t_d = d/V = 12/0.6 = 20 \text{ seconds}$$
$$f_o(t) = f_i(t - 20)$$
$$\frac{F_o(S)}{F_i(S)} = e^{-20S}$$

12.5 The Second-Order Lag Process

General Description. A second order lag process is one that has two capacitance elements, or one capacitance and one inertial element (i.e., inductance, inertance, or mass). Two parameters characterize the step response of a second-order system. One parameter is expressed by the resonant frequency, denoted by ω_o. The other parameter is expressed either by the damping coefficient, denoted by α, or the damping ratio, denoted by ζ. The damping ratio is simply the damping coefficient divided by the resonant frequency ($\zeta = \alpha/\omega_o$).

The time-domain equation and transfer function of a second-order system are given by Equations (12.14) and (12.15), below. Two versions of the time-domain equation are given, one based on the damping coefficient, the other based on the damping ratio. A second version of the transfer function based on the damping ratio is not shown. The equations are arranged in a way that will be useful in later developments, i.e., the coefficient of h is set equal to one.

THE SECOND-ORDER PROCESS

The Time-Domain Equation

$$\left(\frac{1}{\omega_o^2}\right)\frac{d^2h}{dt^2} + 2\left(\frac{\alpha}{\omega_o^2}\right)\frac{dh}{dt} + h = Gm \qquad (12.14)$$

or

$$\left(\frac{1}{\omega_o^2}\right)\frac{d^2h}{dt^2} + 2\left(\frac{\zeta}{\omega_o}\right)\frac{dh}{dt} + h = Gm$$

where:

ω_o = the resonant frequency, radians/second

α = the damping coefficient, 1/second

h = the input signal

m = the output signal

G = a gain constant

$\zeta = \alpha/\omega_o$ = the damping ratio

The Transfer Function

$$\frac{H(S)}{M(S)} = \frac{G}{\left(\frac{1}{\omega_o^2}\right)S^2 + 2\left(\frac{\alpha}{\omega_o^2}\right)S + 1} \qquad (12.15)$$

The step response of a second-order system is divided into two regions depending on the value of the damping ratio, ζ. If the damping ratio is greater than one ($\alpha > \omega_o$), the response is overdamped as shown in Figure 12.14–a. If the damping ratio is less than one ($\alpha < \omega_o$), the response is underdamped as shown in Figure 12.14–c. The two regions are

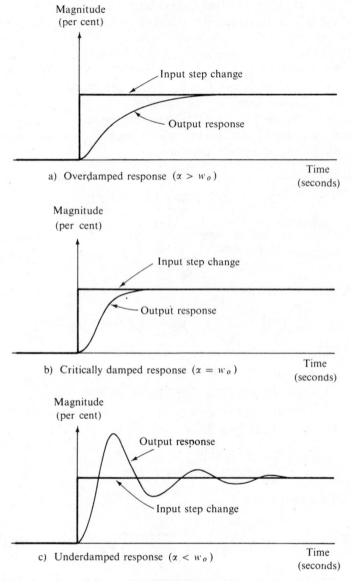

a) Overdamped response ($\alpha > w_o$)

b) Critically damped response ($\alpha = w_o$)

c) Underdamped response ($\alpha < w_o$)

FIGURE 12.14

THE STEP RESPONSE OF A SECOND-ORDER PROCESS

separated by the critically damped response which occurs when the damping ratio is equal to one ($\alpha = \omega_o$). The critically damped response is shown in Figure 12.14–b.

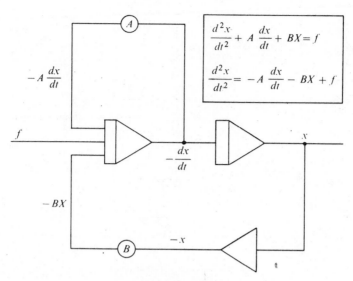

$$\frac{d^2x}{dt^2} + A\frac{dx}{dt} + BX = f$$

$$\frac{d^2x}{dt^2} = -A\frac{dx}{dt} - BX + f$$

FIGURE 12.15

THE ANALOG-COMPUTER DIAGRAM OF A SECOND-ORDER SYSTEM

The analog-computer diagram of a second-order system is shown in Figure 12.15. Coefficient A is equal to 2α and coefficient B is equal to ω_o^2.

The spring-mass-damping system shown in Figure 12.16 will be used to illustrate the development of the time-domain equation of a second-order system. Three forces act on the mass (M); the force exerted by the spring (F_s), the force exerted by the damping piston (F_r), and the externally applied force (F). The sum of these three forces will cause the mass to accelerate according to Newton's law of motion.

$$\sum F = F + F_s + F_r = M\frac{d^2x}{dt^2}$$

The spring force (F_s) is equal to minus the spring constant (K) multiplied by the distance the spring is compressed or extended (x). The negative sign indicates that the direction of the force is down when x is positive and up when x is negative.

$$F_s = -Kx$$

The damping force is equal to minus the dashpot resistance times the velocity of the dashpot piston. The piston velocity is represented mathematically by the derivative of x with respect to time (i.e., dx/dt). The negative sign indicates that the force is down when the piston velocity

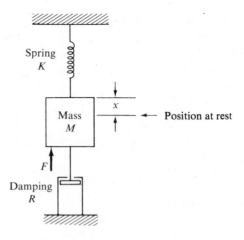

$$M \frac{d^2 x}{d t^2} + R \frac{dx}{dt} + Kx = F$$

FIGURE 12.16

A SPRING-MASS-DAMPING SECOND-ORDER SYSTEM

is positive (moving up) and the force is up when the piston velocity is negative.

$$F_r = - R\frac{dx}{dt}$$

Substituting $-Kx$ for F_s and $-R \, dx/dt$ for F_r into the first equation and rearranging the terms gives the time-domain equation (12.16).

$$F - Kx - R\frac{dx}{dt} = M \frac{d^2x}{dt^2}$$

$$\frac{M}{K}\frac{d^2x}{dt^2} + \frac{R}{K}\frac{dx}{dt} + x = \frac{1}{K} F \qquad \textbf{(12.16)}$$

where:

M = the mass of the moving body, kilograms

R = the dashpot resistance, newton-seconds/meter

K = the spring constant, newtons/meter

F = the input force, newtons

x = the position of the moving body, meters

Equation (12.16) is the time-domain equation of the spring-mass-damping second-order system. In Equation (12.16), F is the input signal and x is the output signal. The damping coefficient (α) is equal to $R/(2M)$, and the resonant frequency (ω_o) is equal to $\sqrt{K/M}$.

Example 12.9 A spring-mass-damping, second-order process A spring-mass-damping system consists of a 10-kilogram mass, a spring constant of 1000 newtons/meter, and a damping resistance of 20 newtons-seconds per meter. Determine the time-domain equation, the transfer function, the resonant frequency (ω_o), the damping coefficient (α), and the type of damping the system has.

Solution:

a. The time-domain equation is given by Equation (12.16).

$$\frac{10}{1000}\frac{d^2x}{dt^2} + \frac{20}{1000}\frac{dx}{dt} + x = \frac{1}{1000}F$$

or

$$0.01\frac{d^2x}{dt^2} + 0.02\frac{dx}{dt} + x = 0.001\,F$$

b. The transfer function is

$$\frac{X(S)}{F(S)} = \frac{0.001}{0.01S^2 + 0.02S + 1}$$

c. From Equation (12.14), $1/\omega_o^2 = 0.01$

$$\omega_o^2 = 100$$
$$\omega_o = 10$$

d. From Equation (12.14), $2\alpha/\omega_o^2 = 0.02$

$$\alpha = 0.01\,\omega_o^2$$
$$\alpha = 0.01(100)$$
$$\alpha = 1 \quad 1/\text{second}$$

e. $\alpha = 1$ and $\omega_o = 10$; therefore $\alpha < \omega_o$, and the step response is underdamped (see Figure 12.14).

Example 12.10 An electrical series RLC, second-order process
An *electrical series RLC circuit* is illustrated in Figure 12.17. Also a second-order system this is an analog of the mechanical system just mentioned. The time-domain equation is

$$LC\frac{d^2e_o}{dt^2} + RC\frac{de_o}{dt} + e_o = e_i \qquad \textbf{(12.17)}$$

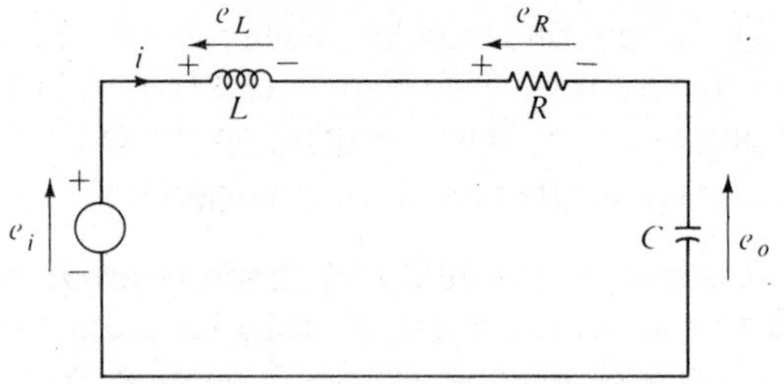

$$i = C \frac{de_o}{dt}$$

$$e_R = Ri = RC \frac{de_o}{dt}$$

$$e_L = L \frac{di}{dt} = L \frac{d}{dt}\left(C \cdot \frac{de_o}{dt}\right) = LC \frac{d^2 e_o}{dt}$$

$$e_L + e_R + e_o = e_i$$

$$\boxed{LC \frac{d^2 e_o}{dt^2} + RC \frac{de_o}{dt} + e_o = e_i}$$

FIGURE 12.17

AN ELECTRICAL SECOND-ORDER SYSTEM

where:

$$e_i = \text{the input signal, volts}$$
$$e_o = \text{the output signal, volts}$$
$$L = \text{the inductance, henrys}$$
$$C = \text{the capacitance, farads}$$
$$R = \text{the resistance, ohms}$$

An electrical series RLC circuit consists of a 0.022-henry inductor, a 10-microfarad capacitor, and a 200-ohm resistance. Determine the time-domain equation, the transfer function, the resonant frequency, the damping coefficient, and the type of damping.

Solution:

a. The time-domain equation is given by Equation (12.17).

$$LC = (0.022)(10^{-5}) = 22 \times 10^{-8}$$
$$RC = (200)(10^{-5}) = 20 \times 10^{-4}$$

$$22 \times 10^{-8} \frac{d^2 e_o}{dt^2} + 20 \times 10^{-4} \frac{de_o}{dt} + e_o = e_i$$

b. The transfer function is given by Equation (12.15).

$$\frac{E_o(S)}{E_i(S)} = \frac{1}{22 \times 10^{-8}S^2 + 20 \times 10^{-4}S + 1}$$

c. From Equation (12.14),

$$1/\omega_o{}^2 = 22 \times 10^{-8}$$
$$\omega_o = 2,130 \text{ radians/second}$$

d. From Equation (12.14),

$$2\alpha/\omega_o{}^2 = 20 \times 10^{-4}$$
$$\alpha = (10 \times 10^{-4})(4.55 \times 10^6) = 4.55 \times 10^3$$
$$\alpha = 4,550 \quad 1/\text{seconds}$$

e. $\alpha = 4,550$ and $\omega_o = 2,130$; therefore $\alpha > \omega_o$ and the system is overdamped.

Example 12.11 The non-interacting, second-order process Both the liquid flow and the electrical non-interacting, second-order processes are illustrated in Figure 12.18. The time-domain equation of the liquid system is given by Equation (12.18).

$$\tau_1\tau_2\frac{d^2h_2}{dt^2} + (\tau_1 + \tau_2)\frac{dh_2}{dt} + h_2 = \left(\frac{R_2}{\rho g}\right)m \qquad \textbf{(12.18)}$$

where:

$\tau_1 = R_1C_1$

$\tau_2 = R_2C_2$

R = liquid resistance, newton-seconds/meter^5

C = liquid capacitance, cubic meters/newton per meter^2

h_2 = the level in tank 2, meters

m = the input flow rate to tank 1, cubic meters/second

The *non-interacting electrical system* is illustrated in Figure 12.18–b. The time-domain equation is

$$\tau_1\tau_2\frac{d^2e_o}{dt^2} + (\tau_1 + \tau_2)\frac{de_o}{dt} + e_o = e_i \qquad \textbf{(12.19)}$$

where:

$\tau_1 = R_1C_1$

$\tau_2 = R_2C_2$

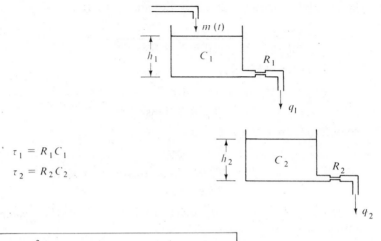

$$\tau_1 = R_1 C_1$$
$$\tau_2 = R_2 C_2$$

$$\tau_1 \tau_2 \frac{d^2 h_2}{dt^2} + (\tau_1 + \tau_2)\frac{dh_2}{dt} + h_2 = (\frac{R_2}{\rho g})m$$

a) A non-interacting liquid process

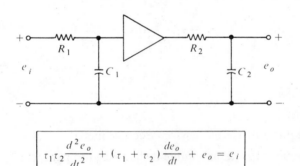

$$\tau_1 \tau_2 \frac{d^2 e_o}{dt^2} + (\tau_1 + \tau_2)\frac{de_o}{dt} + e_o = e_i$$

b) A non-interacting electrical circuit

FIGURE 12.18

TWO-CAPACITY, NON-INTERACTING PROCESSES

$R =$ the electrical capacitance, ohms

$C =$ the electrical capacitance, farads

$e_o =$ the output signal, volts

$e_i =$ the input signal, volts

A liquid, non-interacting, two-capacity system has time constants τ_1 and τ_2 of 520 seconds and 960 seconds. The liquid is water and the value of R_2 is 1.6×10^6 newton-seconds/meter⁵. Determine the time-domain equation and transfer function.

Solution:

a. The time-domain equation is given by Equation (12.18).

$$\tau_1\tau_2 = (520)(960) = 5 \times 10^5 \text{ seconds}^2$$
$$\tau_1 + \tau_2 = 520 + 960 = 1480 \text{ seconds}$$
$$\rho = 1000 \text{ kilograms/cubic meter (Appendix A)}$$
$$g = 9.81 \text{ meters/second}^2$$
$$R_2/(\rho g) = 1.6 \times 10^6/9.81 \times 10^3 = 163$$
$$5 \times 10^5 \frac{d^2h}{dt^2} + 1.48 \times 10^3 \frac{dh}{dt} + h = 163m$$

b. The transfer function is

$$\frac{H(S)}{M(S)} = \frac{163}{5 \times 10^5 S^2 + 1.48 \times 10^3 S + 1}$$

Example 12.12 The interacting, second-order process The liquid-flow and the electrical-interacting, second-order processes are illustrated in Figure 12.19. The liquid-interacting system is illustrated in Figure 12.19–a. The time-domain equation is given by Equation (12.20).

$$\tau_1\tau_2\frac{d^2P_o}{dt^2} + \left[\tau_1 + \tau_2 + \tau_1\left(\frac{R_2}{R_L}\right) + \tau_2\left(\frac{R_1}{R_2}\right)\right]\frac{dP_o}{dt} +$$
$$\left[1 + \frac{R_2}{R_L} + \frac{R_1}{R_L}\right]P_o = P_i \qquad \textbf{(12.20)}$$

where τ_1, τ_2, R, C, and m are defined for Equation (12.18).

$$P_i = \text{the inlet pressure, newtons/meter}^2$$
$$P_1 = \text{the pressure in tank 1, newtons/meter}^2$$
$$P_o = \text{the pressure in tank 2, newtons/meter}^2$$

The interacting electrical circuit is shown in Figure 12.19–b. The time-domain equation is given by Equation (12.21).

$$\tau_1\tau_2\frac{d^2e_o}{dt^2} + \left[\tau_1 + \tau_2 + \tau_1\left(\frac{R_2}{R_L}\right) + \tau_2\left(\frac{R_1}{R_2}\right)\right]\frac{de_o}{dt} +$$

$$\left[1 + \frac{R_2}{R_L} + \frac{R_1}{R_L}\right]e_o = e_i \qquad \textbf{(12.21)}$$

where: τ_1, τ_2, R, C, e_i, and e_o are defined in Equation (12.19).

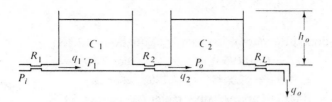

$$\tau_1 \tau_2 \frac{d^2 P_o}{dt^2} + \left[\tau_1 + \tau_2 + \tau_1 \left(\frac{R_2}{R_L}\right) + \tau_2 \left(\frac{R_1}{R_2}\right)\right] \frac{d P_o}{dt} + \left[1 + \frac{R_2}{R_L} + \frac{R_1}{R_L}\right] P_o = P_i$$

a) An interacting liquid process

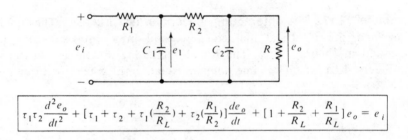

$$\tau_1 \tau_2 \frac{d^2 e_o}{dt^2} + \left[\tau_1 + \tau_2 + \tau_1 \left(\frac{R_2}{R_L}\right) + \tau_2 \left(\frac{R_1}{R_2}\right)\right] \frac{d e_o}{dt} + \left[1 + \frac{R_2}{R_L} + \frac{R_1}{R_L}\right] e_o = e_i$$

b) An interacting electrical circuit

FIGURE 12.19

TWO-CAPACITY, INTERACTING PROCESSES

An electrical, interacting, two-capacity circuit has the following component values.

$$R_1 = 1000 \text{ ohms}$$
$$R_2 = 800 \text{ ohms}$$
$$R_L = 400 \text{ ohms}$$
$$C_1 = 10 \text{ microfarads}$$
$$C_2 = 50 \text{ microfarads}$$

Determine τ_1, τ_2, the time-domain equation, and the transfer function.

Solution:

a. The time-domain equation is given by Equation (12.21).

$$\tau_1 = (1000)(10^{-5}) = 10^{-2} \text{ second}$$
$$\tau_2 = (800)(50 \times 10^{-6}) = 4 \times 10^{-2} \text{ second}$$
$$\tau_1\tau_2 = 4 \times 10^{-4} \text{ second}^2$$
$$\tau_1 R_2/R_L = 2 \times 10^{-2} \text{ second}$$
$$\tau_2 R_1/R_2 = 5 \times 10^{-2}$$
$$R_2/R_L = 2$$
$$R_1/R_L = 10^3/400 = 2.5$$
$$\tau_1 + \tau_2 + \tau_1 R_2/R_L + \tau_2 R_1/R_2 = 0.12$$
$$1 + R_2/R_L + R_1/R_L = 1 + 2 + 2.5 = 5.5$$

The time-domain equation is

$$4 \times 10^{-4}\frac{d^2 e_o}{dt^2} + 0.12\frac{de_o}{dt} + 5.5e_o = e_i$$

or, in the form of Equation (12.14),

$$7.28 \times 10^{-5}\frac{d^2 e_o}{dt^2} + 2.18 \times 10^{-2}\frac{de_o}{dt} + e_o = 0.182e_i$$

b. The transfer function is

$$\frac{E_o(S)}{E_i(S)} = \frac{0.182}{7.28 \times 10^{-5}S^2 + 2.18 \times 10^{-2}S + 1}$$

Example 12.13 A dc motor, second-order process The armature-controlled dc motor in Figure 12.20 is another example of a second-order system. A brief development of the time-domain equation is given in Figure 12.20, and a detailed discussion of the armature-controlled dc motor is included in Section 10.3. The time-domain equation (12.22) is given below.

$$\tau_m\tau_a\frac{d^2 w}{dt^2} + (\tau_m + \tau_a)\frac{dw}{dt} + \left(1 + \frac{K_e K_t}{RB}\right)w = \left(\frac{K_t}{RB}\right)e \quad \textbf{(12.22)}$$

where:

$\tau_m = J/B$, the mechanical time constant, seconds

$\tau_a = L/R$, the armature time constant, seconds

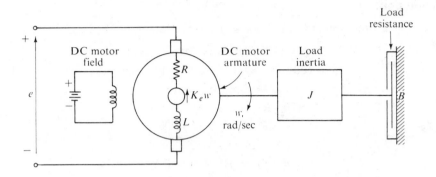

Basic equations

$$L\frac{di}{dt} + Ri + K_e w = e$$

$$T = K_t i$$

$$T = J\frac{dw}{dt} + Bw$$

$$i = \frac{1}{K_t} T = \frac{1}{K_t}[J\frac{dw}{dt} + Bw]$$

$$L\frac{d}{dt}[\frac{1}{K_t}(J\frac{dw}{dt} + Bw)] + R[\frac{1}{K_t}(J\frac{dw}{dt} + Bw)] + K_e w = e$$

$$JL\frac{d^2w}{dt^2} + (BL + RJ)\frac{dw}{dt} + (RB + K_e K_t)w = K_t e$$

divide by BR and define τ_m and τ_a as follows:

$$\tau_m \triangleq \frac{J}{B} = \text{the mechanical time constant}$$

$$\tau_a \triangleq \frac{L}{R} = \text{the armature time constant}$$

$$\tau_m \tau_a \frac{d^2w}{dt^2} + (\tau_m + \tau_a)\frac{dw}{dt} + (1 + \frac{K_e K_t}{RB})w = (\frac{K_t}{RB})e$$

FIGURE 12.20

AN ARMATURE-CONTROLLED DC MOTOR

J = the moment of inertia of the load, kilograms-meter^2

B = the damping resistance, newton-meters/radian per second

L = the armature inductance, henrys

R = the armature resistance, ohms

K_t = the torque constant, newton-meters/ampere

e = the armature voltage, volts

w = the motor speed, radians/second

K_e = the emf constant, volts/radian per second

An armature-controlled dc motor has the following characteristics.

$$J = 6.2 \times 10^{-4} \text{ kilogram-meter}^2$$

$$B = 1 \times 10^{-4} \text{ newton-meter/radian per second}$$

$$L = 0.020 \text{ henry}$$

$$R = 1.2 \text{ ohms}$$

$$K_e = 0.043 \text{ volt/radian per second}$$

$$K_t = 0.043 \text{ newton-meter/ampere}$$

Determine the mechanical time constant, the electrical time constant, the time-domain equation, and the transfer function.

Solution:

a.   The time-domain equation is given by Equation (12.22).

$$\tau_m = J/B = 6.2 \times 10^{-4}/10^{-4} = 6.2 \text{ seconds}$$

$$\tau_a = L/R = 0.020/1.2 = 0.0167 \text{ second}$$

$$\tau_m\tau_a = 0.104$$

$$\tau_m + \tau_a = 6.2 + 0.0167 = 6.22$$

$$1 + K_eK_t/(RB) = 1 + 15.4 = 16.4$$

$$K_t/(RB) = 358$$

The time-domain equation is

$$0.104\frac{d^2w}{dt^2} + 6.22\frac{dw}{dt} + 16.4w = 358e$$

or, in the form of Equation (12.14),

$$6.35 \times 10^{-3}\frac{d^2w}{dt^2} + 0.379\frac{dw}{dt} + w = 21.8e$$

b.   The transfer function is

$$\frac{W(S)}{E(S)} = \frac{21.8}{6.35 \times 10^{-3}S^2 + 0.379S + 1}$$

12.6   The First-Order Lag Plus Dead-Time Process

The first-order lag plus dead-time process is a series combination of a first-order lag process and a dead-time element. The step response, illustrated in Figure 12.21, is characterized by the first-order lag time constant (τ) and the dead-time lag (τ_d). The first order lag plus dead-time characteristic is frequently used as a first approximation of the model of more complex processes for the purpose of analysis and design. The values of τ_d and τ can be determined from a step-response test of the process as indicated in Figure 12.21. An example of a first-order lag plus dead-time salt brine process is shown in Figure 12.22. A mixing tank is used to continuously blend salt and water to form a brine solution. The inlet water flow rate is regulated to match the outlet flow of brine solution. The salt flow rate is regulated to maintain the desired salt concentration in the mixing tank. The salt flow rate, $m(t)$, is the input signal of the process. The salt concentration in the tank, $C_o(t)$, is the output signal of the process. The belt conveyor is the dead-time element and the mixing tank provides the first-order lag characteristic. In this example, the dead-time element preceded the first-order lag element. However, the response is the same for processes in which the first order lag characteristic precedes the dead-time element.

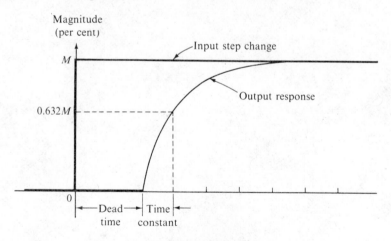

FIGURE 12.21

THE STEP RESPONSE OF A FIRST-ORDER LAG PLUS DEAD-TIME PROCESS

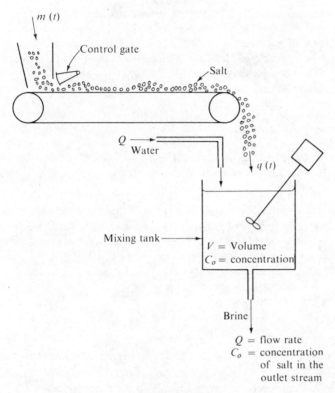

FIGURE 12.22

A FIRST-ORDER LAG PLUS DEAD-TIME BLENDING PROCESS

EXERCISES

12–1. A liquid tank process similar to Figure 12.3 has the following parameters and conditions.

> tank height $= 5.2$ meters
> tank diameter $= 1.8$ meters
> level at time $t_o = h(t_o) = 1.7$ meters
> output flow rate $= q = 0.009$ cubic meter/second
> input flow range $= 0$ to 0.02 cubic meter/second
> average input flow rate between t_o and $t_o + 50$ seconds $=$
> $m_{avg} = 0.011$ cubic meter/second

Determine the time-domain equation, the transfer function, the integral time, and the level at time $t_o + 50$ seconds.

12–2.   An oil tank similar to Figure 12.7 has a diameter of 1.5 meters and a
height of 3.6 meters. The outlet at the bottom has a length of 0.8
meter and a diameter of 0.020 meter. The oil temperature is 15° C
and the outlet flow is laminar. Determine each of the following.
a.   The capacitance of the tank.
b.   The resistance of the outlet.
c.   The RC time constant.
d.   The time domain equation.
e.   The transfer function.

12–3.   An electrical circuit similar to Figure 12.8–c has a capacitance value
of 10 microfarad. Determine the resistance value that will result in a
time constant equal to 0.083 second.

12–4.   An oil-bath thermal system similar to Figure 12.9 has an inside diameter of 0.75 meter and a height of 0.6 meter. The inside film coefficient
is 58 watts/meter^2—Kelvin. The wall resistance and outside film coefficient are negligible. Determine each of the following.
a.   The thermal resistance.
b.   The thermal capacitance.
c.   The time constant.
d.   The time-domain equation.
e.   The transfer function.

12–5.   A nitrogen (N_2) gas process similar to Figure 12.10 has the following
parameters.

> pressure vessel volume = 1.1 cubic meters
> temperature = 540 Kelvin
>
> gas flow resistance = $1.8 \times 10^5 \dfrac{\text{newton-second}}{\text{kilogram-meter}^2}$

Determine each of the following.
a.   The capacitance of the pressure vessel
b.   The time constant
c.   The time-domain equation
d.   The transfer function

12–6.   A blending system similar to Figure 12.11 has a flow rate of 0.006
cubic meter per second. Determine the tank volume that will result
in a time constant of 10 minutes. Then determine the time-domain
equation and transfer function using the 10-minute time constant.

12–7.   A dead-time process similar to Figure 12.13–a consists of an 8.2-
meter long belt conveyor with a belt velocity of 0.44 meter per
second. Determine the dead-time lag, the time-domain equation, and
the transfer function.

12–8.   A spring-mass-damping system consists of a 25-kilogram mass, a
spring constant of 1450 newtons/meter, and a damping resistance of

42 newton- seconds per meter. Determine the time-domain equation, the transfer function, the resonant frequency (ω_o), the damping coefficient (α), and the type of damping.

12–9. An electrical series RLC circuit consists of a 0.04-henry inductor, a 4-microfarad capacitor, and a 100-ohm resistor. Verify that the system is underdamped. The damping of the circuit may be increased by adding a second resistor in series with the 100-ohm resistor. Determine the value of the second resistor that will result in a critically damped circuit. Determine the time-domain equation and the transfer function of the critically damped circuit.

12–10. A liquid non-interacting, two-capacity system has time constants τ_1 and τ_2 of 300 seconds and 1200 seconds. The liquid is water, and the value of R_2 is 2.2 $\times$ 10^6 newton-seconds/meter.[5] Determine the time domain equation and transfer function.

12–11. A electrical, interacting, two-capacity circuit has the following component values.

$$R_1 = 100 \text{ ohms}$$
$$R_2 = 300 \text{ ohms}$$
$$R_L = 50 \text{ ohms}$$
$$C_1 = 0.1 \text{ microfarad}$$
$$C_2 = 0.8 \text{ microfarad}$$

Determine τ_1, τ_2, the time-domain equation, and the transfer function.

12–12. An armature-controlled dc motor has the folowing characteristics.

$$J = 3.2 \times 10^{-3} \text{ kilogram-meter}^2$$
$$B = 2 \times 10^{-3} \text{ newton-meter/radian per second}$$
$$L = 0.075 \text{ henry}$$
$$R = 1.2 \text{ ohms}$$
$$K_e = 0.22 \text{ volt/radian per second}$$
$$K_t = 0.22 \text{ newton-meter/ampere}$$

Determine the mechanical-time constant, the electrical-time constant, the time-domain equation, the transfer function, the resonant frequency (ω_o), the damping coefficient (α), and the type of damping (i.e., overdamped, critically damped, or underdamped).

12–13. A first-order lag plus dead-time process similar to Figure 12.22 has the following parameter values.

distance traveled on the belt = 7.6 meters
belt speed = 1.1 meter/second
tank volume = 12.2 cubic meters
water flow rate = 0.01 cubic meter/second

Determine the dead-time lag (T_d) and the first-order time constant (τ).

12-14. The response of a control system is very much like the response of a second-order system. A closed-loop system may be overdamped, critically damped, or underdamped. The damping coefficient (α) determines the type of damping present in a system. Notice that, in Equation (12.14), α is part of the coefficient of the first derivative of the output (dx/dt).

$$\frac{d^2x}{dt^2} + 2\alpha\frac{dx}{dt} + \omega_o{}^2x = \frac{1}{M}F$$

The first derivative is the rate of change (or velocity) of the output. Consider a dc motor position control system which is underdamped. A speed sensor measures the rate of change of the output (i.e., dx/dt). Explain how the signal from the speed sensor could be used to increase the damping coefficient of the system.

12–15. A blending and heating system is illustrated in Figure 15.9. The thermal system is a first-order lag and the blending system is a first-order lag plus dead time. Determine the time constant and dead-time lag of the blending system and the time constant of the thermal system. Also determine the transfer function of the thermal system. The system parameters are

Production rate: 0.00018 cubic meters/second
Mixing tank diameter: 0.55 meters
Height of liquid in the mixing tank: 0.65 meters
Inside thermal film coefficient: $h_i = 810$ watts/meter^2-Kelvin
Outside thermal film coefficient: $h_o = 1200$ watts/meter^2-Kelvin
Tank wall thickness: 0.7 centimeters
Tank wall material: steel
Density of liquid in the tank: 1008 kilograms/meter^3
Specific heat of the liquid: 4060 joules/kilogram-Kelvin
Diameter of the mixing tank outlet pipe: 2 centimeters
Distance from tank outlet to the analyzer feed pipe inlet:
    1.5 meters
Concentration analyzer feed pipe flow rate: 0.2 meters/second
Length of the concentration analyzer feed pipe: 2 meters

13

Frequency
Response

13.1 Introduction

A control system consists of components connected by signal paths. Each component may change the size and the timing of the signal. Usually, the amount of the change in size and timing depends on the frequency of the signal. Figure 13.1 is an example of the graphs of the input and output signals of a linear control system component.

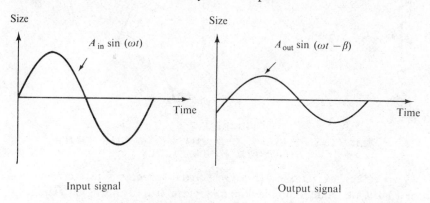

FIGURE 13.1

THE RESPONSE OF A LINEAR CONTROL SYSTEM COMPONENT

411

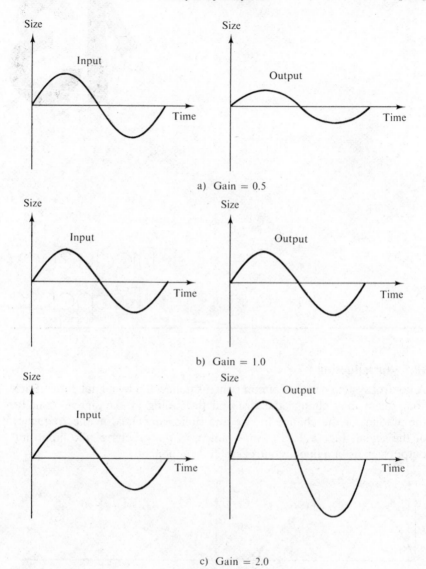

a) Gain = 0.5

b) Gain = 1.0

c) Gain = 2.0

FIGURE 13-2

*EXAMPLES OF INPUT AND OUTPUT SIGNALS WITH
DIFFERENT GAIN VALUES*

Two values are required to represent the response illustrated in Figure 13.1. One value expresses the change in size of the signal. The second value expresses the change in timing of the signal. The change in size of the signal is illustrated in Figure 13.2, and the change in timing is illustrated in Figure 13.3.

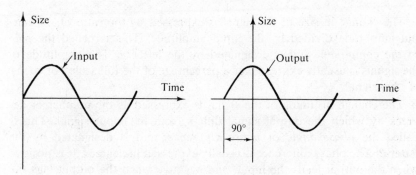

a) Phase angle = + 90° , a 90° lead

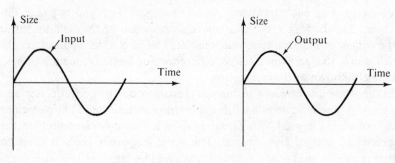

b) Phase angle = 0°

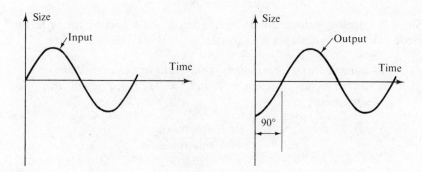

c) Phase angle = −90°, − a 90° lag

FIGURE 13.3

*EXAMPLES OF INPUT AND OUTPUT SIGNALS WITH
DIFFERENT PHASE ANGLES*

The change in size of the signal is expressed by the ratio of the output amplitude divided by the input amplitude. This is called the *gain of the component* and is designated by the letter *g*. The amplitude of the signals is usually expressed as a percentage of the full scale (or span) of the signal.

The change in timing of the signal is expressed as the number of degrees by which the output signal differs from the input signal. This is called the *phase angle* or *phase difference,* and is designated by the letter β. The phase difference is usually expressed in degrees. It is positive when the output leads the input, and negative when the output lags the input.

The frequency response of a component is the set of values of *g* and β which occur when a sinusoidal input signal is varied over a range of frequencies. For each frequency, there is a gain (*g*) and a phase difference (β) which give the characteristic response of the component *at that frequency*. The frequency response may be plotted as a pair of graphs which show the gain and phase difference for each frequency. These graphs are known as *Bode diagrams*.

In the Bode diagram, logarithmic scales are used for the frequency and the gain. A linear scale is used for the phase difference. This particular choice of scales is used to facilitate graphical methods of combining the responses of several components. The same frequency scale is used on both graphs to make comparisons easier. The gain (*g*) is often expressed in decibels. In this book, the letter *m* will be used to represent the gain in decibels.

$$m \; (db) = 20 \log_{10} g$$

Since the decibel is logarithmic unit, a linear scale is used for *m* in the Bode diagram. Figure 13.4 is an example of a Bode diagram.

The gain (g) of a component is the ratio of the amplitude of the output signal over the amplitude of the input signal at a given frequency.

$$g = \frac{\text{output amplitude}}{\text{input amplitude}}$$

The gain is often expressed in decibels.

$$m \; (db) = 20 \log_{10} g$$

The phase angle *(β) is the number of degrees by which the output signal differs from the input signal. The phase angle is posi-*

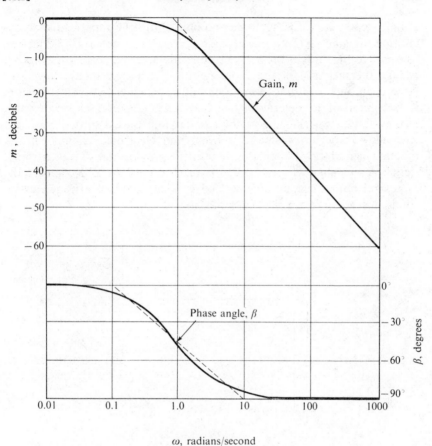

FIGURE 13.4
AN EXAMPLE OF A BODE DIAGRAM

tive if the output leads the input, and negative if the output lags
the input.

The frequency response *of a component is the set of values of
g and β (or m and β) which occur when a sinusoidal input signal is varied over a range of frequencies.*

13.2 Determination of the Frequency Response

There are two ways to determine the frequency response of a component: experimentally and analytically. In the experimental method,
a sinusoidal input signal is applied to an actual component. The output

signal is compared in magnitude and phase with the input signal to determine the gain (g) and phase angle (β). This is not a simple matter. The measuring equipment must be carefully selected so that it does not change the response of the component. Frequency-response testing of large components is time-consuming and expensive. In some systems it is not practical to determine the frequency response experimentally. However, the experimental method is frequently used to determine the frequency response of small electrical and electronic components. A function generator or oscillator is used to provide the sinusoidal input signal. A two-channel oscilloscope is used to measure and compare the input and output signals. Figure 13.5 is a schematic of a frequency-response test setup for an electronic component.

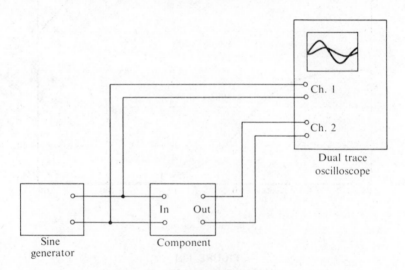

FIGURE 13.5
SCHEMATIC OF A FREQUENCY-RESPONSE TEST SETUP

The frequency response of a component can be determined analytically from its transfer function, which was introduced in Chapter Four. A brief review will be helpful in understanding the relationship between the transfer function and the frequency response; refer to Chapter Four for further details.

In the *time domain,* the input-output relationship of a component is defined by a differential equation. Methods of operational calculus are required to determine the gain (g) and phase angle (β) from the differential equation. The Laplace transform allows us to transform the differential equation into an algebraic equation in the *frequency domain.* The transfer function is obtained by solving the algebraic equation for the ratio of the output over the input. For example, the time domain equation and transfer function of a first-order lag are given below.

Time Domain Equation:

$$\tau \frac{dh}{dt} + h = m \tag{13.1}$$

$$\text{Transfer Function} = \frac{H}{M} = \frac{1}{\tau S + 1} \tag{13.2}$$

where:

τ = the time constant

h = time-domain output signal

H = frequency-domain output signal

m = the time-domain input signal

M = the frequency-domain input signal

S = the frequency parameter

When the input signal (m) is a sinewave with a radian frequency of ω, then the frequency parameter, S, is equal to $j\omega$ where $j = \sqrt{-1}$.

$$S = j\omega \text{ when } m = A \sin (\omega t)$$

The frequency response of a component is given by its transfer function when S is replaced by $j\omega$. For each value of ω, the transfer function may be reduced to a single complex number. The magnitude of this complex number is the gain (g) of the component at the frequency ω. The angle of the complex number is the phase difference (β) at the frequency ω.

Complex numbers can be represented by a point on a two-dimensional plane called the *complex plane*. The x-axis is called the *real axis,* and the y-axis is called the *imaginary* or *j-axis*. The graph of the complex number N is shown in Figure 13.6. The complex number may be defined in rectangular form by its real coordinate (a) and its imaginary coordinate (b); i.e., $N = a + jb$ where j identifies the imaginary coordinate. In this form, a is called the *real part of the complex number,* and b is called the *imaginary part*. The complex number may also be defined in polar form by its distance from the origin (c) and the angle (θ). In the polar form, c is called the *magnitude of the complex number,* and θ is called the *angle of the complex number;* i.e., $N = c \lfloor \theta$. A complex number may be converted from one form to the other by using the appropriate conversion equations below.

Rectangular to Polar Conversion	Polar to Rectangular Conversion
$c = \sqrt{a^2 + b^2}$	$a = c \cos \theta$
$\theta = \tan^{-1}\left(\dfrac{b}{a}\right)$	$b = c \sin \theta$

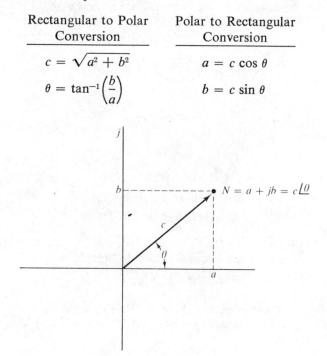

FIGURE 13.6
THE COMPLEX PLANE

The frequency response of a component is obtained by substituting $j\omega$ for S in the transfer function. With S replaced by $j\omega$, the transfer function is a complex number which can be converted to the polar form. The magnitude of the complex number is the gain (g) of the component at the frequency ω. The angle of the complex number is the phase difference (β) of the component at the frequency ω.

Example 13.1 The first-order lag defined by Equation (13.2) has a time constant (τ) equal to 0.1 second. Determine the frequency response when the frequency (ω) is equal to 10 radians/second.

Solution:

Substitute $j10$ for S and 0.1 for τ in Equation (13.2).

$$\frac{H}{M} = \frac{1}{\tau S + 1} = \frac{1}{(0.1)(j10) + 1}$$

$$\frac{H}{M} = \frac{1}{1 + j} = \frac{1}{\sqrt{2} \; \underline{|45°}}$$

$$\frac{H}{M} = \frac{1}{\sqrt{2}} \; \underline{|-45°}$$

$$\frac{H}{M} = 0.707 \; \underline{|-45°}$$

Therefore, the gain, $g = 0.707$. The phase difference, $\beta = -45°$.

13.3 Overall Frequency Response of Several Components

A control system consists of several components connected in series. The design and analysis of a control system require a knowledge of the transfer function and the frequency response of the group of components considered as a unit. In other words, the overall transfer function and frequency response must be known.

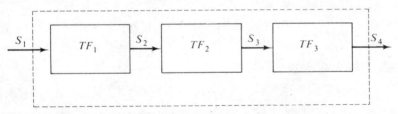

FIGURE 13.7
THREE COMPONENTS CONNECTED IN SERIES

Consider the three components in Figure 13.7 with transfer functions TF_1, TF_2, and TF_3. Signal S_1 is the input signal of component No. 1. Signal S_2 is the output signal of component No. 1 and the input signal of component No. 2. Signal S_3 is the output signal of component No. 2 and the input signal of component No. 3. Signal S_4 is the output signal of component No. 3. Also, signal S_1 is the input signal, and S_4 is the output signal of the three components as a unit. By definition, the transfer function of a component or group of components is the ratio of the output signal over the input signal.

Thus,

Transfer function of component No. 1 $= TF_1 = S_2/S_1$
Transfer function of component No. 2 $= TF_2 = S_3/S_2$
Transfer function of component No. 3 $= TF_3 = S_4/S_3$
Transfer function of the group of components $= TF = S_4/S_1$

But,

$$S_4/S_1 = (S_2/S_1)(S_3/S_2)(S_4/S_3)$$

and

$$TF = (TF_1)(TF_2)(TF_3) \tag{13.3}$$

Equation (13.3) states that the overall transfer function of several components in series is equal to the product of the transfer functions of the individual components.

The frequency response of a component or group of components is obtained by substituting $j\omega$ for S in the transfer function. The result is a complex number which may be converted to the polar form. The magnitude of the complex number is the frequency-response gain (g). The angle of the complex number is the frequency-response angle (β). Let the frequency responses be represented by the following complex numbers.

$g_1 \;\underline{|\beta_1}$ = the frequency response of component No. 1

$g_2 \;\underline{|\beta_2}$ = the frequency response of component No. 2

$g_3 \;\underline{|\beta_3}$ = the frequency response of component No. 3

$g \;\;\underline{|\beta}$ = the frequency response of the components as a unit

Then, from Equation (219),

$$g\,\underline{|\beta} = (g_1 \;\underline{|\beta_1})(g_2 \;\underline{|\beta_2})(g_3 \;\underline{|\beta_3}) \tag{13.4}$$

However, the product of two complex numbers in polar form is a complex number whose magnitude is the product of the magnitudes of the two complex numbers, and whose angle is the sum of the angles of the two complex numbers.

$$(g_1 \;\underline{|\beta_1})(g_2 \;\underline{|\beta_2})(g_3 \;\underline{|\beta_3}) = (g_1 g_2 \;\underline{|\beta_1 + \beta_2})(g_3 \;\underline{|\beta_3})$$
$$= g_1 g_2 g_3 \;\underline{|\beta_1 + \beta_2 + \beta_3}$$

and

$$g \;\underline{|\beta} = g_1 g_2 g_3 \;\underline{|\beta_1 + \beta_2 + \beta_3} \tag{13.5}$$

Equation (13.5) states that the overall gain of several components in series is equal to the product of the gains of the individual components. It further states that the overall phase angle of several components in series is equal to the sum of the phase angles of the individual components. In other words, the gains are multiplied and the phase angles are added.

The Bode diagram scales were selected so that graphical addition can be used to combine two or more frequency-response graphs. The logarithmic gain scale is used so that multiplication of two gain terms can be accomplished graphically by adding the logarithms of the two gain terms; i.e.

$$\log g = \log (g_1 \cdot g_2 \cdot g_3) = \log g_1 + \log g_2 + \log g_3$$

The decibel scale is also a logarithmic scale, and the addition of two decibel values is equivalent to multiplying the corresponding gain values.

$$20 \log g = 20 \log (g_1 \cdot g_2 \cdot g_3) = 20 \log g_1 + 20 \log g_2 + 20 \log g_3$$
$$m(\text{db}) = m_1(\text{db}) + m_2(\text{db}) + m_3(\text{db}) \qquad \textbf{(13.6)}$$

The linear phase-angle scale was selected so that two or more phase angles can be added graphically.

The overall transfer function of several components in series is equal to the product of the transfer functions of the individual components.

The overall gain of several components in series is equal to the product of the gains of the individual components.

The overall phase angle of several components in series is equal to the sum of the phase angles of the individual components.

The overall frequency response of several components can be determined on a Bode diagram graphically by adding the decibel gains and by adding the phase angles of the individual components.

Example 13.2   A thermal process and a temperature measuring means are connected in series in a control system. Determine the overall transfer function from the following individual transfer functions.

$$\text{thermal process transfer function} = \frac{1}{6.42 \times 10^3 S + 1}$$

$$\text{measuring means transfer function} = \frac{1}{78S + 1}$$

Solution:

The overall transfer function is the product of the individual transfer functions.

$$\text{overall transfer function} = \left(\frac{1}{6.42 \times 10^3 S + 1}\right)\left(\frac{1}{78S + 1}\right)$$
$$= \frac{1}{(6.42 \times 10^3 S + 1)(78S + 1)}$$

Example 13.3 Three components are connected in series. The transfer functions are given below. Determine the overall transfer function (TF).

$$TF_1 = \frac{1 + \tau_1 S}{1 + \tau_2 S}$$

$$TF_2 = \frac{1}{AS^2 + BS + 1}$$

$$TF_3 = \frac{1}{\tau_3 S + 1}$$

Solution:

The overall transfer function is the product of the individual transfer functions.

$$TF = (TF_1)(TF_2)(TF_3)$$

$$TF = \left(\frac{1 + \tau_1 S}{1 + \tau_2 S}\right)\left(\frac{1}{AS^2 + BS + 1}\right)\left(\frac{1}{\tau_3 S + 1}\right)$$

$$TF = \frac{(1 + \tau_i S)}{(1 + \tau_2 S)(AS^2 + BS + 1)(\tau_3 S + 1)}$$

Example 13.4 The following gains and phase angles were measured at a frequency of 1 cycle per second. Determine the overall gain (g) and phase angle (β) if the components are connected in series.

	Gain	Phase Angle
Final control element	1.5	$-5°$
Process	2.0	$-170°$
Measuring means	0.9	$-15°$

Solution:

The gain (g) is the product of the individual gains and the phase angle (β) is the sum of the individual phase angles.

$$g = (1.5)(2.0)(0.9) = 2.7$$
$$\beta = (-5°) + (-170°) + (-15°) = -190°$$

13.4 Frequency Response of a Proportional Component

The transfer function of a proportional component is simply a constant which indicates the gain between the input and the output. The phase difference is zero at all frequencies, and the gain is the same at all frequencies. A potentiometer used as a position sensor and a proportional controller are examples of a proportional component. The transfer function of a proportional component is given by Equation (13.7). The Bode diagram of a proportional component with a gain g of 2.51 ($m = 8.0$ db) is given in Figure 13.8.

$$\text{Proportional component transfer function} = \frac{H}{M} = g \qquad \textbf{(13.7)}$$

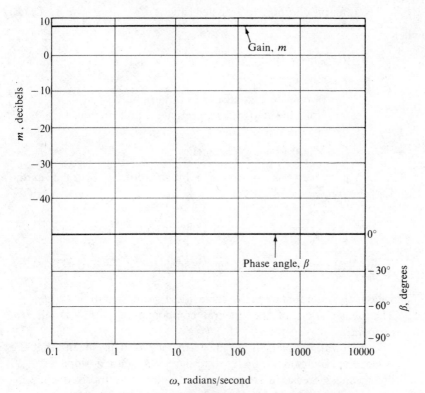

ω, radians/second

FIGURE 13.8

THE BODE DIAGRAM OF A PROPORTIONAL COMPONENT

PROPORTIONAL COMPONENT BODE DIAGRAM

The Bode diagram of a proportional component consists of horizontal straight-line gain and phase graphs. The gain graph is determined by the gain of the component. The gain is usually expressed in decibels (appendix D contains a convenient decibel conversion table). The phase angle is 0° for all values of ω.

13.5 Frequency Response of an Integral Component

The integral or ramp process (explained in Chapter Twelve) and the integral control mode (explained in Chapter Nine) are both examples of an integral component or, simply, an integrator. The transfer function has the following general form.

$$\text{Integral component transfer function} = \frac{H}{M} = \frac{1}{AS} \qquad \textbf{(13.8)}$$

where:

H = the output signal

M = the input signal

A = a constant equal to the time required for a unit input signal to produce a unit change in the output signal

S = the frequency parameter

The frequency response is obtained by substituting $j\omega$ for S in the transfer function.

$$\frac{1}{A(j\omega)} = -j\left(\frac{1}{A\omega}\right) = \left(\frac{1}{A\omega}\right) \underline{/-90°}$$

$$\text{Integral component frequency response} = \left(\frac{1}{A\omega}\right) \underline{/-90°} \qquad \textbf{(13.9)}$$

The following general statements are evident from Equation (13.9).

1. The phase angle of the integral component is −90° at all frequencies.
2. The gain varies inversely as the frequency ω. This means that the Bode diagram gain graph is a straight line with a slope of −20 db/decade increase in frequency (i.e., each time the frequency increases by a factor of 10, the gain is reduced by 20 db).
3. The gain is 0 db ($G = 1$) when $\omega = 1/A$. This is a convenient point for constructing the Bode diagram of the integral component.

INTEGRAL COMPONENT BODE DIAGRAM

The Bode diagram of an integral component consists of straight-line gain and phase graphs. The gain graph has a slope of −20 db/decade and passes through the point $\left(\dfrac{1}{A} \text{ rad/sec, } 0 \text{ db}\right)$. The phase graph lies on the −90° line.

The Bode diagram of the integral component is given in Figure 13.9. The gain graph is constructed by locating a point at 0 db and $\omega_o = \dfrac{1}{A}$ rad/sec. The gain line passes through the point just located with a slope of -20 db per decade increase in ω, as shown in Figure 13.9.

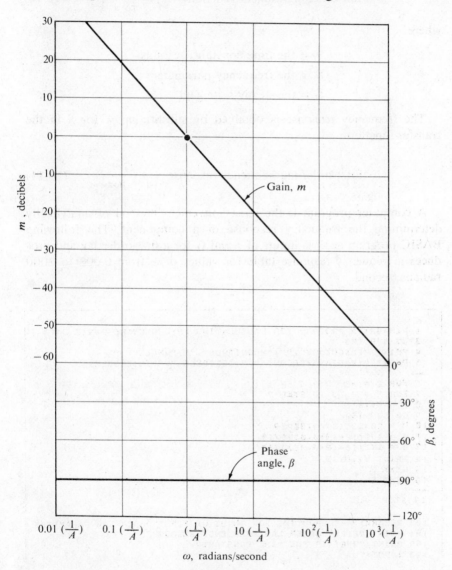

FIGURE 13.9

THE BODE DIAGRAM OF AN INTEGRAL COMPONENT

13.6 Frequency Response of a First-Order Lag Component

The first-order lag component was introduced in Chapter Twelve. The transfer function is given by Equation (13.10) below.

$$\text{First-order lag transfer function} = \frac{H}{M} = \frac{G}{\tau S + 1} \qquad (13.10)$$

where:

$$\tau = \text{the time constant, seconds}$$
$$S = \text{the frequency parameter}$$
$$G = \text{the steady-state gain}$$

The frequency response is obtained by substituting $j\omega$ for S in the transfer function.

$$\text{First-order lag frequency response} = \frac{G}{1 + j\omega\tau} \qquad (13.11)$$

A computer program is the most convenient method of analytically determining the frequency response of a component. The following BASIC program accepts inputs of τ and G for a first-order lag and produces a frequency response table for values of ω from 0.002 to 1000 radians/second.

```
 1 REMARK: A PROGRAM FOR CALCULATING FREQUENCY RESPONSE DATA
 3 GO SUB 100
 4 PRINT "FREQUENCY","","MAGNITUDE","","ANGLE"
 5 PRINT "(RADIANS/SECOND)","(DECIBELS)","","(DEGREES)"

 6 FOR E = -3 TO 2 STEP 1
 7 FOR F = 2 TO 10 STEP 2
 8 W = F*10↑E
 9 GO SUB 150
10 M = 20*LOG(G)/2.30259
11 M = INT(10*(M+0.05))/10
12 B = INT(10*(B+0.05))/10
14 PRINT W,,M,,B
15 NEXT F
16 NEXT E

17 STOP

100 REMARK: A SUBROUTINE FOR INPUT OF FIRST ORDER LAG PARAMETRS
101 PRINT "FIRST ORDER LAG FREQUENCY RESPONSE"
102 PRINT "WHAT IS THE TIME CONSTANT"
103 INPUT T
```

```
104 PRINT "WHAT IS THE VALUE OF G"
105 INPUT G1
106 RETURN

150 REMARK: A SUBROUTINE FOR CALCULATION OF FIRST ORDER LAG DATA
151 R = 1
152 I = W*T
153 GO SUB 400
154 G = G1/C
155 B =-A
156 RETURN

400 REMARK: A SUBROUTINE FOR RECTANGULAR-TO-POLAR CONVERSION
401 C = SQR(R↑2 + I↑2)
402 IF R<=0 THEN 405
403 A = ATN(I/R)*57.2958
404 RETURN
405 IF R = 0 THEN 408
406 A = ATN(I/R)*57.2958 + 180
407 RETURN
408 IF I<0 THEN 411
409 A = 90
410 RETURN
411 A = -90
412 RETURN
```

The above program is divided into a Main Program (lines 3–20) and three subroutines. The Input Subroutine (lines 100–106) prints the title and receives inputs. The Computation Subroutine (lines 150–156) calculates the values of g and β for each value of ω. The Conversion Subroutine (lines 400–415) converts a complex number to the polar form. The Main Program uses the Input Subroutine (line 3), prints a heading (lines 4, 5), sets up a loop to determine the set of ω values (lines 6, 7, and 8), calculates values of g and β (line 9), converts gain to decibels (line 10), rounds off small values of m and β (lines 11–16), and prints each line in the table (line 17).

The following is a sample run of the preceding program. The range of values of ω in the output table can be altered by changing the initial and final values of E in line number 6 of the Main Program.

```
FIRST ORDER LAG FREQUENCY RESPONSE
WHAT IS THE TIME CONSTANT
? 1
WHAT IS THE VALUE OF G
? 1
FREQUENCY                    MAGNITUDE                ANGLE
(RADIANS/SECOND)             (DECIBELS)               (DEGREES)
  2.000000E-3                0                        -.1
  4.000000E-3                0                        -.2
```

6.000000E-3	0	-.3
8.000000E-3	0	-.5
.01	0	-.6
.02	0	-1.1
.04	0	-2.3
.06	0	-3.4
.08	0	-4.6
.1	0	-5.7
.2	-.2	-11.3
.4	-.6	-21.8
.6	-1.3	-31
.8	-2.1	-38.7
1	-3	-45
2	-7	-63.4
4	-12.3	-76
6	-15.7	-80.5
8	-18.1	-82.9
10	-20	-84.3
20	-26	-87.1
40	-32	-88.6
60	-35.6	-89
80	-38.1	-89.3
100	-40	-89.4
200	-46	-89.7
400	-52	-89.9
600	-55.6	-89.9
800	-58.1	-89.9
1000	-60	-89.9

The results from the preceding frequency response table are plotted in Figure 13.10. The BASIC program on pages 426–427 is easily modified to determine the frequency response of any component or group of components. The necessary changes are confined to the Input Subroutine (lines 100–106) and the Calculation Subroutine (lines 150–166). In the remainder of this chapter, the program will be modified to determine the frequency response of a first-order lead, a second-order lag, a dead-time lag, and a three-mode controller.

The Bode-diagram gain graph in Figure 13.10 can be approximated by two straight lines which intersect at a point called the *break point*. The frequency at the break point (ω_b) is equal to the reciprocal of the time constant (τ).

$$\omega_b = \frac{1}{\tau}, \text{ rad/second} \qquad (13.12)$$

FIRST-ORDER LAG BODE DIAGRAM

The Bode diagram of a first-order lag with $G = 1$ can be approximated by straight-line gain and phase-angle graphs. The gain-graph break point is at $\left(\frac{1}{\tau} rad/sec, 0 \ db \right)$. On the left of the break point, the gain is 0 db. On the right, the gain has a *slope of −20 db/decade. The phase-angle break points are located at $\left(\dfrac{0.1}{\tau} \ rad/sec, \ 0°\right)$ and $\left(\dfrac{10}{\tau} \ rad/sec, \ −90°\right)$. A straight line connects the two β break points: On the left of this line, the phase angle is 0°; on the right, the phase angle is −90°.*

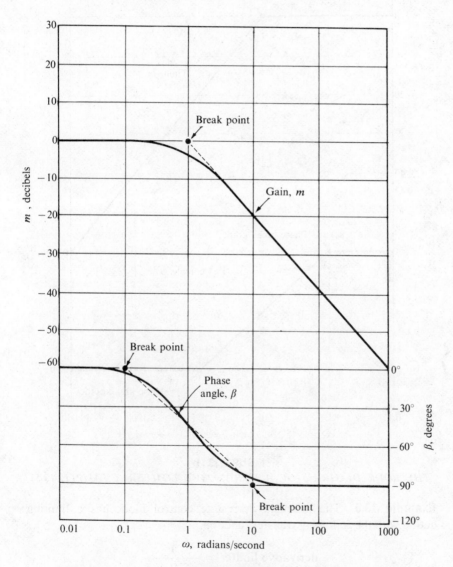

FIGURE 13.10
THE BODE DIAGRAM OF A FIRST-ORDER LAG 1/(1 + jω)

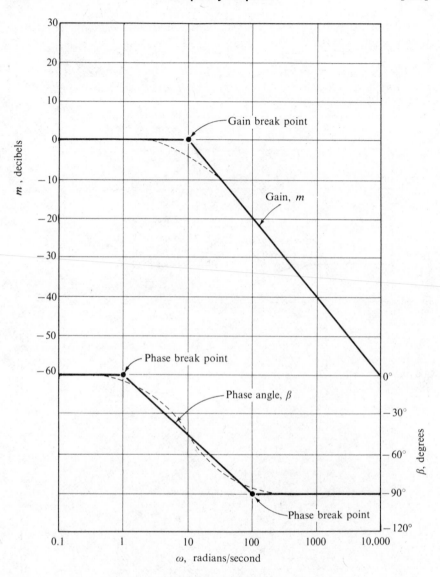

FIGURE 13.11
THE BODE DIAGRAM OF A DERIVATIVE LIMITER (EXAMPLE 13.5)

Example 13.5 The practical derivative control mode has a limiting action with the following transfer function.

$$\text{derivative limiter} = \frac{1}{\alpha T_d S + 1}$$

In Example 9.6, $\alpha T_d = 0.1$. Determine the magnitude and phase-angle break points and sketch the straight-line, Bode-diagram graphs.

Solution:

$$\text{derivative limiter} = \frac{1}{0.1S + 1}$$

$$\omega_b = \frac{1}{0.1} = 10$$

The gain-graph break point is (10 rad/sec, 0 db). The phase-angle break points are (1 rad/sec, 0°) and (100 rad/sec, −90°). The straight-line Bode diagram is given in Figure 13.11.

Example 13.6   The first-order lag flow process of Example (12.3) has the following transfer function.

$$\frac{H}{M} = \frac{1.94 \times 10^3}{5.07 \times 10^3 S + 1}$$

Modify the BASIC program to produce a frequency response table for values of ω from 0.000002 to 0.1 radians/second. Construct a Bode diagram from the frequency response table.

Solution:

Modify line 6 in the BASIC program as follows and run the program.

$$6 \text{ FOR E} = -6 \text{ TO } -2 \text{ STEP } 1$$

The output table from the computer RUN is given below. The Bode diagram is given in Figure 13.12.

FREQUENCY (RADIANS/SECOND)	MAGNITUDE (DECIBELS)	ANGLE (DEGREES)
2.000000E-6	65.8	-.6
4.000000E-6	65.8	-1.2
6.000000E-6	65.8	-1.7
8.000000E-6	65.7	-2.3
1.000000E-5	65.7	-2.9
2.000000E-5	65.7	-5.8
4.000000E-5	65.6	-11.5
6.000000E-5	65.4	-16.9
8.000000E-5	65.1	-22.1
1.000000E-4	64.8	-26.9
2.000000E-4	62.7	-45.4
4.000000E-4	58.7	-63.8
6.000000E-4	55.6	-71.8
8.000000E-4	53.3	-76.2

1.000000E-3	51.5	-78.8
2.000000E-3	45.6	-84.4
4.000000E-3	39.6	-87.2
6.000000E-3	36.1	-88.1
8.000000E-3	33.6	-88.6
.01	31.7	-88.9
.02	25.6	-89.4
.04	19.6	-89.7
.06	16.1	-89.8
.08	13.6	-89.9
.1	11.7	-89.9

13.7   Frequency Response of a First-Order Lead Component

The first order lead transfer function is the reciprocal of the first-order lag transfer function.

$$\text{First order lead transfer function} = \frac{H}{M} = \tau S + 1 \qquad (13.13)$$

The frequency response is obtained by substituting $j\omega$ for S in the transfer function.

$$(\tau)(j\omega) + 1 = 1 + j\omega\tau$$

The BASIC program in Section 13.6 may be modified to produce a frequency response table of a first-order lead component by substituting the following input and calculation subroutines.

```
100 REMARK: A SUBROUTINE FOR INPUT OF FIRST ORDER LEAD PARAMETERS
101 PRINT "FIRST ORDER LEAD FREQUENCY RESPONSE"
102 PRINT "WHAT IS THE FIRST ORDER LEAD TIME"
103 INPUT T
104 PRINT "WHAT IS THE VALUE OF G"
105 INPUT G1
106 RETURN

150 REMARK: A SUBROUTINE FOR CALCULATION OF FIRST ORDER LEAD DATA
151 R = 1
152 I = W*T
153 GO SUB 400
154 G = G1*C
155 B = A
156 RETURN
```

The output of a computer RUN of the BASIC program using the above subroutines is shown next. The output contains the frequency response values of a first-order lead with $\tau = 1$. A comparison of the

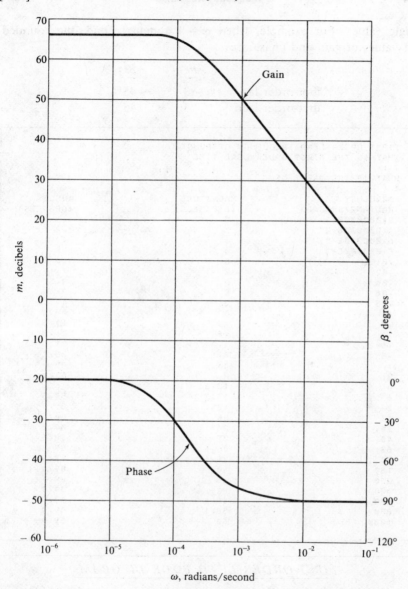

FIGURE 13.12

THE BODE DIAGRAM OF A FIRST-ORDER LAG FLOW PROCESS (EXAMPLE 13.6)

frequency response of a first-order lead with the response of a first-order lag in Section 13.6 ($\tau = 1$, $G = 1$) leads to an interesting conclusion. The only difference is the sign of the magnitude and phase angle values. For example, when $\omega = 1$ radian/second, the rounded off values of gain and phase are:

	m	β
first-order lag	-3	$-45°$
first-order lead	3	45

```
FIRST ORDER LEAD FREQUENCY RESPONSE
WHAT IS THE FIRST ORDER LEAD TIME
? 1
WHAT IS THE VALUE OF G
? 1
FREQUENCY                 MAGNITUDE              ANGLE
(RADIANS/SECOND)          (DECIBELS)             (DEGREES)
  2.000000E-3             0                        .1
  4.000000E-3             0                        .2
  6.000000E-3             0                        .3
  8.000000E-3             0                        .5
   .01                    0                        .6
   .02                    0                       1.1
   .04                    0                       2.3
   .06                    0                       3.4
   .08                    0                       4.6
   .1                     0                       5.7
   .2                      .2                     11.3
   .4                      .6                     21.8
   .6                     1.3                     31
   .8                     2.1                     38.7
  1                       3                       45
  2                       7                       63.4
  4                      12.3                     76
  6                      15.7                     80.5
  8                      18.1                     82.9
 10                      20                       84.3
 20                      26                       87.1
 40                      32                       88.6
 60                      35.6                     89
 80                      38.1                     89.3
100                      40                       89.4
200                      45                       89.7
400                      52                       89.9
600                      55.6                     89.9
800                      58.1                     89.9
1000                     60                       89.9
```

FIRST-ORDER LEAD BODE DIAGRAM

The Bode diagram of a first-order lead can be approximated by straight-line gain and phase-angle graphs. The gain-graph break point is at $\left(\dfrac{1}{\tau} rad/sec, 0\ db\right)$. On the left of the break point, the gain is 0 db. On the right, the gain line has a slope of $+20\ db/$ decade. The phase-angle break points are located at $\left(\dfrac{0.1}{\tau} rad/sec,\right.$ $\left.0°\right)$ and $\left(\dfrac{10}{\tau} rad/sec, +90°\right)$. A straight line connects the two *β break points: on the left of this line, the phase angle is 0°; on the right, the phase angle is +90°.*

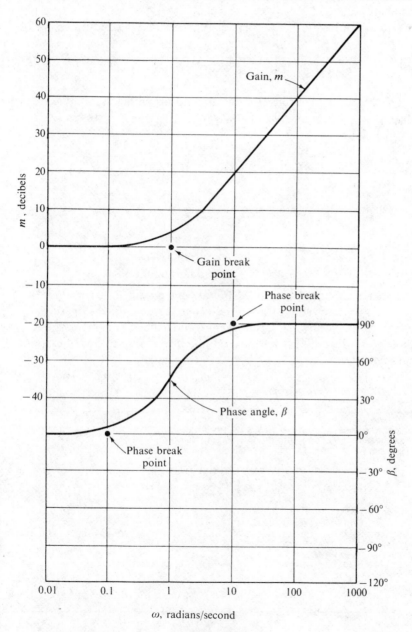

FIGURE 13.13

THE BODE DIAGRAM OF A FIRST-ORDER LEAD $(1 + j\omega)$

13.8 Frequency Response of a Second-Order Lag Component

The second-order lag was also introduced in Chapter Twelve. Its transfer function is characterized by a quadratic equation with S in the denominator.

Equation (13.14) is the transfer function of a second-order lag.

$$\frac{H}{M} = \frac{G}{\dfrac{1}{\omega_o^2} S^2 + 2\dfrac{\alpha}{\omega_o^2} S + 1} \qquad (13.14)$$

The frequency response of the second-order lag is obtained by replacing S in Equation (13.14) by $j\omega$.

$$\frac{G}{\dfrac{1}{\omega_o^2}(j\omega)^2 + 2\left(\dfrac{\alpha}{\omega_o^2}\right)(j\omega) + 1} = \frac{G}{1 - \left(\dfrac{\omega}{\omega_o}\right)^2 + j2\left(\dfrac{\alpha}{\omega_o}\right)\left(\dfrac{\omega}{\omega_o}\right)}$$

$$\text{Second-order lag frequency response} = \frac{G}{1 - \left(\dfrac{\omega}{\omega_o}\right)^2 + j2\,(\zeta)\left(\dfrac{\omega}{\omega_o}\right)}$$

where:

$$\zeta = \alpha/\omega_o = \text{the damping ratio}$$

The BASIC program in 13.6 may be modified to produce a frequency response table of a second-order lag component by substituting either of the following sets of input and calculation subroutines. The choice of subroutines depends on the form desired for the input of parameters.

SUBROUTINES FOR USE WITH ω_o AND ζ

```
100 REMARK: A SUBROUTINE FOR INPUT OF SECOND ORDER LAG PARAMETERS
101 PRINT "SECOND ORDER LAG FREQUENCY RESPONSE"
102 PRINT "WHAT IS THE RESONANT FREQUENCY"
103 INPUT W0
104 PRINT "WHAT IS THE DAMPING RATIO"
105 INPUT R0
106 PRINT "WHAT IS THE VALUE OF G"
107 INPUT G1
108 RETURN

150 REMARK: A SUBROUTINE FOR CALCULATION OF SECOND ORDER LAG DATA
151 R = 1 - (W/W0)↑2
152 I = 2*R0*W/W0
153 GO SUB 400
154 G = G1/C
155 B = -A
156 RETURN
```

SUBROUTINES FOR USE WITH $1/\omega_o{}^2$ AND $2\alpha/\omega_o{}^2$

```
100 REMARK: A SUBROUTINE FOR INPUT OF SECOND ORDER LAG PARAMETERS
101 PRINT "SECOND ORDER LAG FREQUENCY RESPONSE"
102 PRINT "WHAT IS THE VALUE OF 1/W0†2"
103 INPUT A1
104 PRINT "WHAT IS THE VALUE OF 2*ALPHA/W0†2"
105 INPUT B1
106 PRINT "WHAT IS THE VALUE OF G"
107 INPUT G1
108 RETURN

150 REMARK: A SUBROUTINE FOR CALCULATION OF SECOND ORDER LAG DATA
151 R = 1-A1*W†2
152 I = B1*W
153 GO SUB 400
154 G = G1/C
155 B = -A
156 RETURN
```

SECOND-ORDER LAG BODE DIAGRAM

A second-order lag is described by the following general transfer function.

$$\frac{H}{M} = \frac{G}{aS^2 + bs + 1}$$

where:

$G =$ the steady-state gain

$a = 1/\omega_o{}^2 =$ a constant

$b = 2\alpha/\omega_o{}^2 = 2\,\zeta/\omega_o =$ a constant

If $\zeta > 1$, the system is overdamped.
If $\zeta = 1$, the system is critically damped.
If $\zeta < 1$, the system is underdamped.

The overdamped and critically damped second-order lag systems can be expressed in terms of two equivalent first-order lag systems with the following transfer function.

$$\frac{G}{aS^2 + bs + 1} = G\left[\frac{1}{\tau_1 s + 1}\right]\left[\frac{1}{\tau_2 s + 1}\right]$$

where:

$$\tau_1 = (b + \sqrt{b^2 - 4a})/(2a)$$
$$\tau_2 = (b - \sqrt{b^2 - 4a})/(2a)$$

An overdamped second-order system is illustrated in Example 13.7. The underdamped second-order system gain graph can be approximated by a pair of straight lines. The break point is located at $\left(\omega_o = \dfrac{1}{\sqrt{a}} \text{ rad/sec, } 0 \text{ db} \right)$. On the left of the break point, the gain is 0 db. On the right, the gain line has a slope of −40 db per decade. The straight line is accurate except for the region between $0.1\omega_b$ and $10\omega_b$. The phase angle changes from 0° on the right of the break point to −180° on the left. Underdamped second-order systems are illustrated in Figure 13.14.

Example 13.7　　The second-order electrical system in Example 12.9 has the following transfer function.

$$\frac{E_o(S)}{E_i(S)} = \frac{1}{22 \times 10^{-8}S^2 + 20 \times 10^{-4}S + 1}$$

Modify the BASIC program to give values of ω from 20 to 1,000,000 radians/second and construct a Bode diagram.

Solution:

Change line 6 in the program to: 6 FOR E = 1 TO 5 STEP 1. The Table produced by a RUN of the program is given below. The Bode diagram is given in Figure 13.15.

FREQUENCY (RADIANS/SECOND)	MAGNITUDE (DECIBELS)	ANGLE (DEGREES)
20	0	-2.3
40	0	-4.6
60	-.1	-6.8
80	-.1	-9.1
100	-.2	-11.3
200	-.6	-22
400	-2	-39.7
600	-3.6	-52.5
800	-5.2	-61.8
1000	-6.6	-68.7
2000	-12	-88.3
4000	-18.5	-107.5
6000	-22.8	-120
8000	-26.3	-129.3
10000	-29.2	-136.4
20000	-39.6	-155.3
40000	-51.1	-167.2
60000	-58.1	-171.4
80000	-63	-173.5
100000	-66.9	-174.8

200000	-78.9	-177.4
400000	-90.9	-178.7
600000	-98	-179.1
800000	-103	-179.3
1.000000E+6	-106.8	-179.5

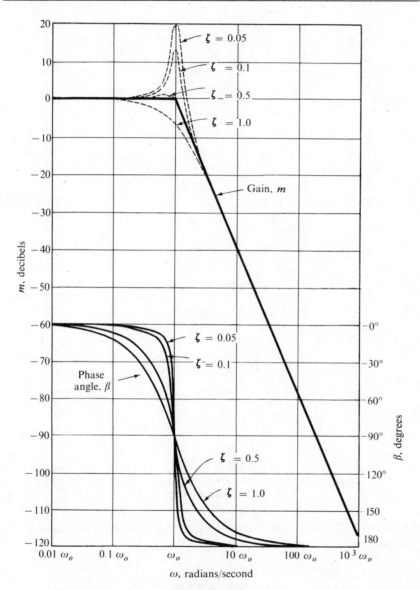

FIGURE 13.14
*THE BODE DIAGRAM OF UNDERDAMPED
SECOND-ORDER SYSTEMS*

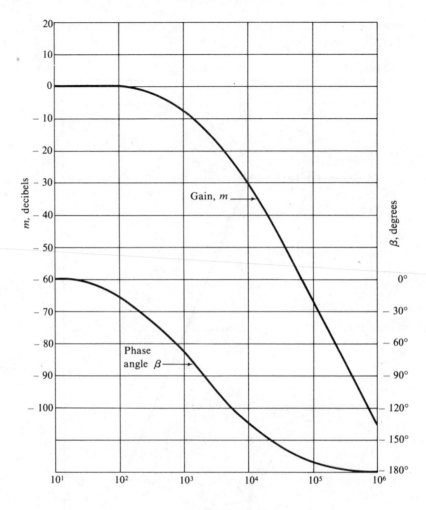

ω, radians/second

FIGURE 13.15

THE BODE DIAGRAM OF AN OVERDAMPED SECOND-ORDER SYSTEM (EXAMPLE 13.7)

Example 13.8 The underdamped second-order system in Example 7.8 has the following transfer function.

$$\frac{X}{F} = \frac{0.001}{0.01S^2 + 0.02S + 1}$$

Determine the break-point coordinates and the damping ratio (α/ω_o). Construct the Bode diagram.

Solution:

$$\omega_o = \sqrt{1/0.01} = 10 \text{ radians/second}$$

$$\frac{2\alpha}{\omega_o^2} = 0.02 \therefore \alpha = 0.01\omega_o^2 = 0.01(100) = 1 \text{ second}^{-1}$$

$$\zeta = \alpha/\omega_o = 1/10 = 0.1$$

The break-point frequency is 10 radians/second, so the BASIC program was modified to give values of ω from 0.2 to 1000 radians/second (line 6 was changed to: FOR $E = -1$ TO 2 STEP 1). The RUN output is shown below. The Bode diagram is given in Figure 13.16.

```
SECOND ORDER LAG FREQUENCY RESPONSE
WHAT IS THE VALUE OF 1/W0↑2
? 0.01
WHAT IS THE VALUE OF 2*ALPHA/W0↑2
? 0.02
WHAT IS THE VALUE OF G
? 0.001
FREQUENCY                MAGNITUDE              ANGLE
(RADIANS/SECOND)         (DECIBELS)             (DEGREES)
  .2                       -60                    -.2
  .4                       -60                    -.5
  .6                       -60                    -.7
  .8                       -59.9                  -.9
  1                        -59.9                  -1.2
  2                        -59.7                  -2.4
  4                        -58.5                  -5.4
  6                        -56.3                  -10.6
  8                        -51.9                  -24
 10                        -46                    -90
 20                        -69.6                  -172.4
 40                        -83.5                  -176.9
 60                        -90.9                  -178
 80                        -96                    -178.5
100                        -99.9                  -178.8
200                        -112                   -179.4
400                        -124.1                 -179.7
600                        -131.1                 -179.8
800                        -136.1                 -179.9
1000                       -140                   -179.9
```

13.9   Frequency Response of a Dead-Time Lag Component

A dead-time lag component delays a signal but does not change the size of the signal. The gain is unity (0 decibels) at all frequencies, and the Bode-diagram gain curve lies on the 0-db line. The phase angle (β) of a dead-time lag can be determined by Equation (13.15).

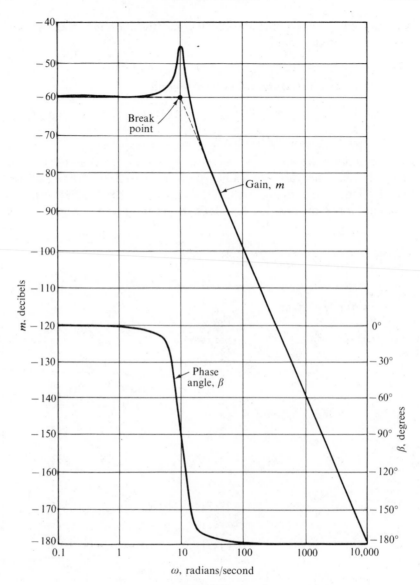

FIGURE 13.16

THE BODE DIAGRAM OF A SECOND-ORDER SYSTEM
(EXAMPLE 13.8)

DEAD-TIME LAG BODE DIAGRAM

The frequency response of a dead time lag is given by the following two relationships.

$$m = 0 \text{ decibels}$$

$$\beta = -57.3\omega T_d \text{ degrees} \qquad \textbf{(13.15)}$$

where:

$$\beta = \text{the phase angle, degrees}$$
$$\omega = \text{the radian frequency, radians/second}$$
$$T_d = \text{the dead-time lag, seconds}$$

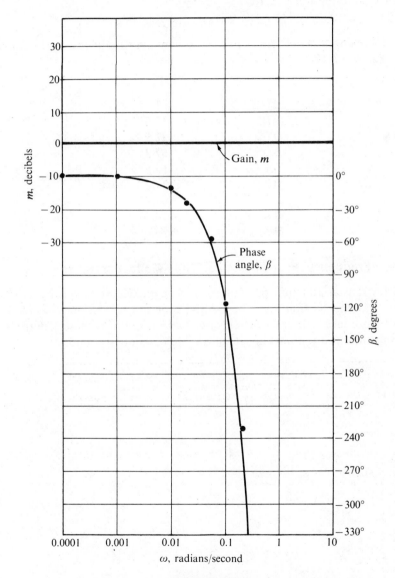

FIGURE 13.17
THE BODE DIAGRAM OF A DEAD-TIME LAG ELEMENT
(EXAMPLE 13.9)

Example 13.9 The belt conveyor in Example 7.7 has a dead-time lag of 20 seconds. Construct the Bode diagram.

Solution:

The gain graph is a straight line at 0 decibels. The phase-angle curve is defined by Equation (13.15).

$$\beta = -1146 \, \omega \text{ degrees}$$

The following table of values is used to construct the phase-angle graph on Figure 13.17.

ω	β
0.001	1.15°
0.01	11.5°
0.02	22.9°
0.05	57.3°
0.1	114.6°
0.2	229.2°
0.3	343.8°
0.4	458.4°
0.5	573.0°

13.10 Frequency Response of a Three-Mode Controller

The transfer function of a three-mode controller is given by Equation (9.22) in Section 9.9. The BASIC program in 13.6 may be modified to produce a frequency response table of a three-mode controller by substituting the following input and calculation subroutines.

```
100 REMARK: A SUBROUTINE FOR INPUT OF THREE MODE CONTROLLER PARAMETERS
101 PRINT "THREE MODE CONTROLLER FREQUENCY RESPONSE"
102 PRINT"GIVE THE VALUES OF THE DERIVATIVE LIMITER COEFFICIENT (A1),"
103 PRINT "DERIVATIVE TIME (T1), INTEGRAL TIME (T2), AND GAIN (K)."
104 PRINT "A1,T1,T2,K";
105 INPUT A1,T1,T2,K
106 RETURN

150 REMARK: A SUBROUTINE FOR CALCULATION OF 3 MODE CONTROLLER DATA
151 R = 1
152 I = T1*W
153 GO SUB 400
154 G = C
155 B = A
156 I = A1*T1*W
157 GO SUB 400
158 G = G/C
159 B = B - A
160 I = T2*W
161 GO SUB 400
162 G = G*K*C/(T2*W)
163 B = B + A - 90
164 RETURN
```

EXERCISES

13–1.   The Bode diagram shown in Figure 13.4 describes the frequency response of an electro-mechanical component. Both the input and the output signals are measured in volts. If the input signal has an amplitude of 10 volts, determine the amplitude and phase difference of the output signal at each of the following frequencies.

a. $\omega = 0.01$ radians/second
b. $\omega = 0.1$ radians/second
c. $\omega = 1$ radian/second
d. $\omega = 10$ radians/second
e. $\omega = 100$ radians/second
f. $\omega = 1000$ radians/second

13–2.   The liquid tank in Example 12.1 is an integral process, with $A = 1.76$ square meters. The Bode diagram is given in Figure 13.9. The input signal (M) is measured in cubic meters/second. The output signal (H) is measured in meters. If the input signal amplitude is 0.01 cubic meter/second, determine the amplitude and phase of the output signal at each of the following frequencies.

a. $0.01/1.76$ radians/second
b. $0.1/1.76$ radians/second
c. $1/1.76$ radians/second
d. $10/1.76$ radians/second
e. $100/1.76$ radians/second

13–3.   The Bode diagram in Figure 13.16 describes the frequency response of the spring-mass-damping system in Example 12.9. The input signal (F) is measured in newtons. The output signal (X) is measured in meters. If the input-signal amplitude is 100 newtons, determine the amplitude and phase of the output signal at each of the following frequencies.

a.   0.1 radians/second
b.   1 radian/second
c.   10 radians/second
d.   100 radians/second
e.   1000 radians/second
f.   10,000 radians/second

13–4.   Construct the Bode diagrams of proportional components with the following gains.

a. $G = 1$ b. $G = 10$ c. $G = 0.1$
d. $G = 6$ e. $G = 0.25$ f. $G = 14$

13–5. Construct the Bode diagram of the sheet-loop process illustrated in Figure 12.4. It is an integral process with the following transfer function.

$$\frac{V_o}{V_i} = \frac{1}{2S}$$

13–6. Construct the Bode diagram of the first-order lag electrical circuit in Example 12.4. The transfer function is given below.

$$\frac{E_o}{E_i} = \frac{1}{0.492S + 1}$$

13–7. Construct the Bode diagram of the first-order lag thermal system in Example 12.5. The transfer function is given below.

$$\frac{\theta_L}{\theta_j} = \frac{1}{6.43 \times 10^3 S + 1}$$

13–8. Construct the Bode diagram of the first-order lag, gas-pressure system of Example 12.6. The transfer function is given below.

$$\frac{P}{M} = \frac{2 \times 10^5}{2.8S + 1}$$

13–9. Construct the Bode diagram of the first-order lag blending system of Example 12.7. The transfer function is given below.

$$\frac{C_o}{C_i} = \frac{1}{10^3 S + 1}$$

13–10. A dead-time lag process has a dead-time lag of 200 seconds. Construct the Bode diagram.

13–11. Construct the Bode diagram of the RLC circuit in Example 12.10. It is an overdamped second-order system with the following transfer function.

$$\frac{E_o}{E_i} = \frac{1}{22 \times 10^{-8} S^2 + 20 \times 10^{-4} S + 1}$$

13–12. Construct the Bode diagram of the liquid non-interacting, two-capacity system in Example 12.11. The two time constants are 520 seconds and 960 seconds. The transfer function is given below.

$$\frac{H}{M} = \frac{163}{5 \times 10^5 S^2 + 1.48 \times 10^3 S + 1}$$

Hint: The denominator is equal to

$$\tau_1 \tau_2 S^2 + (\tau_1 + \tau_2)S + 1$$

which is easily factored.

13-13. Construct the Bode diagram of an electrical, interacting, two-capacity circuit with the following transfer function.

$$\frac{E_o}{E_i} = \frac{0.182}{7.28 \times 10^{-5}S^2 + 1.18 \times 10^{-2}S + 1}$$

13-14. Construct the Bode diagram of the armature-controlled dc motor in Example 12.12. The transfer function is given below.

$$\frac{W}{E} = \frac{21.8}{6.35 \times 10^{-3}S^2 + 0.379S + 1}$$

13-15. Construct the Bode diagram of the proportional-plus-integral controller in Example 9.4. The transfer function is given below.

$$\frac{V}{E} = 2\left[\frac{1 + 50S}{50S}\right]$$

13-16. Construct the Bode diagram of the proportional-plus-derivative controller in Example 9.6. The transfer function is given below.

$$\frac{V}{E} = \frac{0.8(S + 1)}{(0.1S + 1)}$$

13-17. Construct the Bode diagram of the three-mode controller in Example 9.8. The transfer function is given below.

$$\frac{V}{E} = 4\left[\frac{7S + 1}{7S}\right]\left[\frac{0.5S + 1}{0.05S + 1}\right]$$

14

Closed Loop Response
and Stability

14.1 Introduction

The computer methods presented in Chapter Thirteen can be used to construct the overall Bode diagram of all the components in a linear control system, including the controller, final-control element, process, and measuring means. This overall Bode diagram expresses the changes in the amplitude and phase angle between the controller input signal and the measuring means output signal. There are actually two overall Bode diagrams of a control system, one gives the open-loop frequency response of the system; the other gives the closed-loop response. In the design of linear control systems, the *open-loop Bode diagram* is used to determine the optimum controller parameters. When the design is completed, the *closed-loop Bode diagram* is used to express the behavior of the control system.

The *open-loop Bode diagram* expresses the frequency response of the control system when the measuring means output is disconnected from the error detector. Since the measuring means output (Cm) is not used, the output of the error detector is equal to the setpoint signal (i.e., $E = SP$). The open-loop response at a given frequency is obtained by multiplying the gains and adding the phase angles of all components in the system. The open-loop Bode diagram is the graph of open-loop gain and phase angle vs. frequency.

The *closed-loop Bode diagram* expresses the frequency response of the control system when the measuring means output is connected to the error detector. The setpoint is still the input to the system, but the error detector output is the setpoint minus the measuring means output (i.e., $E = SP - C_m$). A computer program can be used to determine the closed-loop frequency response from the open-loop response.

Another method of expressing the behavior of a control system is to compare the error that would result when feedback control is used (closed-loop) with the error when feedback is not used (open-loop). This comparison is expressed as the *error ratio,* which is the ratio of the closed-loop error over the open-loop error. An error ratio less than one means that the closed-loop error is less than the open-loop error. A ratio greater than one means that the closed-loop error is greater than the open-loop error. In other words, if the ratio is less than one, closed-loop control is better than open-loop control. If the ratio is greater than one, closed-loop control does more harm than good—the error is actually greater than it would be without closed-loop control. If the ratio is equal to one, closed-loop control neither increases nor decreases the error.

A third method of expressing the behavior of a control system is to compare the closed-loop error with the setpoint. This comparison is expressed by the *deviation ratio* which is the ratio of the closed-loop error over the setpoint. The deviation ratio is an indication of how accurately a control system can follow a change in the setpoint, and it is used to evaluate follow-up control systems.

The possibility of sustained oscillations always exists in a closed-loop control system. When a system goes into a continuous oscillation, it is said to be *unstable.* Stability refers to the ability of a control system to dampen out any oscillations that result from an upset. The design of a control system involves the determination of controller parameters that will minimize the error produced by a disturbance without making the system unstable.

14.2   Open-Loop Frequency Response

An open-loop control system is illustrated in Figure 14.1. The first block represents the controller. The second block represents the final-control element, process, and measuring means. The symbol $G(S)$ is used to represent the transfer function of the controller [the (S) simply indicates that the transfer function depends on the frequency parameter S]. The

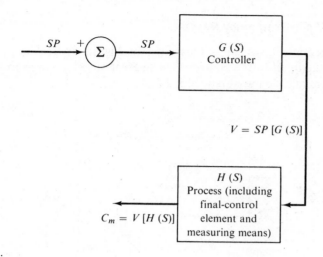

$$\frac{C_m}{SP} = [G\,(S)]\,[H\,(S)]$$

FIGURE 14.1
AN OPEN-LOOP CONTROL SYSTEM

symbol $H(S)$ is used to represent the overall transfer function of the final-control element, process and measuring means.

The open-loop transfer function is obtained by substituting $SP[G(S)]$ for V in the equation for C_m (Figure 14.1) and then dividing the resulting equation by SP.

OPEN-LOOP RESPONSE

Open-loop transfer function $= \dfrac{C_m}{SP} = [G(S)][H(S)]$　　**(14.1)**

The open-loop Bode diagram is constructed with the aid of a computer by multiplying the gains and adding the phase angles of the controller, final-control element, process, and measuring means (see Example 14.1).

The open-loop Bode diagram is constructed graphically by adding the decibel gain and phase graphs of the controller, final-control element, process, and measuring means.

14.3   Closed-Loop Frequency Response

A closed-loop control system is illustrated in Figure 14.2. The error signal (E) is the input to the controller. The error signal is equal to the setpoint (SP) minus the measured value (C_m).

$$E = SP - C_m$$

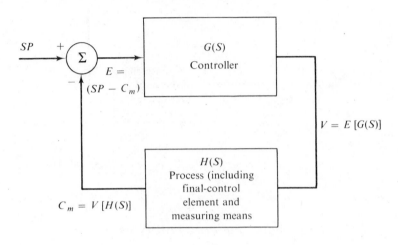

$$\frac{C_m}{SP} = \frac{[G(S)]\,[H(S)]}{[G(S)]\,[H(S)] + 1}$$

FIGURE 14.2

A CLOSED-LOOP CONTROL SYSTEM

The output of the controller (V) is equal to the input (E) times the transfer function $G(S)$.

$$V = E[G(S)] = [SP - C_m][G(S)]$$

The output of the process (C_m) is equal to the input (V) times the transfer function $H(S)$.

$$C_m = V[H(S)] = [SP - C_m][G(S)][H(S)]$$

The last equation may be solved for the closed-loop transfer function $\dfrac{C_m}{SP}$.

$$C_m = SP[G(S)][H(S)] - C_m[G(S)][H(S)]$$
$$C_m + C_m[G(S)][H(S)] = SP[G(S)][H(S)]$$

$$C_m[1 + [G(S)][H(S)]] = SP[G(S)][H(S)]$$

$$\frac{C_m}{SP} = \frac{[G(S)][H(S)]}{1 + [G(S)][H(S)]} = \begin{array}{l}\text{closed loop transfer}\\\text{function}\end{array}$$

where $[G(S)]$ $[H(S)]$ is the open-loop transfer function. In other words, the closed-loop response is equal to the open-loop response divided by 1 plus the open-loop response. The open-loop response values are determined and tabulated in Example 14.1.

Let

$$g_o \; \underline{\big|\beta_o} = \text{the open-loop response at } \omega \text{ rad/sec}$$

$$g_c \; \underline{\big|\beta_c} = \text{the closed-loop response at } \omega \text{ rad/sec}$$

$$a + jb = \text{the rectangular form of } g_o \; \underline{\big|\beta_o}$$

Then

$$g_c \; \underline{\big|\beta_c} = \frac{g_o \; \underline{\big|\beta_o}}{1 + g_o \; \underline{\big|\beta_o}} = \frac{a + jb}{1 + a + jb}$$

CLOSED-LOOP RESPONSE

Closed-loop transfer function $= \dfrac{C_m}{SP} = \dfrac{[G(S)][H(S)]}{1 + [G(S)][H(S)]}$ **(14.2)**

The closed-loop Bode diagram is constructed with the aid of a computer by dividing the open-loop response by 1 plus the open-loop response (see Example 14.1).

The closed-loop Bode diagram is constructed graphically by converting the open-loop gain and phase values to the corresponding closed-loop values. A Nichols Chart is used to convert each set of open-loop gain and phase values into the corresponding closed-loop gain and phase values.

The closed-loop gain and phase graphs may be determined analytically or graphically. The analytical method consists of substituting $j\omega$ for S in the transfer function. For each value of ω, the transfer function reduces to a single complex number. The magnitude of the complex number is the closed-loop gain. The angle of the complex number is the closed-loop phase angle.

The following BASIC program produces a table of the open-loop and the closed-loop frequency response of a first-order lag plus dead-time lag process. The program is similar to the programs in Chapter 13 with the addition of a subroutine to calculate the closed-loop response from the open-loop response (lines 250–264).

The closed-loop gain and phase graphs may be determined graphically from the open-loop gain and phase graph. A Nichols Chart is used

```
 1 REMARK: A PROGRAM FOR CALCULATING FREQUENCY RESPONSE DATA
 3 GO SUB 100
 4 PRINT "FREQUENCY","OPEN LOOP RESPONSE","CLOSED LOOP RESPONSE"
 5 PRINT"(RAD/SEC)","(DECIBELS)","(DEGREES)","(DECIBELS)","(DEGREES)"
 6 FOR E = -4 TO 0 STEP 1
 7 FOR F = 2 TO 10 STEP 2
 8 W = F*10↑E
 9 GO SUB 150
10 M = 20*LOG(G)/2.30259
11 M = INT(10*(M+0.05))/10
12 B = INT(10*(B+0.05))/10
13 GO SUB 250
14 PRINT W,M,B,M0,B0
15 NEXT F
16 NEXT E
17 STOP

100 REMARK: A SUBROUTINE FOR INPUT OF A FIRST ORDER LAG PLUS
101 REMARK: DEAD TIME LAG PROCESS
102 PRINT "FIRST ORDER LAG PLUS DEAD TIME RESPONSE"
103 PRINT "GIVE THE VALUES OF GAIN (G1), TIME CONSTANT (T),"
104 PRINT "AND DEAD TIME LAG (L)"
105 PRINT "G1,T,L";
106 INPUT G1,T,L
107 RETURN

150 REMARK: A SUBROUTINE FOR CALCULATION OF DATA FOR A FIRST ORDER
151 REMARK: LAG PLUS DEAD TIME LAG PROCESS
152 R = 1
153 I = T*W
154 GO SUB 400
155 G = G1/C
156 B = -A - 57.3*W*L
157 RETURN

250 REMARK: A SUBROUTINE FOR CALCULATION OF THE CLOSED LOOP RESPONSE
251 REMARK: OF A SYSTEM FROM THE OPEN LOOP RESPONSE
252 R = 1 + G*COS(B/57.2958)
253 I = G*SIN(B/57.2958)
254 GO SUB 400
255 G0 = G/C
256 B0 = B - A
257 M0 = 20*LOG(G0)/2.30259
258 M0 = INT(10*(M0+0.05))/10
259 B0 = INT(10*(B0+0.05))/10
260 RETURN

400 REMARK: A SUBROUTINE FOR RECTANGULAR-TO-POLAR CONVERSION
401 C = SQR(R↑2 + I↑2)
402 IF R<=0 THEN 405
403 A = ATN(I/R)*57.2958
404 RETURN
405 IF R = 0 THEN 408
406 A = ATN(I/R)*57.2958 + 180
407 RETURN
408 IF I<0 THEN 411
409 A = 90
410 RETURN
411 A = -90
412 RETURN
```

to convert each open-loop gain and phase value into the corresponding closed-loop gain and phase value. Figure 14.3 is an example of a Nichols Chart. The rectangular coordinates are used for the open-loop decibel gain and phase angle. The open-loop decibel gain ranges from -24 to $+24$ on the y-axis. The open-loop phase angle ranges from

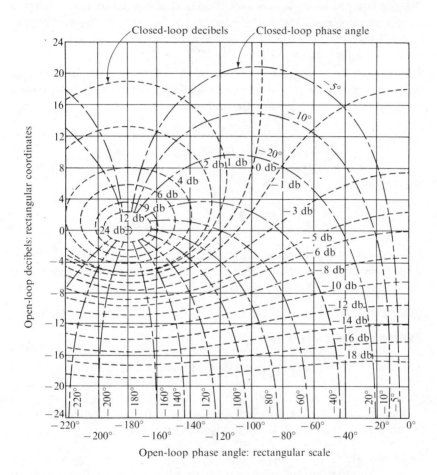

Closed-loop decibels − − − − − − −

Closed-loop phase angle —— - - ——

FIGURE 14.3 [1]

THE NICHOLS CHART

---

[1] Figures 14.3 and 14.4 are based on information from H. M. James, N. B. Nichols and R. S. Phillips, *Theory of Servomechanisms* (New York: McGraw-Hill Book Company, 1947), p. 179.

—220° to 0° on the *x*-axis. The curved coordinates are used for the closed-loop decibel gain and phase angle. They are similar to contour lines on a contour map. The short dashes are used for the closed-loop gain contours. Notice the large gain peak centered at the rectangular co-ordinates (—180°, 0 db). The long-short-long dashes are used for the closed-loop, phase-angle contours. Notice that the phase-angle contours form a spiral staircase with the post at the $(-180°, 0$ db) rectangular coordinates.

Example 14.1 A control system consists of a proportional controller and a first-order lag plus dead-time process. The following values were obtained from a step response test of the process.

$$t_d = \quad 2 \text{ seconds}$$
$$t_1 = 100 \text{ seconds}$$

Therefore,

$$H(S) = [e^{-2S}]\left[\frac{1}{1 + 100S}\right]$$

The controller gain is set at 40, so $G(S) = [40]$. Determine the open-loop transfer function and construct the open-loop and closed loop Bode diagrams of the control system.

Solution:

The open-loop transfer function is

$$\frac{C_m}{SP} = [G(S)][H(S)] = [40][e^{-2S}]\left[\frac{1}{1 + 100S}\right]$$

The solution may be obtained by the computer aided method or by the graphical method. Both methods will be presented.

Graphical Method. The open-loop Bode diagram is obtained by graphi-cally adding the decibel gains and phase angles of the individual com-ponents in the system. The closed-loop frequency response is obtained from the open-loop response by the following two-step procedure.

1. Locate each frequency point on the Nichols Chart. The open-loop decibel gain and phase-angle values are used to locate each point.
2. Determine the closed-loop gain and phase angle of each frequency point. The curved coordinates are used to determine the closed-loop values.

The frequency points for the system of Example 14.1 are plotted on the Nichols Chart in Figure 14.4. Some of the closed-loop lines are deleted to emphasize the frequency points. The closed-loop Bode diagram is given in Figure 14.5.

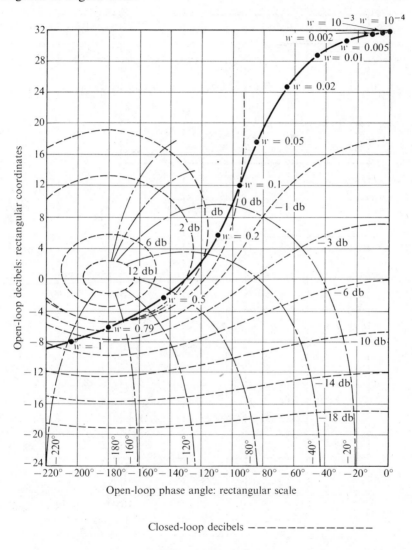

Closed-loop decibels ————————————

Closed-loop phase angle ———-—————--—

FIGURE 14.4

CLOSED LOOP RESPONSE FROM THE NICHOLS CHART
EXAMPLE 14.1

Computer-Aided Method. The output of a RUN of the BASIC program in Section 14.3 is on the next page. The open-loop and closed-loop frequency response graphs are given in Figure 14.5.

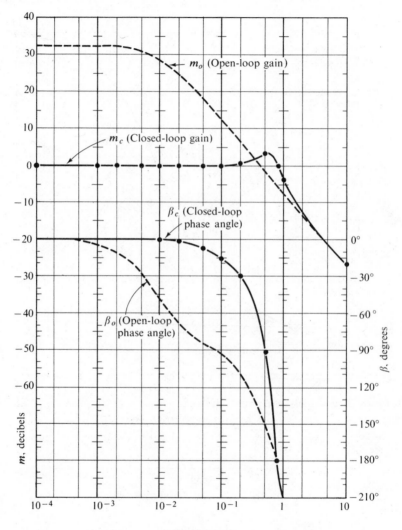

ω, radians/second

FIGURE 14.5

THE OPEN-LOOP AND CLOSED-LOOP BODE DIAGRAMS OF

$$THE\ SYSTEM\ [40][e^{-2S}]\left[\frac{1}{1 + 100S}\right],\ EXAMPLE\ 14.1$$

THE COMPUTER OUTPUT FOR EXAMPLE 14.1: A TABLE OF THE OPEN-LOOP AND CLOSED-LOOP FREQUENCIES OF THE SYSTEM GIVEN BY $[G(S)][H(S)] = [40][e^{-2S}]\left[\dfrac{1}{1 + 100S}\right]$

```
FIRST ORDER LAG PLUS DEAD TIME RESPONSE
GIVE THE VALUES OF GAIN (G1), TIME CONSTANT (T),
AND DEAD TIME LAG (L)
G1,T,L? 40,100,2
```

FREQUENCY (RAD/SEC)	OPEN LOOP RESPONSE		CLOSED LOOP RESPONSE	
	(DECIBELS)	(DEGREES)	(DECIBELS)	(DEGREES)
2.000000E-4	32	-1.2	-.2	0
4.000000E-4	32	-2.3	-.2	-.1
6.000000E-4	32	-3.5	-.2	-.1
8.000000E-4	32	-4.7	-.2	-.1
1.000000E-3	32	-5.8	-.2	-.1
2.000000E-3	31.9	-11.5	-.2	-.3
4.000000E-3	31.4	-22.3	-.2	-.6
6.000000E-3	30.7	-31.7	-.2	-.9
8.000000E-3	29.9	-39.6	-.2	-1.1
.01	29	-46.1	-.2	-1.4
.02	25.1	-65.7	-.2	-2.9
.04	19.7	-80.5	-.2	-5.7
.06	16.4	-87.4	-.2	-8.6
.08	13.9	-92	-.1	-11.5
.1	12	-95.7	-.1	-14.4
.2	6	-110.1	.4	-29.6
.4	0	-134.4	2.2	-67.2
.6	-3.5	-157.8	3.3	-124.5
.8	-6	-181	0	-182
1	-8	-204	-4.3	-218.4
2	-14	-318.9	-15.3	-325.4
4	-20	-548.3	-19.1	-549.2
6	-23.5	-777.5	-23.8	-774.4
8	-26	-1006.7	-26.2	-1009.4
10	-28	-1235.9	-27.6	-1234.9

Example 14.2   Construct the open-loop and the closed-loop Bode diagrams of the system in Example 14.1 if the controller gain is changed to 28.2.

Solution:

The computer output is given below. The open-loop and closed-loop Bode diagrams are given in Figure 14.6.

```
FIRST ORDER LAG PLUS DEAD TIME RESPONSE
GIVE THE VALUES OF GAIN (G1), TIME CONSTANT (T),
AND DEAD TIME LAG (L)
G1,T,L? 28.2,100,2
```

FREQUENCY (RAD/SEC)	OPEN LOOP RESPONSE		CLOSED LOOP RESPONSE	
	(DECIBELS)	(DEGREES)	(DECIBELS)	(DEGREES)
2.000000E-4	29	-1.2	-.3	0
4.000000E-4	29	-2.3	-.3	-.1

6.000000E-4	29	-3.5	-.3	-.1
8.000000E-4	29	-4.7	-.3	-.2
1.000000E-3	29	-5.8	-.3	-.2
2.000000E-3	28.8	-11.5	-.3	-.4
4.000000E-3	28.4	-22.3	-.3	-.8
6.000000E-3	27.7	-31.7	-.3	-1.2
8.000000E-3	26.9	-39.6	-.3	-1.6
.01	26	-46.1	-.3	-2
.02	22	-65.7	-.3	-4
.04	16.7	-80.5	-.3	-8
.06	13.3	-87.4	-.3	-12
.08	10.9	-92	-.3	-16.1
.1	9	-95.7	-.2	-20.2
.2	3	-113.1	-.1	-41.4
.4	-3	-134.4	-.1	-80.6
.6	-6.6	-157.8	-2	-143.4
.8	-9.1	-181	-5.3	-181.5
1	-11	-204	-8.5	-212.8
2	-17	-318.9	-17.9	-323.7
4	-23	-548.3	-22.4	-548.9
6	-26.6	-777.5	-26.8	-775.3
8	-29.1	-1006.7	-29.1	-1008.6
10	-31	-1235.9	-30.8	-1235.2

14.4 Error Ratio and Deviation Ratio

Every closed-loop control system has a maximum frequency limit. A control system cannot respond accurately to setpoint changes above its frequency limit. Disturbances with frequencies above the maximum limit are especially troublesome. The controller cannot reduce the error produced by the disturbance and may actually increase the error. The maximum frequency limit can be determined by comparing the error when feedback control is used with the error when feedback is not used. A convenient means of comparison is the *error ratio,* the ratio of the closed-loop error over the open-loop error.

$$\text{Error ratio} = \frac{\text{closed-loop error magnitude}}{\text{open-loop error magnitude}}$$

A typical error-ratio graph is illustrated in Figure 14.7. The graph is divided into three frequency zones—1, 2, and 3. In zone 1, the error ratio is less than one (zero decibels). This means that the closed-loop error is less than the open-loop error. The controller will reduce errors occurring in zone 1. In zone 2, the error ratio is greater than one (zero decibels). The closed-loop error is actually greater than the open-loop error. The controller is doing more harm than good. In zone 3, the error ratio is equal to one (zero decibels). The closed-loop error is equal to the

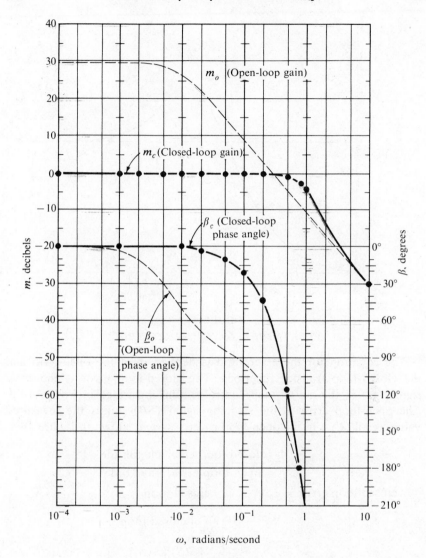

FIGURE 14.6

*THE OPEN-LOOP AND CLOSED-LOOP BODE DIAGRAMS OF
THE SYSTEM* $[28.2][e^{-2S}]\left[\dfrac{1}{1 + 100S}\right]$, *EXAMPLE 14.2*

open-loop error. The presence of the controller neither increases or decreases the error. The maximum frequency limit is the frequency that
divides zone 1 and zone 2.

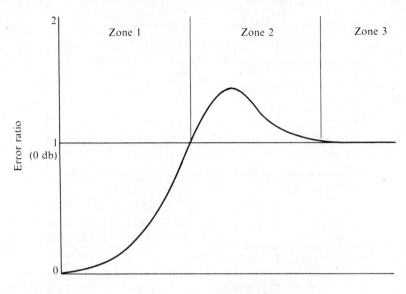

FIGURE 14.7 ²

THE THREE ZONES OF CONTROL

The equation for the error ratio is derived from the open-loop and the closed-loop transfer functions. The closed-loop error is the error signal (E) in the closed-loop control system illustrated in Figure 14.2. The open-loop error is the setpoint signal (SP) minus the measured value signal (C_m) in the open-loop control system shown in Figure 14.1.

$$\text{error ratio} = \frac{\text{closed-loop error magnitude}}{\text{open-loop error magnitude}}$$

$$\text{closed loop error} = (SP - C_m) \text{ closed loop}$$

$$= SP - SP\left[\frac{C_m}{SP}\right] \text{ closed loop}$$

$$= SP - SP\left[\frac{[G(S)][H(S)]}{1 + [G(S)][H(S)]}\right]$$

$$= SP\left[1 - \frac{[G(S)][H(S)]}{1 + [G(S)][H(S)]}\right]$$

$$= SP\left[\frac{1 + [G(S)][H(S)] - [G(S)][H(S)]}{1 + [G(S)][H(S)]}\right]$$

---

² Figure 14.7 is based on information from G. K. Tucker and D. M. Wills, *A Simplified Technique of Control System Engineering* (Minneapolis: Honeywell, Inc., 1962), pp. 92.

$$\text{closed loop error} = SP\left[\frac{1}{1 + [G(S)][H(S)]}\right] \tag{14.3}$$

$$\text{open loop error} = (SP - C_m)\ \text{open loop}$$

$$= SP - SP\left[\frac{C_m}{SP}\right]\ \text{open loop}$$

$$= SP - SP[G(S)][H(S)]$$

$$\text{open loop error} = SP[1 - [G(S)][H(S)]]$$

$$\text{error ratio} = \left|\frac{SP\left[\dfrac{1}{1 + [G(S)][H(S)]}\right]}{SP[1 - [G(S)][H(S)]]}\right|$$

The error ratio is a complex number. The parallel lines indicate that the error ratio is equal to the magnitude of this complex number, and the angle is ignored.

$$\text{Error ratio} = \left|\frac{1}{[1 + [G(S)][H(S)]]\,[1 - [G(S)][H(S)]]}\right| \tag{14.4}$$

Every closed-loop control system has a zone 2: a range of frequencies for which the controller increases the error. The results in Table 14.1 clearly illustrate one frequency which is always in zone 2—the frequency for which the open-loop phase angle is $-180°$. This frequency is designated ω_{-180}. The open loop gain at ω_{-180} is designated by g_{-180}.

TABLE 14.1

ERROR RATIO VERSUS GAIN WHEN $\beta = -180°$

g_{-180}	$1 + g_{-180}$	$1 - g_{-180}$	Error Ratio	
			(ratio)	(db)
1.0	2.0	0	∞	∞
0.9	1.9	0.1	5.26	14.4
0.8	1.8	0.2	2.78	8.9
0.7	1.7	0.3	1.96	5.85
0.6	1.6	0.4	1.56	3.85
0.5	1.5	0.5	1.33	2.50
0.4	1.4	0.6	1.19	1.50
0.3	1.3	0.7	1.10	0.8
0.2	1.2	0.8	1.04	0.35
0.1	1.1	0.9	1.01	0.10
0	1.0	1.0	1.00	0

$$g_{-180} \underline{| -180°} = [G(j\omega_{-180})][H(j\omega_{-180})]$$
$$= -g_{-180} + j0 = -g_{-180}$$

At ω_{-180}, the error ratio is given by

$$\text{Error ratio} = \left| \frac{1}{(1 + g_{-180} \underline{| -180})(1 - g_{-180} \underline{| -180})} \right|$$
$$= \frac{1}{(1 - g_{-180})(1 + g_{-180})}$$

The *deviation ratio*[3] is equal to the ratio of the closed-loop error magnitude over the setpoint magnitude. It is an indication of how accurately a control system can follow a change in setpoint. The deviation ratio is obtained directly from Equation (14.3).

$$\text{closed-loop error} = SP \left[\frac{1}{1 + [G(S)][H(S)]} \right]$$

$$\text{deviation ratio} = \left| \frac{\text{closed-loop error}}{\text{setpoint}} \right| = \left| \frac{1}{1 + [G(S)][H(S)]} \right| \quad \textbf{(14.5)}$$

The deviation ratio may also be determined by dividing the closed-loop response by the open-loop response.

$$\text{closed-loop response} = \left(\frac{C_m}{SP} \right) \text{closed} = \left[\frac{[G(S)][H(S)]}{[G(S)][H(S)] + 1} \right]$$

$$\text{open-loop response} = \left(\frac{C_m}{SP} \right) \text{open} = [G(S)][H(S)]$$

$$\frac{\text{closed-loop response}}{\text{open-loop response}} = \frac{\dfrac{[G(S)][H(S)]}{[G(S)][H(S)] + 1}}{[G(S)][H(S)]}$$

$$= \frac{1}{[G(S)][H(S)] + 1}$$

---

[3] G. K. Tucker and D. M. Wills, *A Simplified Technique of Control System Engineering* (Philadelphia: Honeywell Inc., 1962), p. 91–93.

ERROR RATIO

$$Error\ ratio = \frac{Closed\text{-}loop\ error\ magnitude}{Open\text{-}loop\ error\ magnitude}$$

$$\text{Error ratio} = \frac{1}{(1 + g_o \underline{\big|\beta_o})(1 - g_o \underline{\big|\beta_o})}$$

where:

g_o = Open-loop gain

β_o = Open-loop phase angle

The Three Zones of Control

The Error Ratio gives the relative size of the Error produced by a disturbance if the loop is closed compared with the error if the loop is open.

Zone 1, Error ratio < 0 db, closed-loop control reduces the error

Zone 2, Error ratio > 0 db, closed-loop control magnifies the error

Zone 3, Error ratio = 0 db, closed-loop control has no effect on the error

DEVIATION RATIO

$$\text{Deviation ratio} = \frac{Closed\text{-}loop\ error\ magnitude}{Setpoint\ magnitude}$$

$$\text{Deviation ratio} = \frac{1}{(1 + g_o \underline{\big|\beta_o})}$$

The Three Zones of Control

The Deviation Ratio gives the relative size of the closed-loop error produced by a change in setpoint compared with the size of the change in the setpoint.

Zone 1, Deviation Ratio < 0 db, good control

Zone 2, Deviation Ratio > 0 db, poor control

Zone 3, Deviation Ratio = 0 db, no control

Therefore,

$$\text{deviation ratio} = \left| \frac{\text{closed-loop response}}{\text{open-loop response}} \right|$$

$$\text{deviation ratio} = m_c - m_o \tag{14.6}$$

where:

$$m_c = \text{closed-loop decibel gain}$$

$$m_o = \text{open-loop decibel gain}$$

The following program is a modification of the BASIC program in Section 14.3. It determines the Error Ratio instead of the Closed-loop Response. The changes in the main program are indicated by an asterisk (*). The subroutine at line 300 replaces the one at line 250.

```
   1 REMARK: A PROGRAM FOR CALCULATING FREQUENCY RESPONSE DATA
   3 GO SUB 100
 * 4 PRINT "FREQUENCY","OPEN LOOP RESPONSE","ERROR RATIO"
 * 5 PRINT "(RAD/SEC)","(DECIBELS)","(DEGREES)","(DECIBELS)"
   6 FOR E = -4 TO 0 STEP 1
   7 FOR F = 2 TO 10 STEP 2
   8 W = F*10↑E
   9 GO SUB 150
  10 M = 20*LOG(G)/2.30259
  11 M = INT(10*(M+0.05))/10
  12 B = INT(10*(B+0.05))/10
 *13 GO SUB 300
 *14 PRINT W,M,B,E1
  15 NEXT F
  16 NEXT E
  17 STOP

 100 REMARK: A SUBROUTINE FOR INPUT OF A FIRST ORDER LAG PLUS
 101 REMARK: DEAD TIME LAG PROCESS
 102 PRINT "FIRST ORDER LAG PLUS DEAD TIME RESPONSE"
 103 PRINT "GIVE THE VALUES OF GAIN (G1), TIME CONSTANT (T),"
 104 PRINT "AND DEAD TIME LAG (L)"
 105 PRINT "G1,T,L";
 106 INPUT G1,T,L
 107 RETURN

 150 REMARK: A SUBROUTINE FOR CALCULATION OF DATA FOR A FIRST ORDER
 151 REMARK: LAG PLUS DEAD TIME LAG PROCESS
 152 R = 1
 153 I = T*W
 154 GO SUB 400
 155 G = G1/C
 156 B = -A - 57.3*W*L
 157 RETURN

 300 REMARK: A SUBROUTINE TO CALCULATE THE ERROR RATIO
 301 B1 = B/57.2958
 302 R = 1 + G*COS(B1)
 303 I = G*SIN(B1)
 304 GO SUB 400
 305 C1 = C
 306 R = 1 - G*COS(B1)
 307 I = -G*SIN(B1)
 308 GO SUB 400
```

```
309 E1 = 20*LOG(1/(C1*C))/2.30259
310 E1 = INT(10*(E1+0.05))/10
311 RETURN


400 REMARK: A SUBROUTINE FOR RECTANGULAR-TO-POLAR CONVERSION
401 C = SQR(R↑2 + I↑2)
402 IF R<=0 THEN 405
403 A = ATN(I/R)*57.2958
404 RETURN
405 IF R = 0 THEN 408
406 A = ATN(I/R)*57.2958 + 180
407 RETURN
408 IF I<0 THEN 411
409 A = 90
410 RETURN
411 A = -90
412 RETURN
```

Example 14.3   Construct an Error Ratio graph of the system in Example 14.1. Then calculate the Deviation Ratio and construct a graph of both the Error Ratio and the Deviation Ratio for comparison.

Solution:

The output of the BASIC program to determine the Error Ratio immediately follows. Table 14.2 shows the determination of several values of the Deviation Ratio. The closed-loop decibel gain, m_o, is given in Example 14.1. The graphs are Figures 14.8 and 14.9.

```
FIRST ORDER LAG PLUS DEAD TIME RESPONSE
GIVE THE VALUES OF GAIN (G1), TIME CONSTANT (T),
AND DEAD TIME LAG (L)
G1,T,L? 40,100,2
```

FREQUENCY	OPEN LOOP RESPONSE		ERROR RATIO
(RAD/SEC)	(DECIBELS)	(DEGREES)	(DECIBELS)
2.000000E-4	32	-1.2	-64.1
4.000000E-4	32	-2.3	-64.1
6.000000E-4	32	-3.5	-64
8.000000E-4	32	-4.7	-64
1.000000E-3	32	-5.8	-64
2.000000E-3	31.9	-11.5	-63.7
4.000000E-3	31.4	-22.3	-62.8
6.000000E-3	30.7	-31.7	-61.4
8.000000E-3	29.9	-39.6	-59.8
.01	29	-46.1	-58.1
.02	25.1	-65.7	-50.1
.04	19.7	-80.5	-39.6
.06	16.4	-87.4	-32.9

.08	13.9	-92	-28.2
.1	12	-95.7	-24.5
.2	6	-110.1	-13.6
.4	0	-134.4	-3.1
.6	-3.5	-157.8	2.5
.8	-6	-181	2.5
1	-8	-234	.9
2	-14	-318.9	0
4	-20	-548.3	.1
6	-23.5	-777.5	0
8	-26	-1006.7	0
10	-28	-1235.9	0

TABLE 14.2

COMPARISON OF THE ERROR RATIO AND THE DEVIATION FOR THE SYSTEM IN EXAMPLES 14.1 AND 14.3

w (rad/sec)	m_o	m_c	Deviation Ratio $(m_c - m_o)$	Error Ratio
0.0001	32	-0.2	-32.2 db	-64 db
0.010	29	-0.2	-29.2	-58
0.10	12	-0.2	-12.2	-24.5
1.00	-8	-4.4	+3.6	+1.6
10.0	-28	-27.5	+0.5	0

Example 14.4   Construct the Error Ratio graph of the system in Example 14.2.

Solution:

The Error Ratio graph is given in Figure 14.10.

14.5   Diminishing, Sustained, and Expanding Oscillations

Every closed-loop control system has a tendency to produce an oscillation in the controlled variable. This tendency for self-oscillation is caused by the presence of the feedback signal. The nature of the oscillation is directly related to the gain at the ω_{-180} frequency. This is the frequency at which the open-loop phase angle is equal to $-180°$. If the open-loop gain at ω_{-180} is less than one, the oscillations will diminish in size with each successive cycle. If the open-loop gain at ω_{-180} is equal to one, the oscillation will continue at a constant amplitude. If the open-loop gain at

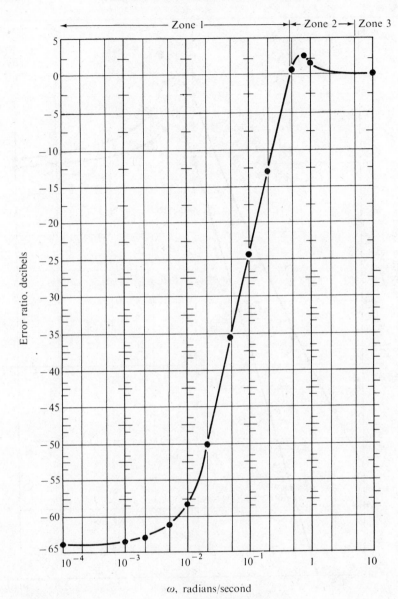

ω, radians/second

FIGURE 14.8

ERROR-RATIO GRAPH, EXAMPLE 14.3

ω_{-180} is greater than one, the oscillation will increase in size with each successive cycle. If the oscillations diminish, the control system is said to be stable. If not, the system is said to be unstable. Therefore, a stable

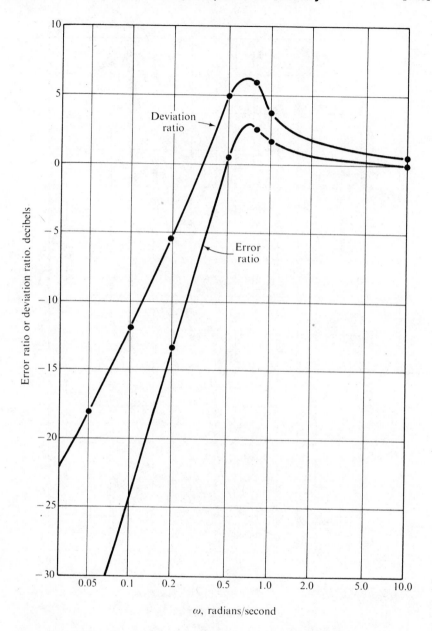

FIGURE 14.9

*A COMPARISON OF THE ERROR RATIO AND THE DEVIATION
RATIO, EXAMPLE 14.3*

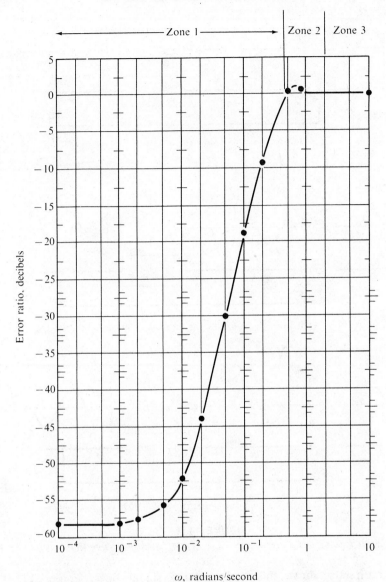

FIGURE 14.10

ERROR-RATIO GRAPH, EXAMPLE 14.4

control system is one in which the open-loop gain is less than one when the open-loop phase angle is $-180°$.

Figure 14.11 illustrates a control system with a sustained oscillation. Graphs of the setpoint, error signal, and controlled-variable signal are

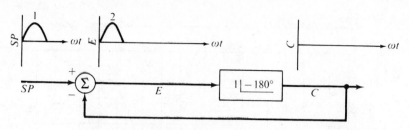

a) Half-wave signal during the interval $0 \leqslant \omega t \leqslant 180°$

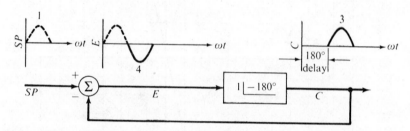

b) Half-wave signal during the interval $180° \leqslant \omega t \leqslant 360°$

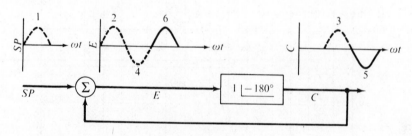

c) Half-wave signal during the interval $360° \leqslant \omega t \leqslant 540°$

FIGURE 14.11

SUSTAINED OSCILLATIONS IN A CLOSED-LOOP CONTROL SYSTEM

located directly above the corresponding signal lines. Figure 14.10 is divided into three, one-half cycle time periods (at the frequency for which the open-loop phase angle is $-180°$). The solid graph lines in each part indicate the value of each signal during the current time interval. The dotted lines indicate the value of each signal during previous intervals. The numbers above the signals indicate the order in which they occur. The sequence is as follows:

Figure	Sequence No.	Explanation
14.11–a	1	The single half-wave pulse 1 is introduced at the setpoint (*SP*).
	2	$E = SP - C = SP - 0 = SP$. Error pulse 2 is identical to setpoint pulse 1.
14.11–b	3	The system block delays the half-wave pulse by $180°$. The gain is 1, so controlled-variable pulse 3 is identical to error pulse 2 except for the $180°$ delay.
14.11–b	4	$E = SP - C = 0 - C = -C$. The error detector inverts the controlled variable. Error pulse 4 is equal to controlled-variable pulse 3 inverted.
14.11–c	5	The system block delays the inverted half-wave pulse by $180°$. Controlled-variable pulse 5 is equal to error pulse 4 except for the $180°$ delay.
14.11–c	6	$E = SP - C = 0 - C = -C$. The error detector inverts the controlled variable. Error pulse 6 is equal to controlled-variable pulse 5 inverted.

The above sequence will continue indefinitely. The single half-wave pulse at the setpoint is required to start the oscillation. Once the oscillation has started, the half-wave pulse will continue to move around the loop as it did in sequence 2–6. The controlled variable is in a state of sustained oscillation because the system gain is 1 and the phase angle is $-180°$.

Figure 14.12 illustrates a control system with a diminishing oscillation. The sequence is the same as in Figure 14.11, except that the system gain is 0.5 instead of 1.0. The half-wave pulse is reduced by one-half each time it passes through the system. The result is an oscillation in which each successive half-cycle is one-half as large as the preceding half-cycle. Each positive half-cycle is one-fourth as large as the preceding positive half-cycle. In other words, the system has a quarter amplitude-decay response to a disturbance when the open-loop gain is 0.5 at a frequency of ω_{-180} radians/second.

Figure 14.13 illustrates a control system with an expanding oscillation. The system gain is 2.0 at ω_{-180}, so the half-wave pulse is doubled each time it passes through the system. The result is an oscillation in which each successive half-wave is twice as large as the preceding half-wave. The oscillations will continue to increase until the limitations of the system are reached. The final-control element will oscillate between

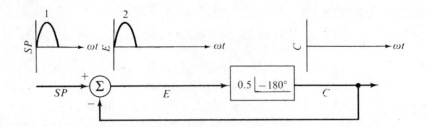

a) Half-wave signal during the interval $0 \leqslant \omega t \leqslant 180°$

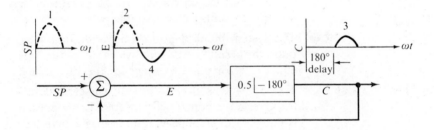

b) Half-wave signal during the interval $180° \leqslant \omega t \leqslant 360°$

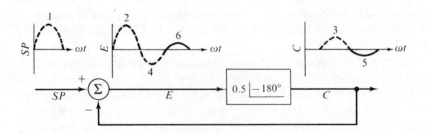

c) Half-wave signal during the interval $360° \leqslant \omega t \leqslant 540°$

FIGURE 14.12

DAMPED OSCILLATIONS IN A CLOSED-LOOP CONTROL SYSTEM

its minimum and maximum values. The controlled variable will oscillate with a large amplitude. Obviously, this is a very undesirable condition. The prevention of such unstable operation is a primary concern in the design of every control system.

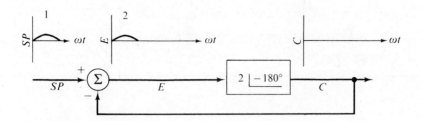

a) Half-wave signal during the interval $0 \leq \omega t \leq 180°$

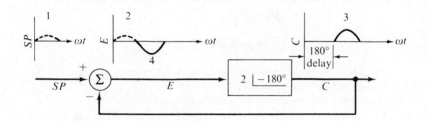

b) Half-wave signal during the interval $180° \leq \omega t \leq 360°$

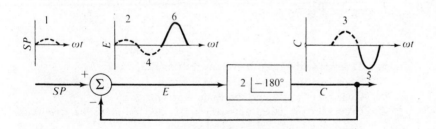

c) Half-wave signal during the interval $360° \leq \omega t \leq 540°$

FIGURE 14.13

EXPANDING OSCILLATIONS IN A CLOSED-LOOP CONTROL SYSTEM

OSCILLATIONS IN CLOSED-LOOP CONTROL SYSTEMS

g_{-180} *is the open loop gain at* ω_{-180}, *the frequency at which the phase angle is* $-180°$.

Expanding oscillations occur in a control system when $g_{-180} > 1$.

> *Sustained oscillations occur when* $g_{-180} = 1$.
>
> *Diminishing oscillations occur when* $g_{-180} < 1$.
>
> *Quarter amplitude decay occurs when* $g_{-180} = 0.5$.

EXERCISES

14–1. A pressure control system has the following transfer functions.

$$G(S) = [5]$$
$$H(S) = [e^{-0.05S}]\left[\frac{2}{1 + 0.5S}\right]$$

   a. Determine the open-loop transfer function.
   b. Construct the open-loop Bode diagram.
   c. Construct the closed-loop Bode diagram.
   d. Construct the deviation-ratio graph and estimate the maximum frequency limit of the control system.

14–2. A thermal control system has a proportional controller with a gain of 1.5. The process is a first-order lag plus dead time with the following transfer function.

$$H(S) = [e^{-200S}]\left[\frac{1}{1 + 1000S}\right]$$

   a. Determine the open-loop transfer function.
   b. Construct the open-loop Bode diagram.
   c. Construct the closed-loop Bode diagram.
   d. Construct the deviation-ratio graph and estimate the maximum frequency limit of the control system.

14–3. A level control system has the following transfer functions.

$$G(S) = [2]$$
$$H(S) = [e^{-2S}]\left[\frac{1}{1 + 50S}\right]$$

   a. Determine the open-loop transfer function.
   b. Construct the open-loop Bode diagram.
   c. Construct the closed-loop Bode diagram.
   d. Construct the deviation-ratio graph and estimate the maximum frequency limit of the control system.

14–4. An armature-controlled dc motor is used in a position control system. The open-loop transfer function of the system is

$$[G(S)][H(S)] = [19]\left[\frac{1}{S}\right]\left[\frac{1}{0.00545S^2 + 0.329S + 1}\right]$$

    a.   Construct the open-loop Bode diagram.

    b.   Construct the closed-loop Bode diagram.

    c.   Construct the deviation ratio graph and estimate the maximum frequency limit.

14–5.   A belt conveyor has a dead-time lag of 20 seconds and a gain of 1.

$$H(S) = [e^{-20S}]$$

    a.   Construct the Bode diagram of the process, $H(S)$. From the Bode diagram, determine the gain of a proportional controller that will result in quarter amplitude-decay response. (Hint: the open-loop gain must be 0.5 or -6 db when the phase angle is $-180°$.)

    b.   Construct the open-loop Bode diagram of the process $H(S)$ and the controller $G(S)$, as determined in step a.

    c.   Construct the closed-loop Bode diagram.

    d.   Construct the deviation-ratio graph and estimate the maximum frequency limit of the control system.

14–6.   A blending system consists of a first-order lag with a time constant of 1000 seconds and a dead-time lag of 20 seconds.

$$H(S) = [e^{-20S}]\left[\frac{1}{1 + 1000S}\right]$$

    a.   Construct the Bode diagram of the process, $H(S)$. From the Bode diagram, determine the gain of a proportional controller that will result in quarter amplitude-decay response.

    b.   Construct the open-loop Bode diagram of the process $H(S)$ and the controller $G(S)$, as determined in step a.

    c.   Construct the closed-loop Bode diagram.

    d.   Construct the deviation-ratio graph and estimate the maximum frequency limit of the control system.

14–7.   A control system consists of the underdamped second-order system in Example 13.8 and a dead-time lag. The transfer function of the process is given below.

$$H(S) = [e^{-0.05S}]\left[\frac{0.001}{0.01S^2 + 0.02S + 1}\right]$$

    a.   Construct the Bode diagram of $H(S)$. Refer to Figure 13.16 and Example 13.8 for further details on the second-order system.

    b.   Determine the gain of a proportional controller that will result in quarter amplitude-decay response.

    c.   Construct the closed-loop Bode diagram of the process and controller.

    d.   Construct the deviation-ratio graph and estimate the maximum frequency limit of the control system.

14–8.   A pressure system consist of a measuring means with a lag coefficient
of 0.2 second, a control valve with two time constants of 0.1 second
each, and a pressure tank with a time constant of 1 second. The
transfer function is given below.

$$H(S) = \left[\frac{1}{1 + 0.2S}\right]\left[\frac{1}{1 + 0.1S}\right]\left[\frac{1}{1 + 0.1S}\right]\left[\frac{10}{1 + S}\right]$$

a.   Construct the Bode diagram of $H(S)$.

b.   Determine the gain of a proportional controller that will result
in quarter amplitude-decay response.

c.   Construct the closed-loop Bode diagram of the process and controller.

d.   Construct the deviation-ratio graph and estimate the maximum
frequency limit.

15

Controller
Design

15.1 Introduction

Controller design consists of selecting the control modes and determining the mode settings which will result in a stable system that meets the control objectives. The control objectives may specify some or all of the following characteristics.

1. The accuracy and speed of response of the measuring means (refer to Chapter Six).
2. The residual error after a load change (refer to Sections 9.4 and 9.6).
3. The response to a step change. Quarter amplitude decay or zero overshoot may be specified (refer to Chapter Two).
4. The maximum frequency limit. The control system is able to follow setpoint changes and minimize disturbances with frequencies less than the maximum limit.
5. The speed of response or rise time. The rise time is the time required for a 63.2 per cent response to a step change.
6. Minimum cost. The initial cost of the hardware is only one part of this objective. Maintenance costs, operating costs, reject product costs, and environmental costs are examples of the other costs

which may be a factor. This objective requires a judgment based on economic and social conditions. Only the first five objectives will be covered in the remainder of this chapter.

Two different approaches are used to design a control system. This does not mean that all design methods use only one approach or the other, but that parts of each approach are found in most design methods.

One approach uses standard, very flexible controllers with one, two, or three control modes. Decisions are often made in selecting the control modes. Controller adjustment formulas are used to determine the mode settings during the startup procedure. The mode adjustments are often "fine tuned" by trial and error. Accurate frequency response data from the process are not required (accurate transfer functions of many industrial processes are very difficult to determine). This method is used when the cost of determining the controller adjustments during startup is less than the cost of the analysis and design required to define the control system before startup. Even the most accurate process-control system design may require fine adjustments in the field.

The second approach uses specialized controllers designed for a specific application. The control modes are specified in the design, and the mode-adjustment range is relatively limited. The open-loop frequency response graphs are used in the design procedure. Accurate frequency-response data of the process are required (accurate transfer functions of most electromechanical systems are relatively easy to determine). This method is used when the cost of determining the controller adjustments during start-up is greater than the cost of the analysis and design required to define the control system before startup.

The frequency response of most industrial processes is difficult to determine or measure. For this reason, process-control design is based more on the first approach. However, frequency-response methods are sometimes used to select the control modes and determine approximate mode settings. Field adjustments are almost always required to determine the exact mode settings.

The second approach has been used very successfully to design servomechanisms. The frequency response of most electromechanical components is relatively easy to determine and measure. Field adjustments are frequently unnecessary.

15.2 Ultimate-Cycle Method

The ultimate-cycle method uses a controller adjustment formula to determine the controller settings. The controller must be installed and the system ready to operate. Two measurements are required. These are the ultimate gain (G_u), and the ultimate period (P_u).

The following procedure is used to determine the ultimate gain and ultimate period. First, set the integral and derivative modes to the least effective setting. Then, start up the process, with the controller gain at a low value. Increase the gain setting until the process starts to oscillate. The last gain setting is the ultimate gain (G_u). The period of the oscillation is the ultimate period (P_u). The controller settings are determined from Table 15.1. The original ultimate cycle method was developed by Nichols and Ziegler.[1] The modified ultimate-cycle method incorporates the quarter amplitude-decay criteria, as shown in Figure 15.1.

TABLE 15.1
THE ULTIMATE-CYCLE METHOD

Control Modes	Original Method	Modified Method
Proportional Control	$G = 0.5G_u$	Adjust the gain to obtain quarter amplitude-decay response to a step change in setpoint (see Figure 15.1).
Proportional-Plus-Integral Control	$T_i = \dfrac{P_u}{1.2}$ (min) $G = 0.45G_u$	$T_i = P_u$ (min) Adjust the gain to obtain quarter amplitude-decay response to a step change in setpoint.
Proportional-Plus-Integral-Plus-Derivative Control	$T_i = \dfrac{P_u}{2.0}$ (min) $T_d = \dfrac{P_u}{8}$ (min) $G = 0.6G_u$	$T_i = \dfrac{P_u}{1.5}$ (min) $T_d = \dfrac{P_u}{6}$ (min) Adjust the gain to obtain quarter amplitude-decay response to a step change in setpoint.

Example 15.1 A process control system is tested at startup. The derivative mode is turned off, and the integral mode is set at the lowest setting. The gain is gradually increased until the controlled variable starts to oscillate. The gain setting is 2.2 and the period of oscillation is 12 minutes. Use the original ultimate-cycle method to determine the three-mode controller settings.

---

[1] N. B. Nichols and J. G. Ziegler, "Optimum Settings for Automatic Controllers," *ASME Transactions,* 64, no. 8 (1942), pp. 759–768.

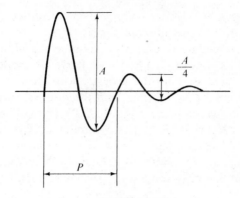

FIGURE 15.1

QUARTER AMPLITUDE-DECAY RESPONSE

Solution:

$$T_i = \frac{P_u}{2.0} = \frac{12}{2} = 6 \text{ minutes}$$

$$T_d = \frac{Pu}{8} = \frac{12}{8} = 1.5 \text{ minutes}$$

$$G = 0.6G_u = (0.6)(2.2) = 1.32$$

15.3 Process-Reaction Method

The process reaction method uses a controller adjustment formula to determine the controller settings. This method assumes that the process can be approximated by a first-order lag plus dead-time characteristic. An open-loop step response test of the process is required.

The open-loop response test consists of operating the process open loop on manual control until the measured variable remains constant. Then a step change in the final-control element is made. The response of the measured variable is illustrated in Figure 15.2.

Three variables are required. Two are obtained from the reaction graph. The third ($\triangle P$) is the per cent change in the final-control element position. The two variables from the reaction graph are the effective lag (L), and the slope of the tangent line (N). The controller settings are determined from the following formulas.

<u>The Process Reaction Method</u>

L = effective lag, minutes

$$N = \frac{\text{change in the measured variable}}{t}, \text{ per cent/min}$$

$\triangle P$ = change in the final-control element, per cent

---

[2] *Ibid.*

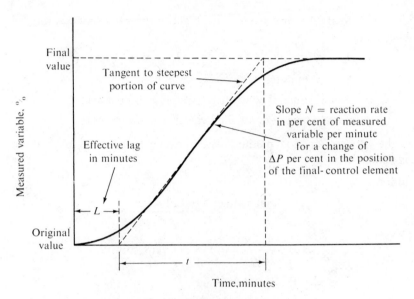

FIGURE 15.2
A TYPICAL PROCESS-REACTION GRAPH

1. Proportional Control

$$G = \frac{\triangle P}{NL}$$

2. Proportional-plus-Integral Control

$$G = 0.9\frac{\triangle P}{NL}$$

$$T_i = 3.33L \text{ minutes}$$

3. Proportional-plus-Integral-plus-Derivative Control

$$G = 1.2\frac{\triangle P}{NL}$$

$$T_i = 2L \text{ minutes}$$

$$T_d = 0.5L \text{ minutes}[3]$$

Example 15.2 During startup, the final control element of a process control system is maintained constant until the controlled variable levels out near the operating value. Then a 10 percent change is produced in the final-control element position. A response graph similar to Figure

------

[3] *Ibid.*

15.2 is obtained. An analysis of the reaction graph produces the following values.

$$L = 5 \text{ minutes}$$
$$t = 10 \text{ minutes}$$

Initial value of controlled variable = 40 per cent

Final value of controlled variable = 48 per cent

Use the process-reaction method to determine the settings of a three-mode controller.

Solution:

$$N = \frac{\text{final value} - \text{initial value}}{t}$$

$$N = \frac{48 - 40}{10} = 0.8 \text{ per cent/minute}$$

$$\triangle P = 10 \text{ per cent}$$
$$L = 5 \text{ minutes}$$

$$G = 1.2\frac{\triangle P}{NL} = 1.2\frac{(10)}{(0.8)(5)} = 3$$

$$T_i = 2L = 2(5) = 10 \text{ minutes}$$
$$T_d = 0.5L = (0.5)(5) = 2.5 \text{ minutes}$$

15.4 Frequency-Response Method

The frequency-response method is a design procedure based on the open-loop Bode diagram of the process to be controlled. The objective of the design is to obtain a stable control system that provides accurate control for a wide range of frequencies. Stable control means that the open-loop gain must be less than one at the $-180°$ phase-angle frequency. Control over a wide range of frequencies means that the $-180°$ phase angle occurs at as high a frequency as possible. Accurate control means that the open-loop gain is as large as possible below the $-180°$ phase-angle frequency.

The stability of the control system is a primary concern of the design procedure. The relationship between stability and the open-loop frequency response was demonstrated in Section 14.5. This relationship may be stated in two ways.

1. A control system is stable only if the open-loop gain is less than one (or zero db) at the frequency for which the phase angle is $-180°$.

2. A control system is stable only if the phase angle is less than $-180°$ at the frequency for which the gain is one (or zero db).

Two methods of viewing the stability requirement suggest a two-way margin of safety. The gain at the $-180°$ frequency must be a safe distance below one (or zero db), and the phase angle at the zero db frequency must be a safe distance above $-180°$. The first safety margin is called the *gain margin*. The second safety margin is called the *phase margin*. Standard practice uses a gain margin of 6 db (Quarter Amplitude Decay) and a phase margin of $40°$. The gain and phase margins are illustrated in Figure 15.3.

Minimum gain margin = 6 db
Minimum phase margin = $40°$

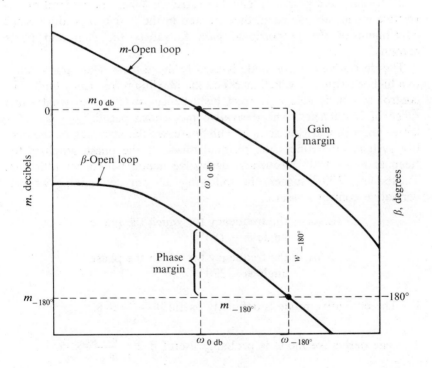

ω, radians/second

FIGURE 15.3
*A BODE DIAGRAM ILLUSTRATING THE GAIN AND
PHASE MARGINS*

GAIN AND PHASE MARGINS

The gain margin must not be less than 6 db and the phase margin must not be less than 40°

The design procedure consists of using the three control modes to modify the open-loop Bode diagram in such a way that the control objectives are satisfied. The Bode diagrams of the three control modes are given in Figures 15.4, 15.5, and 15.6.

The proportional-control gain is used to raise or lower the open-loop gain graph without changing the phase graph. For proportional control, the design procedure consists of the adjustment of the proportional gain until the gain and phase margins are satisfied. When the integral or derivative modes are included, the last step in the design procedure is the adjustment of the proportional gain to satisfy the gain and phase margins.

The derivative control mode is used to move the $-180°$ phase angle to a higher frequency which increases the maximum frequency limit. The control system is able to correct disturbances over a somewhat wider range of frequencies. The derivative mode also permits the use of a larger proportional-gain setting. This increases the accuracy of the control system and reduces the residual offset. If the phase graph is too steep at the $-180°$ frequency, derivative action will be of little use. Tucker and Wills[4] suggest the following rule for determining whether derivative control is useful.

$$\omega_{-180} = \text{the frequency for which the phase angle is } -180°$$

$$\omega_{-270} = \text{the frequency for which the phase angle is } -270°$$

1. The derivative mode is definitely useful if $\dfrac{\omega_{-270}}{\omega_{-180}} > 5$.

2. The derivative mode is probably useful if $2 < \dfrac{\omega_{-270}}{\omega_{-180}} < 5$.

3. The derivative mode is of little use if $\dfrac{\omega_{-270}}{\omega_{-180}} < 2$.

---

[4] G. K. Tucker and D. M. Wills, *A Simplified Technique of Control System Engineering* (Philadelphia: Honeywell Inc., 1962), pp. 63.

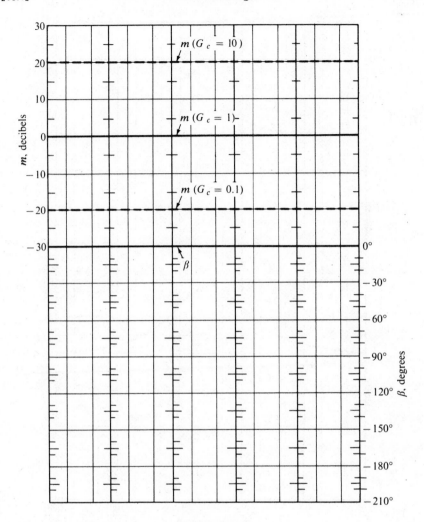

FIGURE 15.4

THE BODE DIAGRAM OF A PROPORTIONAL-CONTROL MODE,
$$G(S) = G_c \,|\, 0°$$

The derivative control mode illustrated in Figure 15.5 has a derivative limiter coefficient (α) equal to 0.1. The limiter break point frequency (ω_L) is equal to $1/a$, or ten times the derivative break-point frequency.

$$\omega_L = \frac{1}{\alpha T_d} = \frac{1}{\alpha}\omega_d = \frac{1}{0.1}\omega_d = 10\omega_d$$

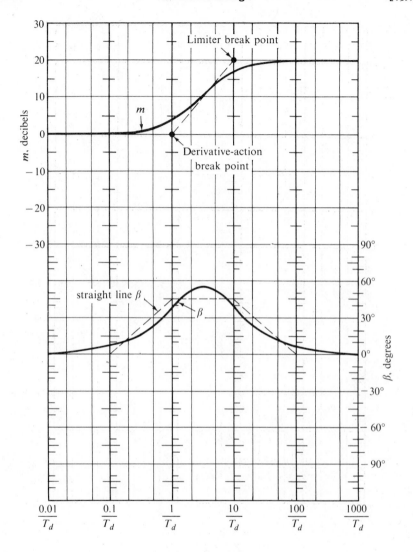

FIGURE 15.5

THE BODE DIAGRAM OF A DERIVATIVE-CONTROL MODE,
$(T_dS + 1)/(0.1T_dS + 1)$

Notice the increase in the gain graph at high frequencies. This increase is called the *derivative amplitude.* It is an unwanted side effect of the derivative control mode. In Figure 15.5, the derivative amplitude is 20 db, or a gain of 10. The phase-lead graph reaches a maximum value half way between the two magnitude break points. When α is equal to 0.1, the maximum phase lead is about $55°$. Other phase-lead values are given in Table 15.2.

TABLE 15.2
GAIN AND PHASE VALUES FOR THE DERIVATIVE CONTROL MODE
$(T_d S + 1) \, / \, (0.1 \, T_d S + 1)$

ω	m	β
$0.1/T_d$	0 db	$5.4°$
$0.2/T_d$	0.2 db	$10.2°$
$0.5/T_d$	1.0 db	$23.6°$
$1.0/T_d$	3.0 db	$39.0°$
$2.0/T_d$	6.8 db	$52.2°$
$3.16/T_d$	10.0 db	$55°$
$5/T_d$	13.2 db	$52.2°$
$10/T_d$	17 db	$39°$
$20/T_d$	20 db	$23.6°$
$50/T_d$	20 db	$10.2°$
$100/T_d$	20 db	$5.4°$

The derivative mode break point is located in such a way that the derivative action produces the maximum improvement in the control system. The positive phase angle of derivative action is used to increase the phase margin. However, the gain of the derivative action reduces the gain margin, which nullifies part of the improvement. A compromise is required between the benefit produced by the phase lead and the harm produced by the gain of the derivative mode.

One approach is to locate the maximum derivative phase lead at the zero-db frequency on the open-loop Bode diagram.[5] This places the zero-db frequency ($\omega_{(0 \text{ db})}$) at the geometric mean between the derivative break-point frequency (ω_d) and the limiter break-point frequency $\left(\omega_L = \dfrac{1}{\alpha}\omega_d \right)$.

$$\omega_{\max} = \omega_{(0 \text{ db})} = \sqrt{\omega_d \omega_L} = \sqrt{\frac{1}{\alpha}\omega_d}$$

$$\omega_d = \sqrt{\alpha}\ \omega_{(0 \text{ db})} \qquad\qquad \textbf{(15.1)}$$

This method produces the maximum possible increase in the phase margin. However, a relatively large decrease in the gain margin is an undesirable but unavoidable side effect.

---

[5] John L. Bower and Peter M. Schultheiss, *Introduction to the Design of Servomechanisms* (New York: John Wiley & Sons, 1958), p. 182–189.

An alternate method is to locate the first $+50°$ point on the derivative phase graph at the $-180°$ frequency on the open-loop Bode diagram.[6] When $\alpha = 0.1$, this results in a derivative break frequency equal to one-half the $-180°$ frequency.

$$\omega_d = 0.5\omega_{-180°} \tag{15.2}$$

This method tends to minimize the harmful effect of the derivative gain without seriously reducing the benefit of the derivative phase lead. It applies only when α is less than 0.13 (the phase angle does not reach $+50°$ for values of α greater than 0.13).

The value of α determines the maximum value of the derivative-action phase angle. This maximum phase angle occurs when the frequency is equal to the geometric mean between the derivative break-point frequency (ω_d) and the limited break-point frequency (ω_L). Equations (15.3) and (15.4) give the relationship between α, ω_{max}, and β_{max}.

$$\omega_d = \sqrt{\alpha}\,\omega_{max} \tag{15.3}$$

$$\alpha = \frac{1 - \sin \beta_m}{1 + \sin \beta_m} \tag{15.4}$$

The derivative control mode is sometimes used to obtain a specific increase in the phase margin. Equation (15.4) is used to determine the value of α required to produce the desired increase. A slightly larger value of α must be used to compensate for the negative effect of the derivative gain. After α has been determined, Equation (15.1) is used to determine ω_d and $T_d = 1/\omega_d$.

The integral control mode is used to increase the gain at the low and intermediate frequencies. This increases the accuracy of the control system and eliminates the residual offset. An undesirable side effect of the integral mode is that it reduces the $-180°$ frequency. This cancels part of the benefit of the derivative control mode. It also explains why a lower gain setting is required when the integral mode is used. If the gain graph slopes upward toward the low frequency end, the integral mode may not be required. If the slope of the graph is zero toward the low frequency end, then the integral control mode will be useful.

The integral mode is applied after the derivative mode. The integral break point is placed in such a way that the integral mode contributes about 5 per cent of the phase angle at the $-180°$ frequency. This is

---

[6] Tucker and Wills, *A Simplified Technique,* p. 70.

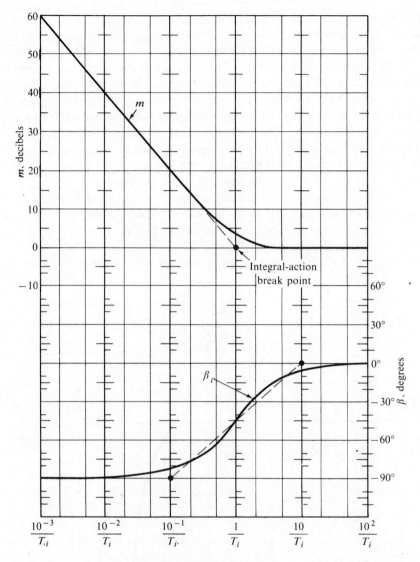

ω, radians/second or radians/minute

FIGURE 15.6
THE BODE DIAGRAM OF AN INTEGRAL-CONTROL MODE,
$(T_iS + 1)/(T_iS)$

accomplished by setting the integral break frequency (ω_i) equal to about one-fifth of the $-170°$ open-loop frequency.

$$\omega_i = 0.2\omega_{-170°} \tag{15.5}$$

However, Equation (15.5) does not always produce a satisfactory result. One exception occurs with first-order lag plus dead-time processes when the first-order time constant is greater than about ten times the dead-time lag. If Equation (15.5) is used in this situation, the combined process plus derivative plus integral phase graph dips below $-140°$ at a very low frequency. As the frequency increases, the phase graph rises above $-140°$ and then drops rapidly below $-140°$ as the dead time takes over. The problem becomes evident when the phase margin criteria are applied in the proportional mode design. The system gain must be reduced to 0 decibels at the lowest $-140°$ frequency. The result is an extremely overdamped, sluggish system.

Figure 15.7 illustrates the problem as it occurs in the design of a process with a first-order time constant of 100 seconds and a dead-time lag of 2 seconds. Step 1 of the design procedure results in a derivative action time constant, T_d, of 2.67 seconds ($\omega_d = 0.375$ radians/second). The phase graph of the process plus derivative mode is shown

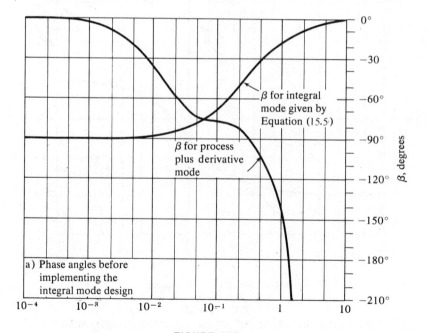

FIGURE 15.7-a
INTEGRAL MODE DESIGN USING EQUATION (15.5):
THE PHASE-ANGLE GRAPHS OF THE PROCESS-PLUS-
DERIVATIVE MODE AND THE INTEGRAL MODE

in Figure 15.7–a. The $-170°$ frequency is about 1.25 radians/second. From Equation (15.5),

$$\omega_i = 0.2(1.25) = 0.25 \text{ radians/second}$$

Figure 15.7–a also shows the phase graph of the integral mode with $\omega_i = 0.25$ radians/second. The combined process plus derivative plus integral graph is shown in Figure 15.7–b. The lowest $-140°$ frequency is about 0.015 radians/second. The phase margin criteria specify a system gain of 0 decibels or less at the $-140°$ frequency. The result is a proportional mode that reduces the system gain to 0 decibels at 0.015 radians/second as shown in Figure 15.7–b. The gain margin of the system is 52 decibels, a clear indication of the problem presented by the dip in the phase graph between 0.015 and 0.15 radians/second.

The preceding design can be improved by moving the integral break point to a lower frequency. Figure 15.8–a shows the phase angle of

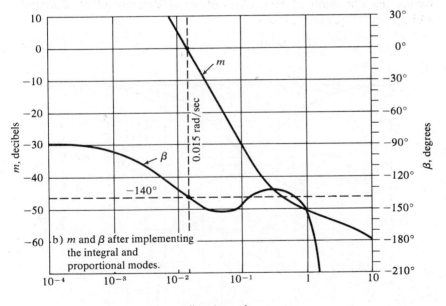

FIGURE 15.7-b
*INTEGRAL MODE DESIGN USING EQUATION (15.5):
THE MAGNITUDE AND PHASE-ANGLE GRAPHS AFTER
IMPLEMENTING THE INTEGRAL AND PROPORTIONAL MODES*

an integral mode with $\omega_i = 0.08$ radians/second. The dotted line shows the integral mode as determined by Equation (15.5). Figure 15.8–b shows the phase graph of the process plus derivative plus improved integral mode. The dip in the phase graph has been raised above $-140°$ and the $-140°$ frequency is about 1 radian/second (compared with 0.015 radians/second in the previous design). The phase margin criteria now result in a gain of 0 decibels at 1 radian/second as shown in Figure 15.8–b. The dotted line in Figure 15.8–b is the gain of the first design. The improved design gain is 40 decibels (100 times) greater than the gain in the design based on Equation (15.5).

The improved design is optimized by setting the integral break point frequency equal to the frequency at which the process plus derivative phase angle is $-82°$.

$$\omega_i = \omega_{-82°} \tag{15.6}$$

The improvement can be incorporated in the design procedure by selecting the minimum of the values of ω_i as determined by Equation (15.5) and Equation (15.6).

$$\omega_i = \text{Min}(\omega_{-82°} \text{ or } 0.2\omega_{-170°}) \tag{15.7}$$

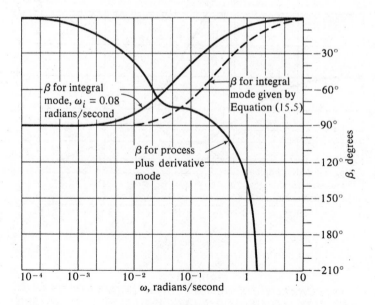

FIGURE 15.8-a

IMPROVED INTEGRAL MODE DESIGN: THE PHASE ANGLE GRAPHS OF THE PROCESS PLUS DERIVATIVE MODE AND THE INTEGRAL MODE

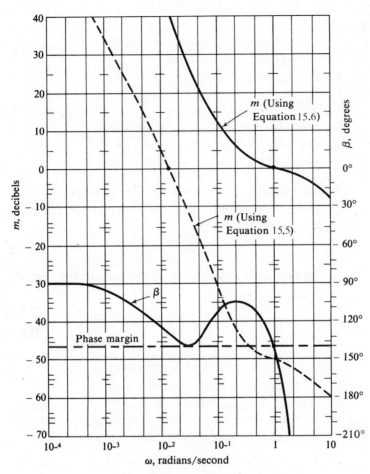

FIGURE 15.8-b

IMPROVED INTEGRAL MODE DESIGN: THE MAGNITUDE AND PHASE ANGLE GRAPHS AFTER IMPLEMENTING THE INTEGRAL AND PROPORTIONAL MODES

CONTROLLER DESIGN

Comment: The simple rules presented here do not apply in all design situations. The intelligent use of a trial-and-error procedure may be necessary to achieve a satisfactory design.

Step 1—Derivative mode

(If the derivative mode is not used, this step is not required.)

Effects:

1. Increase the maximum frequency limit.
2. Increase the phase margin.
3. Decrease the gain margin.

Usefulness:

1. Definitely useful if $\omega_{-270°}/\omega_{-180°} > 5$.
2. Probably useful if $2 < \omega_{-270°}/\omega_{-180°} < 5$.
3. Not useful if $\omega_{-270°}/\omega_{-180°} < 2$.

Design Method 1: Increasing the Phase Margin

1. Determine α from the desired increase in the phase margin (β_m).

$$\alpha = \frac{1 - \sin \beta_m}{1 + \sin \beta_m} \qquad \textbf{(15.4)}$$

2. Determine T_d from α and the open-loop, zero-db frequency ($\omega_{0\ db}$).

$$\omega_d = \sqrt{\alpha}\ \omega_{(0\ db)} \qquad \textbf{(15.1)}$$

$$T_d = 1/\omega_d \qquad \textbf{(15.8)}$$

Design Method 2: General Applications

1. Use a predetermined value of α such as 0.1.
2. Determine T_d such that the first $+50°$ derivative angle is at the open-loop $-180°$ frequency. For $\alpha = 0.1$, this results in the following value for T_d.

$$\omega_d = 0.5\omega_{-180°}$$

$$T_d = 1/\omega_d.$$

Step 2—Integral Mode

(If the integral mode is not used, this step is not required.)

Effects:

1. Increase the low-frequency gain.
2. Reduce the maximum frequency limit.
3. Reduce the phase margin.

Usefulness:

1. Definitely useful if the open-loop gain graph is level at the low-frequency end.

> 2. Not useful if the open-loop gain graph continues to slope upward at the low-frequency end.

Design Method:

Determine T_i such that the integral action break point frequency is the minimum of $\omega_{-82°}$ or $0.2\omega_{-170°}$.

$$\omega_i = \text{Min}[\omega_{-82°} \text{ or } 0.2\omega_{-170°}] \qquad \textbf{(15.7)}$$

$$T_i = 1/\omega_i \qquad \textbf{(15.9)}$$

Step 3—Proportional Mode

Effects:
1. Move the gain graph up or down.
2. Does not affect the phase graph.

Usefulness: Almost always useful.

Design Method:

Determine the controller gain (M_c) such that the *gain margin* and the *phase margin* are both satisfied. The gain margin is satisfied if: $M_c + m_{-180°} \leq -6$ db.

$$M_c \leq -m_{-180°} - 6 \qquad \textbf{(15.10)}$$

The phase margin is satisfied if: $M_c + m_{-140°} = 0$ db.

$$M_c = -m_{-140°} \qquad \textbf{(15.11)}$$

Both the gain margin and the phase margin are satisfied if the controller gain is determined such that:

$$M_c = \text{Min}[-m_{-140°} \text{ or } (m_{-180°} - 6)] \qquad \textbf{(15.12)}$$
$$G_c = 10^{(m_c/20)} \qquad \textbf{(15.13)}$$

The following BASIC program can be used to design a three-mode controller for a process with any combination of one or more first order lags, second order lags, and dead time lags. In its present form, the program prints a frequency response table for each step in the design procedure. This will afford the opportunity to examine the effect of each step on the system frequency response. The program can be modified to provide additional information, such as the values of $\omega_{-270°}/\omega_{-180°}$, $\omega_{-170°}$, $\omega_{-82°}$, $m_{-140°}$, and $m_{-180°}$. It could also be modified to make some or all of the design decisions and to plot the Bode diagrams. The present form was selected to illustrate the design procedure. Sample outputs from the program are presented in Example 15.3.

```
 1 REMARK: A PROGRAM FOR CALCULATING FREQUENCY RESPONSE DATA
 3 GO SUB 100
 4 PRINT "FREQUENCY","OPEN LOOP RESPONSE","ERROR RATIO"
 5 PRINT "(RAD/SEC)","(DECIBELS)","(DEGREES)","(DECIBELS)"
 6 FOR E = -5 TO 0 STEP 1
 7 FOR F = 2 TO 10 STEP 2
 8 W = F*10↑E
 9 GO SUB 150
10 M = 20*LOG(G)/2.30259
11 M = INT(10*(M+0.05))/10
12 B = INT(10*(B+0.05))/10
13 GO SUB 300
14 PRINT W,M,B,E1
15 NEXT F
16 NEXT E
17 STOP

100 REMARK: A SUBROUTINE FOR INPUT OF CONTROL SYSTEM PARAMETERS
101 PRINT "CONTROL SYSTEM DESIGN BY THE FREQUENCY RESPONSE METHOD"
102 PRINT "GIVE THE VALUES OF OVERALL PROCESS GAIN (G1), DERIVATIVE"
103 PRINT "TIME (T1), INTEGRAL TIME (T2), AND CONTROLLER GAIN (K)."
104 PRINT "IF NOT USED, SET T1=0, T2=0, AND K=1."
105 PRINT "G1,T1,T2,K";
106 INPUT G1,T1,T2,K
107 PRINT "GIVE THE NUMBER OF FIRST ORDER LAGS.";
108 INPUT N1
109 IF N1 = 0 THEN 115
110 PRINT "GIVE THE FIRST ORDER TIME CONSTANTS."
111 FOR J = 1 TO N1
112 PRINT J,
113 INPUT T(J)
114 NEXT J
115 PRINT "GIVE THE NUMBER OF 2ND ORDER LAGS.";
116 INPUT N2
117 IF N2 = 0 THEN 123
118 PRINT"GIVE THE SECOND ORDER RESONANT FREQUENCIES, DAMPING RATIOS."
119 FOR J = 1 TO N2
120 PRINT J,
121 INPUT W0(J),R0(J)
122 NEXT J
123 PRINT "GIVE THE NUMBER OF DEAD TIME LAGS.";
124 INPUT N3
125 IF N3 = 0 THEN 131
126 PRINT "GIVE THE DEAD TIME LAGS"
127 FOR J = 1 TO N3
128 PRINT J,
129 INPUT L(J)
130 NEXT J
131 RETURN

150 REMARK: A SUBROUTINE FOR CALCULATION OF CONTROL SYSTEM DATA
151 G = K*G1
152 B = 0
153 IF N1 = 0 THEN 161
154 FOR J = 1 TO N1
155 R = 1
156 I = W*T(J)
157 GO SUB 400
158 G= G/C
159 B = B - A
160 NEXT J
161 IF N2 =0 THEN 169
```

```
162 FOR J = 1 TO N2
163 R = 1 - W↑2/W0(J)↑2
164 I = 2*W*R0(J)/W0(J)
165 GO SUB 400
166 G = G/C
167 B = B - A
168 NEXT J
169 IF N3 = 0 THEN 173
170 FOR J = 1 TO N3
171 B = B - L(J)*W*57.2958
172 NEXT J
173 IF T1 = 0 THEN 183
174 R = 1
175 I = T1*W
176 GO SUB 400
177 G = G*C
178 B = B + A
179 I = 0.1*T1*W
180 GO SUB 400
181 G = G/C
182 B = B - A
183 IF T2 = 0 THEN 189
184 R = 1
185 I = T2*W
186 GO SUB 400
187 G = G*C/(T2*W)
188 B = B+A-90
189 GO SUB 300
190 RETURN

300 REMARK: A SUBROUTINE TO CALCULATE THE ERROR RATIO
301 B1 = B/57.2958
302 R = 1 + G*COS(B1)
303 I = G*SIN(B1)
304 GO SUB 400
305 C1 = C
306 R = 1 - G*COS(B1)
307 I = -G*SIN(B1)
308 GO SUB 400
309 E1 = 20*LOG(1/(C1*C))/2.30259
310 E1 = INT(10*(E1+0.05))/10
311 RETURN

400 REMARK: A SUBROUTINE FOR RECTANGULAR-TO-POLAR CONVERSION
401 C = SQR(R↑2 + I↑2)
402 IF R<=0 THEN 405
403 A = ATN(I/R)*57.2958
404 RETURN
405 IF R = 0 THEN 408
406 A = ATN(I/R)*57.2958 + 180
407 RETURN
408 IF I<0 THEN 411
409 A = 90
410 RETURN
411 A = -90
412 RETURN
```

The blending and heating system shown in Figure 15.9 will be used to illustrate the controller design procedure. The system specifications are as follows.

Specified Parameters—Blending and Heating System
1. Product
    Production rate: 2.5×10^{-4} cubic meters/second (15 liters/minute)
    Composition: 90% water and 10% syrup
    Density: $\rho = 1005$ kilograms/cubic meter
    Viscosity: $\mu = 0.1$ newton-seconds/square meter
    Specific heat: $S_h = 4170$ joules/kilogram-Kelvin

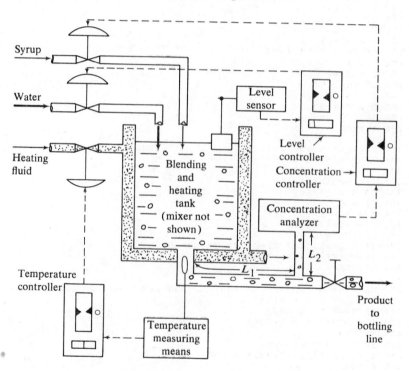

FIGURE 15.9
A LIQUID BLENDING AND HEATING SYSTEM

2. Mixing tank
   Dimensions: 0.75-meter diameter, 1-meter height
   Operating level: 0.75-meter height
   Mixing temperature: 60° C
   Heating fluid temperature: 100° C
   Outlet pipe diameter: 0.0351 meters
   Distance from outlet to temperature probe: 0.26 meters
   Wall thickness: 0.01 meters
   Wall material: steel
3. Concentration measuring means
   Model: first-order lag plus dead time
   Time constant: $T_1 = 200$ seconds
   Dead-time lag: $L = 15$ seconds
   Input range: 0 to 20%
   Output range: 4 to 20 milliamperes
   Gain: 0.8 milliamperes/% syrup
4. Level measuring means
   Model: first-order lag plus dead time
   Time constant: $T_1 = 2$ seconds
   Dead-time lag: $L = 0.5$ seconds
   Input range: 0 to 1 meter
   Output range: 4 to 20 milliamperes
   Gain: 16 milliamperes/meter
5. Temperature measuring means
   Model: Overdamped second-order lag
   Time constants: $T_1 = 50$ seconds, $T_2 = 240$ seconds
   Input range: 50 to 70° C
   Output range: 4 to 20 milliamperes
   Gain: 0.8 milliamperes/°C
6. Control valves
   Model: Underdamped second-order lag
   Water valve: $\omega_o = 10.2$ radians/second, $\zeta = 0.75$
   Syrup valve: $\omega_o = 2.4$ radians/second, $\zeta = 0.90$
   Heating valve: $\omega_o = 21.6$ radians/second, $\zeta = 0.8$
   Input range: 4 to 20 milliamperes
7. Concentration process
   Model: first-order lag
   Gain: control valve and process: 39.06%/milliampere

8. Level process
   Model: first-order lag
   Gain, control valve, and process: 1.9 meters/milliampere
9. Thermal process
   Model: first-order lag plus dead time
   Gain, control valve, and process: 10°C/milliampere

Example 15.3 Design a three-mode controller for the control of concentration in the blending and heating system, Figure 15.9.

Solution:

1. Overall process gain, K

$$K = 0.8 \times 39.06 = 31.25$$

2. Concentration process time constant, T_2

Liquid volume, $V = \pi D^2 h / 4 = \pi (0.75)^2 (0.75)/4$

$$V = 0.3313 \text{ cubic meters}$$

Liquid flow rate $= 2.5 \times 10^{-4}$ cubic meters/second
Time constant, $T_2 = V/Q = 0.3313/2.5 \times 10^{-4} = 1325$ seconds

3. Summary of the Input Parameters—Concentration Control System
   Overall process gain: 31.25
   Two first-order lags: $T_1 = 200$ seconds, $T_2 = 1325$ seconds
   One second-order lag: $\omega_o = 2.4$ radians/second, $\zeta = 0.9$
   One dead-time lag: $L = 15$ seconds

The design of a control system involves four runs of the design program. The output of the first run is on the next page. The derivative time, integral time, and controller gain were not used. The output table gives the open-loop frequency response of the concentration process, control valve, and measuring means (the last six lines of the frequency response table were deleted for brevity). The values of $\omega_{-180°}$ and $\omega_{-270°}$ can be obtained directly from the output table by linear interpolation or they may be read from a Bode graph of the results.

```
CONTROL SYSTEM DESIGN BY THE FREQUENCY RESPONSE METHOD
GIVE THE VALUES OF OVERALL PROCESS GAIN (G1), DERIVATIVE
TIME (T1), INTEGRAL TIME (T2), AND CONTROLLER GAIN (K).
IF NOT USED, SET T1=0, T2=0, AND K=1.
G1,T1,T2,K? 31.25,0,0,1
GIVE THE NUMBER OF FIRST ORDER LAGS.? 2
```

```
GIVE THE FIRST ORDER TIME CONSTANTS.
  1              ? 200
  2              ? 1325
GIVE THE NUMBER OF 2ND ORDER LAGS.? 1
GIVE THE SECOND ORDER RESONANT FREQUENCIES, DAMPING RATIOS.
  1              ? 2.4,0.9
GIVE THE NUMBER OF DEAD TIME LAGS.? 1
GIVE THE DEAD TIME LAGS
  1              ? 15
FREQUENCY      OPEN LOOP RESPONSE              ERROR RATIO
(RAD/SEC)      (DECIBELS)    (DEGREES)         (DECIBELS)
 2.000000E-5    29.9          -1.8             -59.8
 4.000000E-5    29.9          -3.5             -59.8
 6.000000E-5    29.9          -5.3             -59.7
 8.000000E-5    29.8          -7               -59.7
 1.000000E-4    29.8          -8.8             -59.6
 2.000000E-4    29.6          -17.3            -59.2
 4.000000E-4    28.8          -32.9            -57.6
 6.000000E-4    27.7          -45.9            -55.4
 8.000000E-4    26.5          -56.5            -53
 1.000000E-3    25.3          -65.2            -50.7
 2.000000E-3    20.2          -92.9            -40.5
 4.000000E-3    13.1          -121.6           -26.4
 6.000000E-3    7.9           -138.4           -15.8
 8.000000E-3    3.8           -149.8           -6.5
 .01            .4            -158.1            2
 .02            -10.9         -181.9            .7
 .04            -22.7         -207.9            0
 .06            -29.7         -228.7            0
 .08            -34.7         -248.1            0
 .1             -38.6         -266.9            0
 .2             -50.6         -358.8            0
 .4             -62.8         -540.1            0
 .6             -70           -720.8            0
 .8             -75.3         -901.2            0
```

By linear interpolation:

$$\frac{\omega_{-180°} - \omega_{-158.1°}}{\omega_{-181.9°} - \omega_{-158.1°}} = \frac{180 - 158.1}{181.9 - 158.1}$$

$$\omega_{-180°} = \omega_{-158.1°} + (\omega_{-181.9°} - \omega_{-158.1°})\frac{(180 - 158.1)}{(181.9 - 158.1)}$$

$$\omega_{-180°} = 0.01 + (0.02 - 0.01)(180 - 158.1)/(181.9 - 158.1)$$

$$\omega_{-180°} = 0.0192 \text{ radians/second}$$

$$\omega_{-270°} = 0.1 + (0.2 - 0.1)(270 - 266.9)/(358.8 - 266.9)$$

$$\omega_{-270°} = 0.103 \text{ radians/second}$$

$$\omega_{-270°}/\omega_{-180°} = 0.103/0.0192 = 5.36$$

Therefore, the derivative mode is definitely useful.

$$T_d = 2/\omega_{-180°} = 2/0.0192 = 104 \text{ seconds}$$

The above value of T_d was used in the second run of the design program shown below. The output table from the second run gives the open-loop frequency response of the process, control value, measuring means, and derivative mode (the integral time and Controller gain are not used). The gain graph is flat at the low-frequency end, so the integral mode is useful.

```
CONTROL SYSTEM DESIGN BY THE FREQUENCY RESPONSE METHOD
GIVE THE VALUES OF OVERALL PROCESS GAIN (G1), DERIVATIVE
TIME (T1), INTEGRAL TIME (T2), AND CONTROLLER GAIN (K).
IF NOT USED, SET T1=0, T2=0, AND K=1.
G1,T1,T2,K? 31.25,104,0,1
GIVE THE NUMBER OF FIRST ORDER LAGS.? 2
GIVE THE FIRST ORDER TIME CONSTANTS.
   1            ? 200
   2            ? 1325
GIVE THE NUMBER OF 2ND ORDER LAGS.? 1
GIVE THE SECOND ORDER RESONANT FREQUENCIES, DAMPING RATIOS.
   1            ? 2.4,0.9
GIVE THE NUMBER OF DEAD TIME LAGS.? 1
GIVE THE DEAD TIME LAGS
   1            ? 15
FREQUENCY      OPEN LOOP RESPONSE             ERROR RATIO
(RAD/SEC)      (DECIBELS)    (DEGREES)        (DECIBELS)
   2.000000E-5   29.9         -1.7            -59.8
   4.000000E-5   29.9         -3.3            -59.8
   6.000000E-5   29.9         -5              -59.7
   8.000000E-5   29.8         -6.6            -59.7
   1.000000E-4   29.8         -8.2            -59.6
   2.000000E-4   29.6         -16.2           -59.2
   4.000000E-4   28.8         -30.7           -57.6
   6.000000E-4   27.7         -42.7           -55.4
   8.000000E-4   26.5         -52.2           -53.1
   1.000000E-3   25.4         -59.8           -50.8
   2.000000E-3   20.4         -82.4           -43.9
   4.000000E-3   13.8         -101.4          -27.9
   6.000000E-3   9.4          -110            -19.5
   8.000000E-3   6.1          -114.8          -13.6
    .01          3.6          -118            -9.4
    .02          -3.8         -129.3          -1.3
    .04          -10.8        -154             .4
    .06          -15.1        -179.7           .3
    .08          -18.5        -204.7           .1
    .1           -21.4        -228.6          0
    .2           -31.5        -335.9          0
    .4           -43         -528             0
    .6           -53.1       -712.6           0
    .8           -55.4       -895             0
```

$$\omega_{-82°} = 0.001 + (0.002 - 0.001)(82 - 59.8)/(82.4 - 59.8)$$

$$\omega_{-82°} = 0.002 \text{ radians/second}$$

$$\omega_{-170°} = 0.04 + (0.06 - 0.04)(170 - 154)/(179.7 - 154)$$
$$\omega_{-170°} = 0.0525 \text{ radians/second}$$
$$\omega_i = \text{Min}(\omega_{-82°} \text{ or } 0.2\omega_{-170°})$$
$$= \text{Min}(0.002 \text{ or } 0.0105)$$
$$\omega_i = 0.002 \text{ radians/second}$$
$$T_i = 1/\omega_i = 500 \text{ seconds}$$

The value of $T = 500$ seconds was used in the third run of the design program shown on the next page. The output table from the third run gives the open-loop frequency response of the process, control valve, measuring means, derivative mode, and integral mode (the controller gain was not used). All that remains is the determination of the controller gain to satisfy the gain margin and the phase margin. By linear interpolation,

$$m_{-140°} = -3.8 - (10.8 - 3.8)(140 - 135)/(156.9 - 135)$$
$$m_{-140°} = -5.40 \text{ decibels}$$
$$m_{-180°} = -10.8 - (15.1 - 10.8)(180 - 156.9)/(181.6 - 156.9)$$
$$m_{-180°} = -14.82 \text{ decibels}$$
$$M_c = \text{Min}(-m_{-140°} \text{ or } (-m_{-180°} - 6))$$
$$M_c = \text{Min}(5.40 \text{ or } (14.82 - 6))$$
$$M_c = \text{Min}(5.40 \text{ or } 8.82)$$
$$M_c = 5.40 \text{ decibels}$$
$$G_c = 10^{5.4/20} = 1.86$$

The above value of G_c was used in the fourth run of the design program shown immediately following the third run. The fourth run of the design program gives the open-loop frequency response of the system including all three control modes. Figure 15.10 shows the final open-loop Bode diagram of the complete system. Figure 15.11 is the error ratio graph of the complete system.

Design Conclusion—Concentration Control System

1. Controller derivative action time constant, T_d: 104 seconds
2. Controller integral action time constant, T_i: 500 seconds
3. Controller gain, G_c: 1.86
4. Maximum frequency limit: 0.035 radians/second (0.0056 Hertz) (Maximum frequency limit occurs where the error ratio becomes positive).

```
CONTROL SYSTEM DESIGN BY THE FREQUENCY RESPONSE METHOD
GIVE THE VALUES OF OVERALL PROCESS GAIN (G1), DERIVATIVE
TIME (T1), INTEGRAL TIME (T2), AND CONTROLLER GAIN (K).
IF NOT USED, SET T1=0, T2=0, AND K=1.
G1,T1,T2,K? 31.25,104,500,1
GIVE THE NUMBER OF FIRST ORDER LAGS.? 2
GIVE THE FIRST ORDER TIME CONSTANTS.
   1            ? 200
   2            ? 1325
GIVE THE NUMBER OF 2ND ORDER LAGS.? 1
GIVE THE SECOND ORDER RESONANT FREQUENCIES, DAMPING RATIOS.
   1            ? 2.4,0.9
GIVE THE NUMBER OF DEAD TIME LAGS.? 1
GIVE THE DEAD TIME LAGS
   1            ? 15
FREQUENCY     OPEN LOOP RESPONSE              ERROR RATIO
(RAD/SEC)     (DECIBELS)    (DEGREES)         (DECIBELS)
  2.000000E-5   69.9         -91.1            -139.8
  4.000000E-5   63.9         -92.2            -127.7
  6.000000E-5   60.3         -93.2            -120.7
  8.000000E-5   57.8         -94.3            -115.6
  1.000000E-4   55.9         -95.4            -111.7
  2.000000E-4   49.6         -100.5           -99.3
  4.000000E-4   43           -109.4           -85.9
  6.000000E-4   38.6         -116             -77.1
  8.000000E-4   35.1         -120.4           -70.3
  1.000000E-3   32.4         -123.3           -64.7
  2.000000E-3   23.4         -127.4           -46.8
  4.000000E-3   14.8         -127.9           -29.6
  6.000000E-3   9.8          -128.5           -19.9
  8.000000E-3   6.4          -128.9           -13.3
   .01          3.7          -129.3           -8.8
   .02          -3.8         -135             -.7
   .04          -10.8        -156.9            .5
   .06          -15.1        -181.6            .3
   .08          -18.5        -206.1            .1
   .1           -21.4        -229.7            0
   .2           -31.5        -336.5            0
   .4           -43          -528.2            0
   .6           -50.1        -712.8            0
   .8           -55.4        -895.1            0
```

```
CONTROL SYSTEM DESIGN BY THE FREQUENCY RESPONSE METHOD
GIVE THE VALUES OF OVERALL PROCESS GAIN (G1), DERIVATIVE
TIME (T1), INTEGRAL TIME (T2), AND CONTROLLER GAIN (K).
IF NOT USED, SET T1=0, T2=0, AND K=1.
G1,T1,T2,K? 31.25,104,500,1.86
GIVE THE NUMBER OF FIRST ORDER LAGS.? 2
GIVE THE FIRST ORDER TIME CONSTANTS.
   1            ? 200
   2            ? 1325
GIVE THE NUMBER OF 2ND ORDER LAGS.? 1
GIVE THE SECOND ORDER RESONANT FREQUENCIES, DAMPING RATIOS.
   1            ? 2.4,0.9
```

```
GIVE THE NUMBER OF DEAD TIME LAGS.? 1
GIVE THE DEAD TIME LAGS
 1              ? 15
FREQUENCY      OPEN LOOP RESPONSE           ERROR RATIO
(RAD/SEC)      (DECIBELS)    (DEGREES)      (DECIBELS)
 2.000000E-5   75.3          -91.1          -150.6
 4.000000E-5   69.3          -92.2          -138.5
 6.000000E-5   65.7          -93.2          -131.4
 8.000000E-5   63.2          -94.3          -126.4
 1.000000E-4   61.2          -95.4          -122.5
 2.000000E-4   55            -100.5         -110.1
 4.000000E-4   48.3          -109.4         -96.7
 6.000000E-4   43.9          -116           -87.9
 8.000000E-4   40.5          -120.4         -81.1
 1.000000E-3   37.8          -123.3         -75.5
 2.000000E-3   28.8          -127.4         -57.6
 4.000000E-3   20.2          -127.9         -40.3
 6.000000E-3   15.2          -128.5         -30.5
 8.000000E-3   11.7          -128.9         -23.6
 .01           9.1           -129.3         -18.5
 .02           1.6           -135           -4.9
 .04           -5.4          -156.9          1.7
 .06           -9.7          -181.6          1
 .08           -13.1         -206.1          .3
 .1            -16           -229.7          0
 .2            -26.2         -336.5          0
 .4            -37.7         -528.2          0
 .6            -44.8         -712.8          0
 .8            -50           -895.1          0
```

Example 15.4   Design a three-mode controller for the control of temperature in the blending and heating system, Figure 15.9.

Solution:

1. Overall process gain, K

$$K = 0.8 \times 10 = 8$$

2. Thermal time constant, T_3

    Film coefficients, h (use Equation 11.40).

$$h_i = h_o = h = 2.26(T_w + 34.4)\sqrt{T_d}$$
$$T_w = 60°C$$
$$T_d = 100 - 60 = 40°C$$
$$h = 2.26(60 + 34.4)\sqrt{40}$$
$$h = 1349 \text{ watts/square meter-Kelvin}$$

Heating area, A

$$A = \pi D^2/4 + \pi Dh = \pi D(D/4 + h)$$
$$A = \pi(0.75)(0.75/4 + 0.75)$$
$$A = 2.21 \text{ square meters}$$

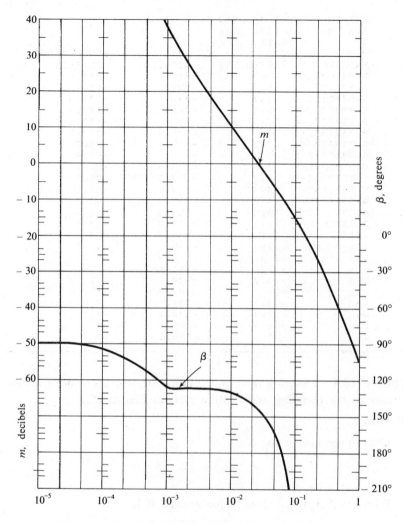

FIGURE 15.10
*OPEN-LOOP BODE DIAGRAM FOR THE CONCENTRATION
CONTROL SYSTEM, EXAMPLE 15.3*

Resistance, R

$$R = (1/h_o + x/K = 1/h_i)/A$$
$$R = (1/1349 + 0.01/45 + 1/1349)/2.21$$
$$R = 7.71 \times 10^{-4} \text{ Kelvin/watt}$$

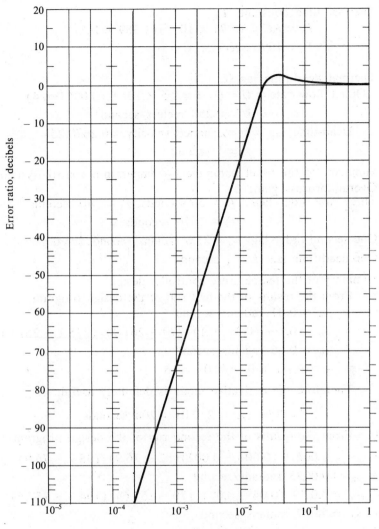

FIGURE 15.11

ERROR RATIO GRAPH FOR THE CONCENTRATION
CONTROL SYSTEM, EXAMPLE 15.3

Thermal capacitance, C

$\quad C = mS_h$

$\quad m = \rho h \pi D^2 / 4 = (1005)(0.75)\pi(0.75^2)/4$

$\quad m = 333$ kilograms

$\quad C = (333)(4170) = 1.389 \times 10^6$ joule/Kelvin

Thermal time constant, T_3
$$T_3 = RC = (7.71 \times 10^{-4})(1.389 \times 10^6)$$
$$T_3 = 1070 \text{ seconds}$$

3. Thermal dead-time lag, L
    Outlet flow rate $= U = Q/A = 2.5 \times 10^{-4}/(0.0351^2\pi/4)$
    $$U = 0.258 \text{ meters/second}$$
    Dead-time lag $= L = \text{distance/velocity} = 0.26/0.258$
    $$L = 1 \text{ second}$$

4. Summary of the input parameters—temperature control system
    Overall process gain: 8
    Three first-order lags: $T_1 = 50$ seconds, $T_2 = 240$ seconds,
    $$T_3 = 1070 \text{ seconds}$$
    One second-order lag: $\omega_o = 21.6$ radians/second, $\zeta = 0.8$
    One dead-time lag: $L = 1$ second

5. Summary of the temperature controller design
    a. From the results of the first run of the design program:
        $$\omega_{-180°} = 0.01 \text{ radians/second}$$
        $$\omega_{-270°} = 0.1 + (0.2 - 0.1)(270 - 261.9)/(275.1 - 261.9)$$
        $$\omega_{-270°} = 0.16 \text{ radians/second}$$
        $$\omega_{-270°}/\omega_{-180°} = 0.16/0.01 = 16$$

        Therefore the derivative mode is definitely useful.

        $$T = 2/\omega_{-180°} = 2/0.01 = 200 \text{ seconds}$$

    b. From the results of the second run of the design program:
        $$\omega_{-82°} = 0.002 + (0.004 - 0.002)(82 - 76.9)/(98.2 - 76.9)$$
        $$\omega_{-82°} = 0.0025 \text{ radians/second}$$
        $$\omega_{-170°} = 0.02 + (0.04 - 0.02)(170 - 157.6)/(194.4 - 157.6)$$
        $$\omega_{-170°} = 0.027 \text{ radians/second}$$

        The low-frequency gain is level, therefore the integral mode is useful.

        $$\omega_i = \text{Min}(0.0025 \text{ or } 0.2 \times 0.027)$$
        $$= \text{Min}(0.0025 \text{ or } 0.0054)$$
        $$\omega_i = 0.0025 \text{ radians/second}$$
        $$T_i = 1/\omega_i = 1/0.0025 = 400 \text{ seconds}$$

    c. From the results of the third run of the design program:
        $$m_{-140°} = -2.2 - (4.7 - 2.2)(140 - 136.6)/(141.1 - 136.6)$$

$$m_{-140°} = -4.1 \text{ decibels}$$
$$m_{-180°} = -13.6 - (25.3 - 13.6)(180 - 164.7)/(198 - 164.7)$$
$$m_{-180°} = -19 \text{ decibels}$$
$$M_c = \text{Min}(4.1 \text{ or } (19 - 6)) = \text{Min}(4.1 \text{ or } 13)$$

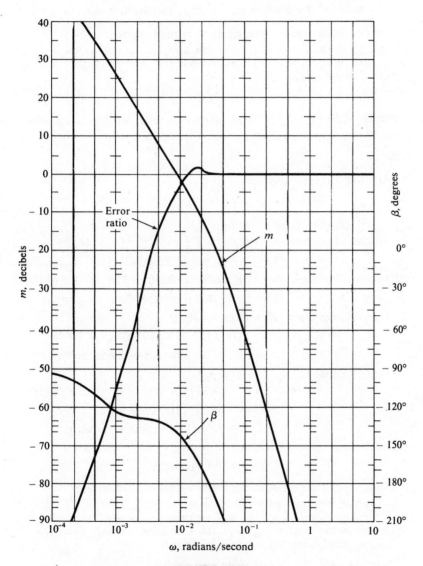

FIGURE 15.12

*OPEN-LOOP BODE DIAGRAM AND ERROR RATIO GRAPH FOR
THE TEMPERATURE CONTROL SYSTEM, EXAMPLE 15.4*

$M_c = 4.1$ decibels

$G_c = 10^{4.1/20} = 1.60$

d. From the results of the fourth run of the design program: Figure 15.12 shows the open-loop Bode diagram and the error ratio of the final design of the temperature control system.

6. Design conclusion—temperature control system
   Controller derivative action time constant, T_d: 200 seconds
   Controller integral action time constant, T_i: 400 seconds
   Controller gain, 1.6
   Maximum frequency limit: 0.014 radians/second (0.0022 Hertz)

Example 15.5 Design a three-mode controller for the control of level in the blending and heating system, Figure 15.9.

Solution:

1. Overall process gain, K

$$K = 16 \times 1.9 = 30.4$$

2. Level process time constant, T_2
   Outlet flow rate, $U = Q/A = 2.5 \times 10^{-4}/(0.0351^2\pi/4)$

$$U = 0.258 \text{ meters/second}$$

   Reynolds number, $\text{Re} = \rho V D/u = (1005)(0.258)(0.0351)/0.01$

$$\text{Re} = 910, \text{ therefore the flow is laminar}$$

   Liquid resistance, $R = P/Q = \rho g h/Q$

$$R = (1005)(9.81)(0.75)/2.5 \times 10^{-4}$$

$$R = 30 \times 10^6 \text{ newtons-seconds/meter}^5$$

   Liquid capacitance, $C = A/(\rho g)$

$$A = \pi(0.75)^2/4 = 0.4418 \text{ square meters}$$

$$C = 0.4418/(1005)(9.81)$$

$$C = 4.48 \times 10^{-5} \text{ meters}^5/\text{newton}$$

   Level time constant, $T_2 = (30 \times 10^6)(4.48 \times 10^{-5})$

$$T_2 = 1344 \text{ seconds}$$

3. Summary of the input parameters—level control system
   Overall process gain: 30.4
   Two first-order lags: $T_1 = 2$ seconds, $T_2 = 1344$ seconds
   One second-order lag: $\omega_o = 10.2$ radians/second, $\zeta = 0.75$
   One dead-time lag: $L = 0.5$ seconds

4. Summary of the level controller design
   a. Step 1: $\omega_{-270°}/\omega_{-180°} = 2.7/0.84 = 3.21$

$$T_d = 2/0.84 = 2.38 \text{ seconds}$$

b.   Step 2: $\omega_i = \text{Min}(0.0054 \text{ or } 0.2 \times 1.65) = 0.0054 \text{ rad/sec}$

$$T_i = 1/0.0054 = 185 \text{ seconds}$$

c.   Step 3: $M_c = \text{Min}(32.3 \text{ or } 37.41 - 6) = 31.41$

$$G_c = 10^{31.41/20} = 37.20$$

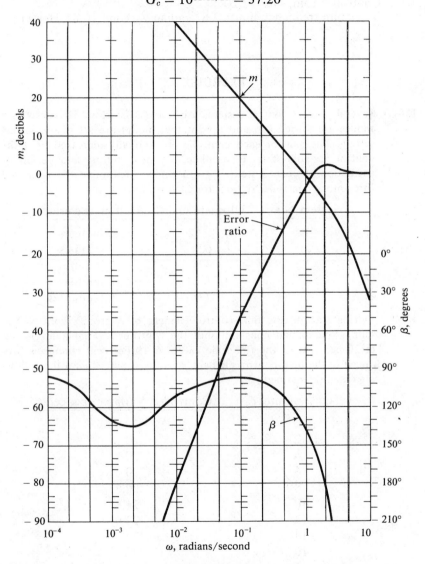

FIGURE 15.13

*OPEN-LOOP BODE DIAGRAM AND ERROR RATIO GRAPH
FOR THE LEVEL CONTROL SYSTEM, EXAMPLE 15.5*

d.　Step 4:　Figure 15.13 shows the open-loop Bode diagram and error ratio graph of the final design.

5.　Design conclusion—level control system

Controller derivative action time constant, T_d: 2.38 seconds
Controller integral action time constant, T_i: 185 seconds
Controller gain, G_c: 37.20
Maximum frequency limit: 3.83 radians/second (0.61 Hertz)

EXERCISES

15–1.　Several process control systems are tested at startup. The derivative modes are turned off, and the integral modes are set at the lowest setting. The gain of each controller is gradually increased until the control variable starts to oscillate. The gain setting and period of oscillation of each system is given below. Determine the two- and three-mode controller settings for each system.

System No.	Ultimate Gain	Ultimate Period
1	0.42	20 minutes
2	6.3	6 minutes
3	0.8	2 minutes
4	1.2	18 minutes
5	2.0	0.5 minutes

15–2.　During startup, the final-control element of a process control system is maintained constant until the controlled variable levels out at 20 per cent of the full-scale value. A 10 per cent change is produced in the final-control element position at time 0, and the following data are obtained.

Time, minutes	Controlled Variable, per cent
0	20.0
2	20
4	20
6	20
8	20.3
10	21.8
12	23.6
14	26
16	29
18	31.5
20	33.4

Time, minutes	Controlled Variable, per cent
22	35
24	36
26	37
28	37.8
30	38.2
32	38.7
34	39
36	39.3
38	39.5
40	39.6

Construct the process reaction curve and use the process reaction method to determine the settings of a two- and a three-mode controller.

15–3. A dc motor position control system has the following open-loop transfer function.

$$\frac{C_m}{SP} = \frac{K_A 16.1}{S(3.32 \times 10^{-3} \, S^2 + 0.201S + 1)}$$

where K_A is the gain of the amplifier. Use the frequency-response method to design a proportional-plus-derivative controller with a gain margin of 5 db and a phase margin of 40°.

15–4. The Amplidyne position control system shown in Figure 10.17 has the following open-loop transfer function.

$$\frac{C_m}{SP} = \frac{8500K_A}{(0.02S + 1)(0.005S + 1)(0.05S + 1)(0.3S + 1)S}$$

where K_A is the gain of the amplifier. Use the frequency-response method to design a proportional-plus-derivative controller with a gain margin of 5 db and a phase margin of 40°.

15–5. Use the frequency-response method to design a three-mode controller for each of the following processes. The minimum gain margin is 5 db and the minimum phase margin is 40°.
a. The pressure control system in Exercise 14–1.
b. The thermal control system in Exercise 14–2.
c. The level control system in Exercise 14–3.
d. The position control system in Exercise 14–4.
e. The belt conveyor in Exercise 14–5.
f. The blending system in Exercise 14–6.
g. The second-order plus dead-time system in Exercise 14–7.
h. The pressure system in Exercise 14–8.

15–6.   The N. E. W. Chemical Company produces a liquid product which is
sold in one liter containers. The product is a blend of two ingredients
which are mixed and heated in a jacketed kettle similar to the one
shown in Figure 12.9. A pipe line similar to Figure 12.13–b is used to
transport the heated product from the kettle to the bottling line. The
product temperature must be between 60°C and 66°C when it is
bottled. A temperature sensor is used to measure the product temperature at the bottling line, and a control valve is used to regulate the
steam flow to the kettle. The component transfer functions and
important parameters are given below.

$$\underline{\text{Control Valve:}} \quad \frac{M}{V} = \frac{K_v}{(1 + T_v S)(1 + T_v S)}$$

$$\underline{\text{Process:}} \quad \frac{\theta}{M} = \left(\frac{K_p}{1 + T_p S}\right)e^{-T_d S}$$

$$\underline{\text{Measuring Means:}} \quad \frac{\theta_m}{\theta} = \frac{K_m}{1 + T_m S}$$

<u>Parameters</u>

V = Controller output signal, percent of full scale
M = Steam flow rate, kilogram per second
θ = Product temperature at the bottling line, °C
θ_m = Measuring means output signal, percent of full scale
K_v = Gain of the control valve, $\dfrac{\text{kilogram per second}}{\text{percent of full scale}}$
K_p = Gain of the process, $\dfrac{°C}{\text{kilogram per second}}$
K_m = Gain of the measuring means, $\dfrac{\text{percent of full scale}}{°C}$
T_v = Time constant of the control valve, second
T_p = Time constant of the kettle, second
T_d = Dead time lag of the pipe line, second
T_m = Time constant of the measuring means, second
m = Mass of product in the tank, kilogram
S_h = Specific heat of the product, joule/°C—kilogram
A = Area of the heated surface in the kettle meter²
h = Film coefficient of the kettle, watt/meter²—°C
v = Velocity of liquid in the pipe line, meter/second
L = Length of the pipe line, meter

Use the frequency response method to design a three mode controller for each set of data below. The data are obtained from system
specifications, experimental measurements, technical references, or
manufacturer's bulletins. Use the values of M, S_h, A, h, v, and L to
determine the values of T_p and T_d (assume all thermal resistances
except h are negligible).

Data

Parameter	A	B	C
K_v	0.002	0.005	0.001
K_m	0.5	1	2
K_p	500	200	1000
T_v	0.1	0.2	0.1
T_m	50	20	100
m	200	30	100
S_h	2100	2100	1800
A	2.1	1.4	1.8
h	200	90	100
v	0.2	0.1	0.1
L	40	10	20

15–7. The O. L. D. Chemical Company produces a liquid product which is a precise blend of two components, A and B. A blending tank similar to Figure 12.11 is used to thoroughly mix the two ingredients. A final control element adjusts the percentage of component A as the two components are metered into the blender. A pipe line similar to Figure 12.13–b is used to transport the blended product to a composition analyzer in the bottling line. The component transfer functions and important parameters are given below.

$$\underline{\text{Final Control Element:}} \quad \frac{M}{V} = \frac{K_v}{(1 + T_v S)(1 + T_v S)}$$

$$\underline{\text{Process:}} \quad \frac{C}{M} = \left(\frac{K_p}{1 + T_p S}\right) e^{-T_d S}$$

$$\underline{\text{Measuring Means:}} \quad \frac{C_m}{C} = \frac{K_m}{1 + T_m S}$$

Parameters

V = Controller output signal, percent of full scale
M = Percent of component A at the blender input
C = Concentration of component A at the analyzer, parts per 100
C_m = Measuring means output, percent of full scale
K_v = Gain of the final control element, $\dfrac{\text{percent}}{\text{percent of full scale}}$
K_p = Gain of the process, $\dfrac{\text{parts per 100}}{\text{percent}}$
K_m = Gain of the measuring means, $\dfrac{\text{percent of full scale}}{\text{parts per 100}}$

T_v = Time constant of the final control element, second
T_p = Time constant of the blender, second
T_d = Dead time lag of the pipe line, second
T_m =Time constant of the measuring means, second
V = Volume of liquid in the tank, meter³
Q = Product flow rate, meter³/second
v = Velocity of liquid in the pipe line, meter/second
L = Length of the pipe line, meter

Use the frequency response method to design a three mode controller for each set of data below. The data are obtained from system specifications, experimental measurements, technical references, or manufacturer's bulletins. Use the values of V, Q, v, and L to determine the values of T_p, and T_d.

Data

Parameter	A	B	C
K_v	1	2	0.5
K_m	1	0.5	2
K_p	1	1	1
T_v	0.2	0.1	0.1
T_m	200	100	500
V	2	1	5
Q	0.002	0.001	0.005
v	0.05	0.1	0.2
L	5	10	20

REFERENCES

Arthur, K. *Transducer Measurements.* Tektronix, Inc., 1970.

Bateson, R. N. *Motomatic Servo Control Course Manual.* Electro-Craft Corporation, 1971.

Binder, R. C. *Fluid Mechanics.* Prentice-Hall, Inc., 1949.

Bower, J. L. and P. M. Schultheiss. *Introduction to the Design of Servomechanisms.* John Wiley & Sons, Inc., 1964.

Bryan, G. T. *Control Systems for Technicians.* Hart Publishing Company, Inc., 1967.

Buckley, P. S. *Techniques of Process Control.* John Wiley & Sons, Inc., 1964.

Chemical Engineering, Vol. 76 No. 12, June 2, 1969.

Diefenderfer, A. J. *Principles of Electronic Instrumentation*. W. B. Saunders Company, 1972.

Eckman, D. P. *Industrial Instrumentation*. John Wiley & Sons, Inc., 1950.

Erickson, W. H. and N. H. Bryant. *Electrical Engineering Theory and Practice*. John Wiley & Sons, Inc., 1952.

Fundamentals of Industrial Instrumentation, Honeywell, Inc., 1957.

Harrison, H. L. and J. G. Bollinger. *Introduction to Automatic Controls*. International Textbook Company, 1969.

Jennings, B. H. and S. R. Lewis. *Air Conditioning and Refrigeration*. International Textbook Company, 1949.

Morris, N. M. Control Engineering. McGraw-Hill, Ltd., 1968.

Olesten, Nils O. *Numerical Control,* John Wiley & Sons, Inc., 1970.

Principles of Automatic Process Control, Instrument Society of America, 1968. (Text & Film).

Principles of Frequency Response, Instrument Society of America, 1958. (Text & Film).

Shinskey, F. G. *Process Control Systems*. McGraw-Hill, Inc., 1967.

Tucker, G. K. and D. M. Wills. *A Simplified Technique of Control System Engineering*. Minneapolis-Honeywell Regulator Company, 1962.

Wilson, H. S. & Zoss, L. M., "Control Theory Notebook", Instrument Society of America, Reprint from ISA Journal.

Zeines, B. *Automatic Control Systems*. Prentice-Hall Inc., 1972.

Ziegler, J. G. and Nichols, N. B. "Optimum Settings for Automatic Controllers", *ASME Transactions,* Vol. 64, No. 8, Page 759–768, 1942.

APPENDIX A

Properties of Solids

Solid	Density kilograms cubic meter	Thermal Conductivity watt meter—Kelvin	Specific Heat joule kilogram—Kelvin
Aluminum	2700	204	910
Asbestos	2400	0.16	815
Brass	8470	100	370
Cast iron	7400	47	460
Copper	8940	380	400
Glass	2600	1	490
Gold	19,300	294	130
Graphite	2000	5	900
Ice	900	2.25	2000
Insulation	—	0.036	—
Lead	11,340	35	130
Nickel	8900	60	460
Rubber	1500	0.2	2000
Silver	10,500	400	234
Steel	7800	45	500
Wood (typical oak)	740	0.2	2400
Wood (typical pine)	440	0.15	2800

Properties of Liquids* $N = \dfrac{kg-m}{sec^2}$

Liquid	Density kilograms cubic meter	Dynamic Viscosity newton-seconds square meter	Thermal Conductivity watts meter—Kelvin	Specific Heat joules kilogram—Kelvin
Ethyl alcohol	800	0.0013	0.18	2300
Gasoline	740	0.0005	0.14	2100
Glycerine	1260	0.83	0.29	2400
Kerosene	800	0.0024	0.15	2070
Mercury	13,600	0.0015	8.0	140
Oil	880	0.160	0.16	2180
Turpentine	870	0.0015	0.13	1720
Water	1000	0.001	0.6	4190

* At a temperature of 15° Celsius.

SI Units with CGS and English Equivalents

Quantity	SI Units	CGS Equivalent	English Equivalent
Length	1 meter	100 centimeters	3.28 feet 39.37 inches
Mass	1 kilogram	10^3 grams	2.2 pounds
Time	1 second	1 second	1 second
Area	1 square meter	10^4 square centimeters	10.76 square feet 1.55×10^3 square inches
Volume	1 cubic meter	10^6 cubic centimeters 10^3 liters	35.3 cubic feet 6.1×10^4 cubic inches
Density	1 kilogram/cubic meter	10^{-3} gm/cubic cm	0.0624 lb/cubic foot
Velocity	1 meter/second	100 cm/sec	3.28 feet/second 2.24 miles/hour
Force	1 newton	10^5 dynes	0.225 pound-force
Gravitational acceleration g	9.81 meter/sec^2	981 cm/sec^2	32.2 ft/sec^2
Dynamic viscosity	1 newton-sec/meter^2	10 poise (dyne-sec/cm^2)	2.09×10^{-2} lbf-sec/ft^2
Work or energy	1 joule (newton-meter)	10^7 ergs (dyne-cm) 0.239 gram calorie	0.738 foot-lbf 9.48×10^{-4} Btu
Power	1 watt (joule/sec)	10^7 erg/sec	0.738 foot-lbf/sec 1.34×10^{-3} HP
Pressure	1 newton/meter^2 $= 10^{-4}$ newton/cm^2	10 dyne/cm^2	0.0209 lbf/ft^2 1.45×10^{-4} lbf/inch^2

SI Units with CGS and English Equivalents—(Continued)

Quantity	SI Units	CGS Equivalent	English Equivalent
Specific heat	1 joule/kilogram—°Kelvin	2.39×10^{-4} calorie/gm-°C	2.39×10^{-4} Btu/lb-°F
Thermal conductivity	1 watt/meter—°Kelvin	$2.39 \times 10^{-3} \dfrac{\text{calorie}}{\text{cm-sec-°C}}$	$6.94 \dfrac{\text{Btu-in.}}{\text{hr} - \text{ft}^2 - \text{°F}}$ $0.578 \dfrac{\text{Btu}}{\text{hr} - \text{ft} - \text{°F}}$
Torque	1 newton-meter	10^7 dyne-cm	0.738 foot-lbf
Moment of inertia	1 kilogram-meter^2	10^7 gram-cm^2	141.85 ounce-inches 23.7 lb-ft^2
Electric current	1 ampere	1 ampere	1 ampere
Electromotive force	1 volt	1 volt	1 volt
Electric charge	1 coulomb	1 coulomb	1 coulomb
Temperature	1 °Kelvin	°Celsius = °Kelvin $-$ 273	°Farenheit = 1.8(°C) $+$ 32
Entropy	1 joule/°Kelvin	10^7 ergs/°C	5.27×10^{-4} Btu/°F

Properties of Gases*

Gas	Molecular Weight M	Density kilogram cubic meter	Dynamic Viscosity newton-seconds square meter	Thermal Conductivity watts meter—Kelvin	Specific Heat† joules kilogram—Kelvin
Hydrogen (H_2)	2.016	0.0854	8.89×10^{-6}	0.163	14,200
Helium (He)	4.002	0.169	1.97×10^{-5}	0.140	1015
Carbon Monoxide (CO)	28.0	1.19		0.023	1010
Nitrogen (N_2)	28.016	1.19	1.77×10^{-5}	0.024	1030
Air	28.8	1.22	1.81×10^{-5}	0.024	910
Oxygen (O_2)	32.0	1.36		0.024	515
Argon (A)	39.944	1.69	2.2×10^{-5}		906
Carbon Dioxide (CO_2)	44.0	1.88	1.46×10^{-5}	0.014	

* At standard atmospheric conditions, a temperature of 15° Celsius and a pressure of 76 cm of mercury.

† The specific heat is for constant pressure.

Standard Atmospheric Conditions

1. Temperature: 288° Kelvin
    15° Celsius
    59° Farenheit

2. Pressure: 1.013×10^5 newtons per square meter
    1.013×10^6 dynes per square centimeter
    14.7 pounds per square inch
    76 centimeters of mercury
    29.92 inches of mercury
    10.336 meters of water
    34 feet of water

3. Air density: 1.23 kilograms per cubic meter
    1.23×10^{-3} gram per cubic centimeter
    0.07651 pound per cubic foot

Systems of Units[1]

The different systems of units are best understood when applied to Newton's law of motion.

$$f = kma$$

where:

f = force acting on a body

m = mass of the body

a = acceleration of the body

k = a constant whose value depends on the system of units

1. The MKS and SI systems, $k = 1$

$$f(\text{newtons}) = m(\text{kilograms}) \times a(\text{meters/sec}^2)$$

2. The cgs absolute system, $k = 1$

$$f(\text{dynes}) = m(\text{grams}) \times a(cm/\text{sec}^2)$$

3. The fps absolute system, $k = 1$

$$f(\text{poundals}) = m(\text{pounds}) \times a(\text{ft/sec}^2)$$

4. The engineering fps system, $k = \dfrac{1}{g_s} = \dfrac{1}{32.2}$

---

[1] From Lemon and Ference, *Analytical Experimental Physics.* Chicago: University of Chicago Press, 1946, pp. 37–38.

$$f(\text{pounds-force}) = \frac{1}{32.2} \times m(\text{pounds}) \times a(\text{ft/sec}^2)$$

5. The engineering fss system, $k = 1$

$$f(\text{pounds-force}) = m(\text{slugs}) \times a(\text{ft/sec}^2)$$

REFERENCES

The entries in the tables of properties of solids, liquids, and gases were converted to SI units by means of data obtained from the following sources.

Binder, R. C. *Fluid Mechanics*. 2nd edition. Englewood Cliffs, N. J.: Prentice-Hall, Inc., 1950, pp. 6, 27, 62–65.

Carmichael, Colin. *Kent's Mechanical Engineers' Handbook: Design and Production*. 12th edition. New York: John Wiley & Sons, 1950, pp. 1–04, 1–29, 2–57, 2–58, 5–78, 19–06.

Jennings, Burgess H., and Lewis, Samuel R. *Air Conditioning and Refrigeration*. 3rd edition. Scranton, Pa.: International Textbook Company, 1951, pp. 22, 99, 104–114.

Lemon, Harvey B., and Ference, Michael, Jr. *Analytical Experimental Physics*. Chicago: University of Chicago Press, 1946, pp. 37, 38, 134, 142, 177, 178, 181, 188, 219.

Salisbury, J. Kenneth. *Kent's Mechanical Engineers' Handbook: Power*. 12th edition. New York: John Wiley & Sons, 1950, pp. 2–48, 2–59, 3–04, 3–05, 3–06, 3–14, 3–15, 3–16, 3–37, 3–54, 3–58, 5–03, 6–43, 6–44.

APPENDIX B

Laplace Transform Pairs

Entry No.	Time Domain $f(t)$; $t > 0$ (see note 3)	Frequency Domain $F(S)$ (see note 3)
1.	$Kf(t)$	$KF(S)$
2.	K	$\dfrac{K}{S}$
3.	Kt^n	$K\left(\dfrac{n!}{S^{n+1}}\right)$
4.	$Kf(t - a)$	$Ke^{-aS}F(S)$
5.	$K\dfrac{df}{dt}$	KSF (see note 2)
6.	$K\dfrac{d^2f}{dt^2}$	KS^2F (see note 2)
7.	$K\dfrac{d^3f}{dt^3}$	KS^3F (see note 2)
8.	$K\int f(t)dt$	$K\dfrac{1}{S}F$ (see note 2)
9.	Ke^{-at}	$K\left(\dfrac{1}{S + a}\right)$
10.	$Kt^n e^{-at}$	$K\left[\dfrac{n!}{(S + a)^{n+1}}\right]$
11.	$K(1 - e^{-t/T})$	$\dfrac{K}{S(TS + 1)}$
12.	$K \sin \omega t$	$K\left[\dfrac{\omega}{S^2 + \omega^2}\right]$
13.	$K \cos \omega t$	$K\left[\dfrac{S}{S^2 + \omega^2}\right]$
14.	$Kt \sin \omega t$	$K\left[\dfrac{2\omega S}{(S^2 + \omega^2)^2}\right]$
15.	$Kt \cos \omega t$	$K\left[\dfrac{S^2 - \omega^2}{(S^2 + \omega^2)^2}\right]$

Notes:

1. The letters K, T and a represent any numerical constant. The letter e represents 2.71828..., the base of the natural logarithms.

2. This table is intended only for determining the transfer functions of control system components. The simplifying assumption of zero initial conditions was made for entries 5, 6, 7, and 8. This assumption is consistent with the practice of using zero initial conditions for most control system analyses.

3. The symbol f or $f(t)$ is used to represent any signal in the time domain, and F or $F(S)$ is used to represent any signal in the frequency domain. Any other pair of lower- and upper-case letters may be substituted for f and F in working a particular problem.

4. The Laplace transformation which transforms $h(t)$ into $H(S)$ is defined as follows:

$$\zeta h(t) = H(S) = \int_0^\infty h(t)e^{-st}dt$$

5. Example:

Determine the transfer function of a component which is described by the following differential equation.

$$7\frac{d^2h}{dt^2} + 2\frac{dh}{dt} + 5h = y$$

where:

$$x = \text{the output signal}$$
$$y = \text{the input signal}$$

The Laplace transformation is

$$7S^2H + 2SH + 5H = Y$$
$$H(7S^2 + 2S + 5) = Y$$
$$\frac{H}{Y} = \frac{1}{7S^2 + 2S + 5} = \text{the transfer function}$$

APPENDIX C

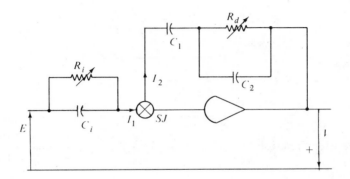

A Proportional-plus-Integral-plus-Derivative Controller

$$E \qquad R_i \qquad C_i \qquad I_1 \qquad SJ \qquad I_2 \qquad C_1 \qquad R_d \qquad C_2 \qquad V$$

$$I_1 = I_2$$

$$I_1 = E\left[SC_i + \frac{1}{R_i} \right] = E\left[\frac{SR_iC_i + 1}{R_i} \right]$$

$$I_2 = V\left[\frac{1}{\dfrac{1}{SC_1} + \dfrac{1}{SC_2 + \dfrac{1}{R_d}}} \right] = V\left[\frac{1}{\dfrac{1}{SC_1} + \dfrac{R_d}{SR_dC_2 + 1}} \right]$$

$$I_2 = V\left[\frac{SC_1}{1 + \dfrac{SR_dC_1}{SR_dC_2 + 1}} \right] = V\left[\frac{SC_1}{\dfrac{SR_dC_2 + 1 + SR_dC_1}{SR_dC_2 + 1}} \right]$$

$$I_2 = V\left[\frac{SC_1(SR_dC_2 + 1)}{SR_d(C_1 + C_2) + 1} \right]$$

529

But

$$I_1 = I_2.$$

$$E\left[\frac{SR_iC_i + 1}{R_i}\right] = V\left[\frac{SC_1(SR_dC_2 + 1)}{SR_d(C_1 + C_2) + 1}\right]$$

$$\frac{V}{E} = \left[\frac{SR_iC_i + 1}{R_i}\right]\left[\frac{SR_d(C_1 + C_2) + 1}{SC_1(SR_dC_2 + 1)}\right]$$

$$\frac{V}{E} = \frac{C_i}{C_1}\left[\frac{SR_iC_i + 1}{SR_iC_i}\right]\left[\frac{SR_d(C_1 + C_2) + 1}{SR_dC_2 + 1}\right]$$

Let

$$T_i = R_iC_i; \qquad T_d = R_d(C_1 + C_2)$$

$$K = C_i/C_1; \qquad \alpha = \frac{C_2}{C_1 + C_2}$$

Then

$$R_dC_2 = \left(\frac{C_2}{C_1 + C_2}\right)R_d(C_1 + C_2) = \alpha T_d$$

Transfer Function $= \dfrac{V}{E} = K\left[1 + \dfrac{1}{T_iS}\right]\left[T_dS + 1\right]\left[\dfrac{1}{\alpha T_dS + 1}\right]$

where:

$$K = C_i/C_1 = \text{the controller gain}$$

$$\left[1 + \frac{1}{T_iS}\right] = \text{the integral model term}$$

$$[T_dS + 1] = \text{the derivative mode term}$$

$$\left[\frac{1}{\alpha T_dS + 1}\right] = \text{the derivative limiter term}$$

Multiply the integral and derivative terms in the above transfer function.

$$\frac{V}{E} = K\left[1 + \frac{T_d}{T_i} + \frac{1}{T_iS} + T_dS\right]\left[\frac{1}{\alpha T_dS + 1}\right]$$

Regroup the term $1 + \dfrac{T_d}{T_i}$ as part of the gain K.

$$\frac{V}{E} = K\left(\frac{T_i + T_d}{T_i}\right)\left[1 + \frac{1}{(T_i + T_d)S} + \left(\frac{T_iT_d}{T_i + T_d}\right)S\right]\left[\frac{1}{\alpha T_dS + 1}\right]$$

The last equation is in the form of the ideal three-mode controller with the addition of the derivative limiter term. There is an interaction of control modes, as follows.

Non-interacting terms:

$$\text{Gain, } K^1 = K\left(\frac{T_i + T_d}{T_i}\right)$$

$$\text{Integral time, } T_i{}^1 = \frac{1}{T_i + T_d}$$

$$\text{Derivative time, } T_d{}^1 = \frac{T_d T_i}{T_d + T_i}$$

$$\alpha^1 = \left(\frac{C_2}{C_1 + C_2}\right)\left(\frac{T_i + T_d}{T_i}\right)$$

With the non-interacting terms, the transfer function is

$$\frac{V}{E} = K^1\left[1 + \frac{1}{T_i{}^1 S} + T_d{}^1 S\right]\left[\frac{1}{\alpha^1 T_d{}^1 S + 1}\right]$$

A Proportional-plus-Derivative Controller

| E | C_i | I_1 | I_2 | C_1 | R_d | C_2 | V |

$$I_1 = I_2$$

$$I_1 = ESC_i$$

$$I_2 = V\left[\frac{SC_1(SR_dC_2 + 1)}{SR_d(C_1 + C_2) + 1}\right] \qquad \text{(see page 1 of this Appendix)}$$

$$ESC_i = V\left[\frac{SC_1(SR_dC_2 + 1)}{SR_d(C_1 + C_2) + 1}\right]$$

$$\frac{V}{E} = \frac{SC_i}{SC_1}\left[\frac{SR_d(C_1 + C_2) + 1}{SR_dC_2 + 1}\right]$$

Let

$$R_d(C_1 + C_2) = T_d$$
$$R_dC_2 = \alpha T_d$$
$$K = C_i/C_1$$

Transfer Function $= K[T_dS + 1]\left[\dfrac{1}{\alpha T_dS + 1}\right]$

where:

$K = C_i/C_1 =$ controller gain

$[T_dS + 1] =$ the derivative term

$\left[\dfrac{1}{\alpha T_dS + 1}\right] =$ the derivative limiter term

A Proportional-plus-Integral Controller

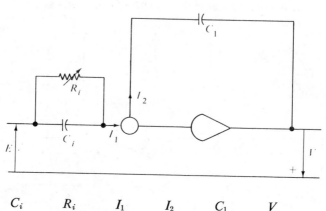

| E | C_i | R_i | I_1 | I_2 | C_1 | V |

$$I_1 = I_2$$

$$I_1 = E\left[SC_i + \frac{1}{R_i}\right] = E\left[\frac{SR_iC_i + 1}{R_i}\right]$$

$$I_2 = VSC_1$$

$$E = \left[\frac{SR_iC_i + 1}{R_i}\right] = VSC_1$$

$$\frac{V}{E} = \frac{SR_iC_i + 1}{SC_1R_i} = \frac{C_i}{C_1}\left[\frac{SR_iC_i + 1}{SR_iC_i}\right]$$

Let

$$T_i = R_i C_i; \qquad K = C_i/C_1$$

$$\text{Transfer Function} = K\left[1 + \frac{1}{T_i S}\right]$$

where:

$$K = C_i/C_1 = \text{controller gain}$$

$$\left[1 + \frac{1}{T_i S}\right] = \text{the integral term}$$

Decibel Conversion Table
NEGATIVE DECIBEL VALUES
GAIN < 1

db	Gain	db	Gain	db	Gain	db	Gain	db	Gain	db	Gain	db	Gain	db	Gain
-.1	.989	-2.6	.741	-5.1	.556	-7.6	.417	-10.2	.309	-15.2	.174	-20.5	.094	-31	.028
-.2	.977	-2.7	.733	-5.2	.550	-7.7	.412	-10.4	.302	-15.4	.170	-21.0	.089	-32	.025
-.3	.966	-2.8	.724	-5.3	.543	-7.8	.407	-10.6	.295	-15.6	.166	-21.5	.084	-33	.022
-.4	.955	-2.9	.716	-5.4	.537	-7.9	.403	-10.8	.288	-15.8	.162	-22.0	.079	-34	.020
-.5	.944	-3.0	.708	-5.5	.531	-8.0	.398	-11.0	.282	-16.0	.158	-22.5	.074	-35	.018
-.6	.933	-3.1	.700	-5.6	.525	-8.1	.394	-11.2	.275	-16.2	.155	-23.0	.071	-36	.016
-.7	.923	-3.2	.692	-5.7	.519	-8.2	.389	-11.4	.269	-16.4	.151	-23.5	.067	-37	.014
-.8	.912	-3.3	.684	-5.8	.513	-8.3	.385	-11.6	.263	-16.6	.148	-24.0	.063	-38	.013
-.9	.902	-3.4	.676	-5.9	.507	-8.4	.380	-11.8	.257	-16.8	.145	-24.5	.060	-39	.011
-1.0	.891	-3.5	.668	-6.0	.501	-8.5	.376	-12.0	.251	-17.0	.141	-25.0	.056	-40	.01
-1.1	.881	-3.6	.661	-6.1	.495	-8.6	.372	-12.2	.245	-17.2	.138	-25.5	.053	-41	.009
-1.2	.871	-3.7	.653	-6.2	.490	-8.7	.367	-12.4	.240	-17.4	.135	-26.0	.050	-42	.008
-1.3	.861	-3.8	.646	-6.3	.484	-8.8	.363	-12.6	.234	-17.6	.132	-26.5	.047	-43	.007
-1.4	.851	-3.9	.638	-6.4	.479	-8.9	.359	-12.8	.229	-17.8	.129	-27.0	.045	-44	.006
-1.5	.841	-4.0	.631	-6.5	.473	-9.0	.358	-13.0	.224	-18.0	.126	-27.5	.042	-45	.006
-1.6	.832	-4.1	.624	-6.6	.468	-9.1	.351	-13.2	.219	-18.2	.123	-28.0	.040	-46	.005
-1.7	.822	-4.2	.617	-6.7	.462	-9.2	.347	-13.4	.214	-18.4	.120	-28.5	.038	-47	.004
-1.8	.813	-4.3	.610	-6.8	.457	-9.3	.343	-13.6	.209	-18.6	.117	-29.0	.035	-48	.004
-1.9	.804	-4.4	.603	-6.9	.452	-9.4	.339	-13.8	.204	-18.8	.115	-29.5	.033	-49	.004
-2.0	.794	-4.5	.596	-7.0	.447	-9.5	.335	-14.0	.200	-19.0	.112	-30.0	.032	-50	.003
-2.1	.785	-4.6	.589	-7.1	.442	-9.6	.331	-14.2	.195	-19.2	.110				
-2.2	.776	-4.7	.582	-7.2	.437	-9.7	.327	-14.4	.191	-19.4	.107				
-2.3	.767	-4.8	.575	-7.3	.432	-9.8	.324	-14.6	.186	-19.6	.105				
-2.4	.759	-4.9	.569	-7.4	.427	-9.9	.320	-14.8	.182	-19.8	.102				
-2.5	.750	-5.0	.562	-7.5	.422	-10.0	.316	-15.0	.178	-20.0	.100				

Decibel Conversion Table

POSITIVE DECIBEL VALUES
GAIN > 1

0.1 to 2.5 db		2.6 to 5.0 db		5.1 to 7.5 db		7.6 to 10.0 db		10.2 to 15.0 db		15.2 to 20.0 db		20.5 to 30 db		31 to 50 db	
db	Gain	db	Gain	db	Gain	db	Gain	db	Gain	db	Gain	db	Gain	db	Gain
.1	1.01	2.6	1.35	5.1	1.80	7.6	2.40	10.2	3.24	15.2	5.75	20.5	10.6	31	35.5
.2	1.02	2.7	1.36	5.2	1.82	7.7	2.43	10.4	3.31	15.4	5.89	21.0	11.2	32	39.8
.3	1.04	2.8	1.38	5.3	1.84	7.8	2.46	10.6	3.39	15.6	6.03	21.5	11.9	33	44.7
.4	1.05	2.9	1.40	5.4	1.86	7.9	2.48	10.8	3.47	15.8	6.17	22.0	12.6	34	50.1
.5	1.06	3.0	1.41	5.5	1.88	8.0	2.51	11.0	3.55	16.0	6.31	22.5	13.3	35	56.2
.6	1.07	3.1	1.43	5.6	1.91	8.1	2.54	11.2	3.63	16.2	6.46	23.0	14.1	36	63.1
.7	1.08	3.2	1.45	5.7	1.93	8.2	2.57	11.4	3.72	16.4	6.67	23.5	15.0	37	70.8
.8	1.10	3.3	1.46	5.8	1.95	8.3	2.60	11.6	3.80	16.6	6.76	24.0	15.8	38	79.4
.9	1.11	3.4	1.48	5.9	1.97	8.4	2.63	11.8	3.89	16.8	6.92	24.5	16.8	39	89.1
1.0	1.12	3.5	1.50	6.0	2.00	8.5	2.66	12.0	3.98	17.0	7.08	25.0	17.8	40	100.0
1.1	1.14	3.6	1.51	6.1	2.02	8.6	2.69	12.2	4.07	17.2	7.24	25.5	18.8	41	112.2
1.2	1.15	3.7	1.53	6.2	2.04	8.7	2.72	12.4	4.17	17.4	7.41	26.0	20.0	42	125.9
1.3	1.16	3.8	1.55	6.3	2.07	8.8	2.75	12.6	4.27	17.6	7.59	26.5	21.1	43	141.3
1.4	1.17	3.9	1.57	6.4	2.09	8.9	2.79	12.8	4.37	17.8	7.76	27.0	22.4	44	158.5
1.5	1.19	4.0	1.58	6.5	2.11	9.0	2.82	13.0	4.47	18.0	7.94	27.5	23.7	45	177.8
1.6	1.20	4.1	1.60	6.6	2.14	9.1	2.85	13.2	4.57	18.2	8.13	28.0	25.1	46	199.6
1.7	1.22	4.2	1.62	6.7	2.16	9.2	2.88	13.4	4.68	18.4	8.32	28.5	26.6	47	224
1.8	1.23	4.3	1.64	6.8	2.19	9.3	2.92	13.6	4.79	18.6	8.51	29.0	28.2	48	251
1.9	1.24	4.4	1.66	6.9	2.21	9.4	2.95	13.8	4.90	18.8	8.71	29.5	29.9	49	282
2.0	1.26	4.5	1.68	7.0	2.24	9.5	2.98	14.0	5.01	19.0	8.91	30.0	31.6	50	316
2.1	1.27	4.6	1.70	7.1	2.27	9.6	3.02	14.2	5.13	19.2	9.12				
2.2	1.29	4.7	1.72	7.2	2.29	9.7	3.06	14.4	5.25	19.4	9.33				
2.3	1.30	4.8	1.74	7.3	2.32	9.8	3.09	14.6	5.37	19.6	9.55				
2.4	1.32	4.9	1.76	7.4	2.35	9.9	3.13	14.8	5.50	19.8	9.77				
2.5	1.33	5.0	1.78	7.5	2.37	10.0	3.16	15.0	5.62	20.0	10.00				

Chapter 1

1–2. Size and timing

1–3. 0.25 milliamps per °C

1–4.

Input Signal	Output Signal
a. Room temperature	Open or closed switch
b. Open or closed switch	Gas flow rate
c. Gas flow rate	Heat
d. Heat	Room temperature

1–6. Read sections 1.3 and 1.4

1–7. (1) Measuring the controlled variable.

(2) Computing the difference between the measured value of the controlled variable and the desired value (the error).

(3) Using the error to generate a control action.

(4) Using the control action to drive the actual value of the controlled variable toward the desired value.

1–8. See Figure 1.6

1–9. E = 4.2 psi

1–13. See Figures 1.7 and 1.8

Chapter 2

2–1. a. (4) b. (3) c. (1) d. (2)

2–2. (a)

2–3. See Figure 2.2

2–5. Critically damped response

Chapter 3

3–4. See Figures 3.2–3.7

3–5. See Figures 3.8–3.12

3–10. b.

3–11. d.

3–13. Continuous

3–14. 0000, 0001, 0010, 0011,
0100, 0101, 0110, 0111,
1000, 1001, 1010, 1011,
1100, 1101, 1110, 1111

3–16.   1.   The computer closes demultiplexer switch No. 1.
2.   The A-D converter converts controlled variable No. 1 to digital input signal No. 1.
3.   The computer compares digital input signal No. 1 with setpoint No. 1 and calculates digital output signal No. 1.
4.   The computer closes multiplexer switch No. 1 and transfers digital output signal No. 1 to D-A converter No. 1.
5.   D-A converter No. 1 holds digital signal No. 1 and converts it to an analog signal which positions control valve No. 1.
6.–10.   Repeat steps 1–5 for controlled variable No. 2 using the No. 2 switches and converters.
11.–15.   Repeat steps 1–5 for controlled variable No. 3 using the No. 3 switches and converters.

As soon as step 15 is completed, the cycle is repeated beginning with step 1.

Chapter 4

4–1. $1745 \dfrac{dh}{dt} + h = 664m$

4–3.   a.   40.33
c.   6.029
e.   0.1222
g.   5521
i.   34.59

4–4.   a. $7.8/S$
c. $109.8/S^4$
e. $-37.2\ S^2 X$
g. $7.3e^{-as}\ G(S)$
i. $\dfrac{4.8\ (S^2 - 377^2)}{(S^2 + 377^2)^2}$
k. $6.2/(S + 8)$

4–6. $\dfrac{H}{M} = \dfrac{1000}{2000S + 1}$

4–7.   b. $\dfrac{H}{M} = \dfrac{1}{100S + 1}$

4–9. $\dfrac{I}{\theta} = \dfrac{0.4}{6S + 1}$

4–11. $\dfrac{\theta}{X} = \dfrac{125}{25S^2 + 26S + 1}$

4–13. $\dfrac{X}{F} = \dfrac{1}{3.2S^2 + 2S + 800}$

4–14. $\dfrac{W}{E} = \dfrac{0.043}{1.6 \times 10^{-5}\ S^2 + 9.66 \times 10^{-4}S + 2.19 \times 10^{-2}}$

4–16. $\dfrac{I}{E} = 1.4 + 0.014S$

Chapter 5

5–1. 500 turns

5–3. b. 4.38 sin 377t
d. -2.12 sin 377t

5–4. b. 141°
d. 64.2°

5–6. b. $N_T = 450$, $X = 32.2$ cm
d. $N_T = 8960$, $X = 641$ cm

5–7. $K_E = 0.0173$ volts per r/min

5–8. b. 147 HZ
d. 293 HZ

5–9. $N = 100$

5–11. $F_{\max} = 109$ newtons
$\triangle R = 1.005$ ohms

5–13. Vapor, Gas, Liquid, Mercury

5–15. predicted $R = 129.6$ ohms

5–17. $a_1 = 5.175 \times 10^{-2}$
$a_2 = 1.05 \times 10^{-5}$
$E_o = 0$
$E = 5.175 \times 10^{-2}T + 1.05 \times 10^{-5}\,T^2$

5–19. $Q_{25} = 0.0755$ g/m
$Q_{75} = 0.1265$ g/m

5–21. a. 2.8×10^5 cc
b. 1.09×10^6 cc

5–25. $F_{\min} = -132.4$ newtons
$F_{\max} = 63.8$ newtons

Chapter 6

6–1. The lag coefficient is the time required for the step response of a sensor to reach 63.2 percent of the final value.

6–3. Worst case error $= 0.03 \times 10^5$ $n/m^2 = 1.5$ percent
Reproducibility $= \pm\, 0.02 \times 10^5$ $n/m^2 = \pm\, 1$ percent

6–5. 40 r/min or 0.8 percent

6–7. 0.1 psi or 0.833 percent

6–9. 0.06 volts or 0.91 percent

6–11. $S_{20} = 0.285$ v/n
$S_{80} = 0.315$ v/n

Chapter 7

7–1. a. 5 d. 192

7–3. a. 6 d. 304

7–5. a. 6 d. 89A

7–9. MICRO . . .

7–11. a. 1.56 volts d. 2.19 volts

7–19. Voltage Gain = 72.96

Chapter 8

8–2. b. 8.7v d. 7.95v

8–3. b. $-9.254v$ d. 29.06v

8–4. b. $k_1 = 1$, $k_2 = 10$, $k_3 = 1$

8–5. b. $y = 10.45v$

8–7. b. $-1.92v$

8–8. $R_2 = 100,000$ ohms

Chapter 9

9–3. Two-position mode

9–7. Proportional plus derivative plus integral modes

9–9. Floating mode

9–11. b. $K = 0.5$ d. $K = 4$ f. $K = 10$
$R_1 = 2 \times 10^5$ $R_1 = 2.5 \times 10^4$ $R_1 = 10^4$

9–13. $R_i = 1.5 \times 10^7$ ohms
$R_1 = 3 \times 10^7$ ohms

$$v = 0.5e + 0.0033 \int_o^t e\, dt + v_o$$

$$\frac{V}{E} = 0.5 \left[\frac{150S + 1}{150S} \right]$$

Chapter 10

10–3. 1/2 inch valve

10–5. $T_{max} = 0.33\, n - m$
$P_{max} = 173$ watts
Regulation $= -35.5$ radians$/s - n - m$
E_{max} 80.42 volts

10–7. $\dfrac{\omega}{E} = \dfrac{6.63}{2.02 \times 10^{-3}\, S^2 + 3.63 \times 10^{-2}\, S + 1}$

$\dfrac{\theta}{E} = \dfrac{6.63}{S(2.02 \times 10^{-3}\, S^2 + 3.63 \times 10^{-2}\, S + 1)}$

10–9. $R_1 = 8{,}000$ ohms
$R_2 = 2{,}000$ ohms
$R_3 = 0.08$ ohms

Chapter 11

11–1. $R = 66.7$ ohms

11–2. a. $R = 13.8$ ohms when $E = 2.5$ volts
b. $R = 18.7$ ohms when $E = 7.5$ volts

11–4. Reynolds number $= 188$, flow is laminar
$R = 2.00 \times 10^9 n - s/m^5$
$Q = 3.42 \times 10^{-4} m^3/s$
$P = 6.85 \times 10^5 n/m^2$

11–7. $R = 0.130$ Kelvin/watt

11–8. a. $R = 0.1075$ Kelvin/watt
b. $R = 5.84 \times 10^{-4}$ Kelvin/watt
c. $R = 3.66 \times 10^{-3}$ Kelvin/watt
d. $R = 1.1 \times 10^{-2}$ Kelvin/watt
e. $R = 3.83 \times 10^{-5}$ Kelvin/watt

11–9. $R = 105 \ n - s/$meter

11–11. $C = 16.7 \times 10^{-6}$ farads

11–12. a. $C = 4.25 \times 10^{-4} m^5/n$
b. $C = 4.83 \times 10^{-4} m^5/n$
c. $C = 5.31 \times 10^{-4} m^5/n$
d. $C = 5.75 \times 10^{-4} m^5/n$
e. $C = 4.88 \times 10^{-4} m^5/n$

11–14. $C = 6.9 \times 10^6$ joules/Kelvin

11–15. $C = 6 \times 10^{-5}$ meters/newton

11–16. $L = 1.6$ henrys

11–17. a. $F = 10.3$ newtons

11–18. $I = 1.93 \times 10^8 \ n - s/m^5$

11–19. $T_d = 7.41 \times 10^{-5} s$

11–21. $T_d = 11.55s$

Chapter 12

12–2. a. $C = 2.05 \times 10^{-4} m^5/n$
b. $R = 3.26 \times 10^7 \ n - s/m^5$
c. $\tau = 6.68 \times 10^3$ sec.

12–3. $R = 8.3 \times 10^3$ ohms

12–5. a. $C = 6.85 \times 10^{-6} \ Kgm - m^2/n$
b. $\tau = 1.23$ sec.

c. $1.23 \dfrac{dp}{dt} + p = 1.8 \times 10^5 m$

d. $\dfrac{P}{M} = \dfrac{1.8 \times 10^5}{1.23S + 1}$

12–9. $R_2 = 100$ ohms

$$16 \times 10^{-8} \dfrac{d^2 e_o}{dt^2} + 8 \times 10^{-4} \dfrac{de_o}{dt} + e_o = e_i$$

$$\dfrac{E_o}{E_i} = \dfrac{1}{16 \times 10^{-8} S^2 + 8 \times 10^{-4} S + 1}$$

12–11. $\tau_1 = 1 \times 10^{-5}$ sec.

$\tau_2 = 2.4 \times 10^{-4}$ sec.

$$2.4 \times 10^{-9} \dfrac{d^2 e_o}{dt^2} + 3.9 \times 10^{-4} \dfrac{de_o}{dt} + 9 e_o = e_i$$

$$\dfrac{E_o}{E_i} = \dfrac{1}{2.4 \times 10^{-9} S^2 + 3.9 \times 10^{-4} S + 9}$$

12–13. $T_d = 6.9$ sec. $\tau = 1220$ sec.

Chapter 13

13–2. b. $0.1 \underline{|-90°}$ meters

c. $0.01 \underline{|-90°}$ meters

e. $0.001 \underline{|-90°}$ meters

13–5.

ω (radian/second)	M (decibels)	β (degrees)
0.05	20	$-90°$
0.5	0	$-90°$
5	-20	$-90°$
50	-40	$-90°$

13–13.

(radians/second)	M (decibels)	(degrees)
1.17	-14.8	0
11.7	-14.8	$-8°$
117	-17.2	$-90°$
1170	-54.8	$-172°$

Chapter 14

14–1.

ω	M_o	β_o	M_c	β_c	$M_c - M_o$
0.1	20	$-3°$	$-.95$	$-.5°$	-21
0.2	20	$-7°$	$-.95$	$-1°$	-21
0.5	20	$-13°$	$-.9$	$-2°$	-20.9
1.0	19	$-29°$	$-.85$	$-3°$	-19.8
2.0	17	$-51°$	$-.8$	$-5°$	-17.8
5.0	12	$-79°$	$-.6$	$-14°$	-12.6
10	6	$-108°$	$.3$	$-31°$	-5.7

| 20 | 0 | $-141°$ | 4 | $-70°$ | 4 |
| 50 | -8 | $-230°$ | -6 | | 2 |

Maximum frequency limit $= 14$ radians/second

14–5.

ω	M_o	β_o	M_c	β_c	$M_c - M_o$
0.0001	-6	$-0.1°$	-9.9	$-0.1°$	-3.9
0.0002	-6	$-0.2°$	-9.9	$-0.2°$	-3.9
0.0005	-6	$-0.6°$	-9.9	$-0.5°$	-3.9
0.001	-6	$-1.1°$	-9.85	$-1°$	-3.85
0.002	-6	$-2.3°$	-9.85	$-2°$	-3.85
0.005	-6	$-5.7°$	-9.85	$-4°$	-3.85
0.01	-6	$-11.5°$	-9.8	$-7°$	-3.8
0.02	-6	$-22.9°$	-9.8	$-15°$	-3.8
0.05	-6	$-57.3°$	-8.6	$-39°$	-2.6
0.10	-6	$-114.6°$	-5	$-83°$	1
0.16	-6	$-180°$	-0.3	$-180°$	5.7
0.20	-6	$-229°$	-5	$-260°$	1

Controller gain $= 0.5$

Maximum frequency limit $= 0.08$ radians/second

Chapter 15

15–1. System No. 2

    Two-mode:

      $T_i = 5$ min

      $G = 2.84$

    Three-mode:

      $T_i = 3$ min

      $T_d = 0.75$ min

      $G = 3.78$

    System No. 4

    Two-mode:

      $T_i = 15$ min

      $G = 0.54$

    Three-mode:

      $T_i = 9$ min

      $T_d = 2.25$ min

      $G = 0.72$

15–3. $K_A = 2.51$

    $T_i = 0.114$ seconds

    Gain margin $= 17$ decibels

    Phase margin $= 40°$

Index